# Logic and Architecture Synthesis

**IFIP – The International Federation for Information Processing**

IFIP was founded in 1960 under the auspices of UNESCO, following the First World Computer Congress held in Paris the previous year. An umbrella organization for societies working in information processing, IFIP's aim is two-fold: to support information processing within its member countries and to encourage technology transfer to developing nations. As its mission statement clearly states,

> IFIP's mission is to be the leading, truly international, apolitical organization which encourages and assists in the development, exploitation and application of information technology for the benefit of all people.

IFIP is a non-profitmaking organization, run almost solely by 2500 volunteers. It operates through a number of technical committees, which organize events and publications. IFIP's events range from an international congress to local seminars, but the most important are:

- the IFIP World Computer Congress, held every second year;
- open conferences;
- working conferences.

The flagship event is the IFIP World Computer Congress, at which both invited and contributed papers are presented. Contributed papers are rigorously refereed and the rejection rate is high.

As with the Congress, participation in the open conferences is open to all and papers may be invited or submitted. Again, submitted papers are stringently refereed.

The working conferences are structured differently. They are usually run by a working group and attendance is small and by invitation only. Their purpose is to create an atmosphere conducive to innovation and development. Refereeing is less rigorous and papers are subjected to extensive group discussion.

Publications arising from IFIP events vary. The papers presented at the IFIP World Computer Congress and at open conferences are published as conference proceedings, while the results of the working conferences are often published as collections of selected and edited papers.

Any national society whose primary activity is in information may apply to become a full member of IFIP, although full membership is restricted to one society per country. Full members are entitled to vote at the annual General Assembly, National societies preferring a less committed involvement may apply for associate or corresponding membership. Associate members enjoy the same benefits as full members, but without voting rights. Corresponding members are not represented in IFIP bodies. Affiliated membership is open to non-national societies, and individual and honorary membership schemes are also offered.

# Logic and Architecture Synthesis

## State-of-the-art and novel approaches

**IFIP Workshop on Logic and Architecture Synthesis, 1994**

Edited by

**Gabrièle Saucier and Anne Mignotte**

*Institut National Polytechnic de Grenoble*
*Grenoble*
*France*

Published by Chapman & Hall on behalf of the
International Federation for Information Processing (IFIP)

CHAPMAN & HALL
London · Glasgow · Weinheim · New York · Tokyo · Melbourne · Madras

**Published by Chapman & Hall, 2–6 Boundary Row, London SE1 8HN, UK**

Chapman & Hall, 2–6 Boundary Row, London SE1 8HN, UK

Black Academic & Professional, Wester Cleddens Road, Bishopbriggs, Glasgow G64 2NZ, UK

Chapman & Hall GmbH, Pappelallee 3, 69469 Weinheim, Germany

Chapman & Hall USA, 115 Fifth Avenue, New York, NY 10003, USA

Chapman & Hall Japan, ITP-Japan, Kyowa Building, 3F, 2-2-1 Hirakawacho, Chiyoda-ku, Tokyo 102, Japan

Chapman & Hall Australia, 102 Dodds Street, South Melbourne, Victoria 3205, Australia

Chapman & Hall India, R. Seshadri, 32 Second Main Road, CIT East, Madras 600 0355, India

First edition 1995



Printed in Great Britain by TJ Press, Padstow, Cornwall

ISBN 0 412 72690 4



A catalogue record for this book is available from the British Library

Printed on permanent acid-free text paper, manufactured in accordance with ANSI/NISO Z39.48-1992 and ANSI/NISO Z39.48-1984 (Permanence of Paper).

# CONTENTS

# Introduction

*Anne Mignotte*
*Laboratoire de l'Informatique du Paralléllisme, URA CNRS N° 1398*
*Ecole Normale Supérieure de Lyon*
*46 Allée d'Italie, 69364 Lyon cedex 07, France*
*Telephone: 33 72 72 85 49, Fax: 33 72 72 80 80*
*E-mail: Anne.Mignotte@ lip.ens-lyon.fr*

Today it is evident that Computer Aided Design (CAD) is essential for ASIC design. This is due to the increasing complexity of integration - a complete paper and pencil design is no longer possible; the fact that CAD tools enable a continuous verification of a circuit's behavior from the highest level down to the electrical level; and the fact that CAD tools make it possible to choose between different implementations of the same application, thereby allowing solution space exploration.

In this book, we will focus on automatic design generation from a behavioral description, in other words, synthesis. Depending on the level of the behavioral input description, one is dealing with "logic synthesis" whose input consists of a set of boolean functions, "controller synthesis" based on a finite state machine input description, "RTL synthesis" starting from a Register Transfer Level description, "high level synthesis" which accepts an algorithmic description or "system level synthesis" which accepts a set of communicating processes as input. A typical design process starts with a higher level of synthesis and works down to logic synthesis producing an optimized structural specification of the circuit. This book, however, follows research topics in a chronological order and therefore starts with the lower level synthesis tasks, working its way up to system level synthesis.

The first four chapters are dedicated to novel approaches in logic synthesis. The main innovations in this area are the use of BDDs (Binary Decision Diagrams) for multilevel logic synthesis and programmable device logic synthesis. Multilevel logic synthesis starts with a set of boolean functions and generates an optimized netlist of cells of the target technology. Programmable devices and in particular, FPGAs (Field Programmable Gate Arrays), are new target technologies implying new synthesis techniques.

Multilevel synthesis optimizes the speed and/or the area of the resulting combinational block. Logic synthesis consists of two main phases: the optimization phase which corresponds to logic minimization and the factorization of the initial logic functions, and the cell assignment phase which allocates the cells of the technology to the optimized functions. Because of the growing complexity of boolean functions to be synthesized, traditional techniques are no longer efficient. The advantage of BDDs is that BDD manipulation does not depend on the boolean function complexity in terms of literals, but only on the number of variables and their order in the BDD representation. Therefore, novel approaches based on BDDs are a good alternative. BDDs can be used for all phases of multi-level logic synthesis: optimization and assignment.

The first chapter concentrates on logic optimization. [1] presents an implicit technique and [2] suggests logic minimization techniques using boolean relations. Boolean relations are more efficient than don't care representations for logic minimization. Their construction is based on the BDD.

Cell assignment [4] is described in the second chapter. In fact, assignment is divided into three steps: decomposition, matching and mapping. The decomposition [5] [6] splits the minimized boolean function into several sub-functions. Matching makes the correspondence between the sub-functions' and the cells' inputs by permutation on the input order. Mapping selects a specific cell of the library to realize each sub-function forming the boolean functions. Another advantage of BDD representation for logic functions is that this allows formal verification of the resulting design [3]. Formal verification is then reduced to the representation of the resulting design and the initial boolean functions with BDDs having the same ordering of variables. One of the main drawbacks of BDDs is that the size of the BDD representation is extremely sensitive to the ordering of variables. Therefore, a considerable amount of research has been carried out on variable ordering for BDD representation. One such approach is based on symmetry [7].

The third and fourth Parts are dedicated to programmable logic synthesis. Programmable devices are enjoying an increase in popularity as they allow rapid prototyping and reprogrammability facilities. They are classified in three main categories: the CPLD (Complex Programmable Logic Device), the multiplexer based FPGA and the look up table based FPGA. The cell granularity of programmable devices is much higher than that of standard cells, but the function complexity implemented with one device may be too low for a complete application. Therefore, a first step in programmable logic synthesis consists of partitioning or clustering [8] and [9].

CPLDs are made up of interconnected PAL (Programmable Array Logic) devices or PLAs (Programmable Logic Arrays). Multiplexer based cells are made up of multiplexer trees. To perform a specific function, these cells are programmed by the application of a specific command on the multiplexers. They are usually modeled by BDDs. A Look Up Table (LUT) can implement any function of k variables. Enumerating all the possible logic functions would give a cell library composed of $2^{2^k}$ basic cells. This is not reasonable and therefore standard techniques for function decomposition used in multilevel synthesis cannot be used for the FPGA technology. Hence, the special structure of FPGA basic cells influences the synthesis process. Logic synthesis techniques such as mapping and function decomposition are presented in [11] for both categories of FPGA. LUT based FPGAs force the synthesis to be based on the number of sub-function inputs. Such a technique is presented in [10]. It is a decomposition technique to detect whether several outputs depend on the same inputs. To conclude on FPGAs, some comparisons of the different FPGA families are given in [12]. These results are analyzed to extract some physical parameters of LUT implementation.

Once the design is defined at the structural level, some modifications are still possible to opitimize it while preserving the global behavior, as shown in Part Five. Retiming is used to speed up the design by moving registers through combinational blocks. One retiming approaches shifts registers to the periphery of the design to obtain larger combinational blocks and to get more freedom for logic optimization [15]. Specific optimization can be done by exploiting the special structure of the data path, as data paths usually have a fixed width for the operands. Data path regularity in terms of number of bits should be extracted from the complete design [14]. Area optimization is then possible by removing redundancies, again using BDDs, to identify duplicate nets [15].

Architectural synthesis starts with a process and produces an optimized design in terms of resources and control steps. The target architecture is usually made up of a data path and a controller which sequences the operations of the data path. Therefore, a control step corresponds to a state of a controller and a resource is a memory resource, a communication resource or a functional unit of the data path. The controller specification is given to controller synthesis tools and the data path is a netlist of macro blocks referring to a specific resource library.

Controller synthesis as described in Part Six starts with a finite state machine and generates several interconnected macro blocks depending on the target architecture. A specific target architecture based on a ROM is presented in [16]. Some of the blocks are described by boolean functions and are given to logic synthesis tools, some others correspond to macro-blocks in a library. The main phase of controller synthesis is the state assignment phase which encodes each state of the finite state machine according to the characteristics of the target architecture and technology. State assignment for programmable devices is presented in [17]. State assignment with false path detection is presented in [20]. The influence of state assignment on fault tolerance is presented in [19]. A formal approach for controller synthesis is proposed in [18].

The controller and the data path are synchronized according to a clocking scheme. An original technique for this synchronization is presented in [22]. An important criterion which is frequently ignored is the power consumption, it should be taken into account during the data path and controller construction [21].

Architectural synthesis is usually decomposed into RTL synthesis and high level synthesis. High level synthesis starts with a process describing an untemporized algorithm and RTL synthesis starts with a temporized process. The main architectural synthesis steps are the selection of the resources in the library and the allocation of the data path resources. High level synthesis also includes a temporization phase, namely the scheduling of the operations into control-steps.

A constant link is then necessary between architectural synthesis and libraries, as described in Part Eight. Specific constraints and criteria of the target library have to be taken into account during architectural synthesis [23] and [25] and on the other hand specific macro block libraries have to be built according to synthesis [24].

The arithmetic operations of the initial specification are linked to operators belonging to a library. The first step of high level synthesis, presented in Part Nine, in particular [26], is the selection of the appropriate operator from among several functional units in the library that are able to realize a given operation in the specification. This step is not always considered locally as specific arithmetic operations have an impact on high level synthesis [27]. Functional units are usually generated based on a structural description. A new approach based on adder synthesis allows solution space exploration [28].

High level synthesis has not yet been accepted in industry and commercialized synthesis tools are still restricted to RTL and logic synthesis, but it is a very active research topic [29]. Different approaches for high level synthesis are possible. Some of them are presented in Part Ten. The classical high level synthesis steps are presented in [31]. An original approach for high level synthesis is the generation of the design from emulation results [30]. Therefore, high level synthesis is restricted to optimizations at the register transfer level. BDDs can be used again to analyze whether or not operations are exclusive [32].

To conclude the high level synthesis topic, two applicative studies are presented in Part Eleven. The first one analyzes existing synthesis tools and the VHDL language for telecom ASIC design [33]. The second one investigates a new research topic: asynchronous circuit design [34].

The higher and more recent synthesis level of co-synthesis is presented in Part Thirteen. It identifies some parts of the initial specification to be implemented in software and some parts to be implemented in hardware [37]. The specification of the hardware describes a complete system. Then, system level synthesis produces a behavioral specification made up of several communicating processes which can be given to lower level synthesis tools to generate specific processors or can be implemented by existing and programmable processors. Therefore, the main step of system level synthesis is mapping. The initial specification is partitioned into processes and each process is assigned to a dedicated processor [38]. A final step, presented in Part Twelve, consists of the allocation of memory elements and buses to realize the communication between the different processes [35] or functional units [36].

PART ONE

# Logic Minimization Based on BDD

1

# Boolean Optimization Using Implicit Techniques

*F. Poirot, G. Tarroux and R. Roane*
*Compass Design Automation*
*505, Route des Lucioles - 06560 Sophia-Antipolis, France.*
*tel: (+33) 92.94.46.21 - fax: (+33) 93.65.39.21*
*e-mail: franck@compass.fr*

**Abstract**

The increasing complexity of industrial designs, and especially with the extensive use of high level languages such as VHDL or Verilog, progressively obsoletes classical Boolean optimization techniques. Thus, the application of Binary Decision Disgrams in Logic Synthesis becomes an attractive alternative to push forward the limit of design complexity. In this paper, we have developed new Optimization techniques only based on implicit techniques, and all classical steps were fully reformulated to provide a powerful system providing better results than other existing techniques.

**Keywords**

Logic Synthesis, Binary Decision Diagrams, Boolean Optimization, Implicit Techniques.

## 1. INTRODUCTION

The generalization of hardware description languages such as VHDL or Verilog and improvements in technologies allow the designer to consider the possibility of defining bigger designs. Also, to make these descriptions more readable and easily debuggable or back-annoted, designers generally introduce local variables or signals to express their logic functions. These descriptions yield to huge redundant and unoptimized Boolean networks which will be under optimized or impractical to process with classical optimization algorithms.

To push forward these limitations, Binary Decision Diagrams have been introduced. BDDs are dense representations of Boolean function based on Shannon decomposition (Bryant 86). If a fixed input order for the successive Shannon decompositions is imposed, they become a canonical representation (ROBDDs) widely used for verification purpose (Madre 90, Touati 91, Jeo 91), testing (Akers 78), or direct technology mapping (Besson 92, Murgai 92).

In this paper, after some definitions, we are preliminary concerned with the description of our implicit Boolean optimization system. This system completely redefines conventional methods and pushes forward the complexity of designs to allow better optimizations. Efficient and novel ideas are presented using dynamic manipulation of implicit structures during the optimization process to allow the consideration of powerful techniques such as Boolean division with a constant control of the memory usage. These techniques have been evaluated on standard MCNC benchmarks and show a strong improvement compared to other published methods.

## 2. DEFINITIONS

### 2.1. Binary Decision Diagrams

Given a Boolean function $F(x_1, x_2, ..., x_n)$ the positive and negative cofactors of f with respect to x are defined by $F_x = F(1, x_2, ..., x_n)$ and $F_{\bar{x}} = F(0, x_2, ..., x_n)$. The Shannon expansion with respect to variable x is given by $F = x.F_x + \bar{x}.F_{\bar{x}}$. A Binary Decision Diagram is a Direct Acyclic Graph (DAG) representing a Boolean function. The construction of a BDD is based on the successive Shannon decompositions of the function according to a splitting variable as shown in figure 1. Each non terminal node n rooted by a variable x, corresponds to a function F(n) and is the origin of two arcs $F(n)_0$ and $F(n)_1$. The graph rooted at $F(n)_0$ represents the negative cofactor $F(n)_{\bar{x}}$ and is called the *Else Edge* of n. The graph rooted at $F(n)_1$ represents the positive cofactor $F(n)_x$ and is called the *Then* Edge of n.

For an Ordered BDD (OBDD) a global ordering "<" over the set of variables is imposed. On any path from a root to a terminal node the variables occur in the given order. The graph may have redundant vertices and duplicated subgraphs. They are eliminated by repeatedly applying transformations rules (Bryant 86). The BDD shown in figure 1 is in fact an OBDD with the order a<b<c<d.

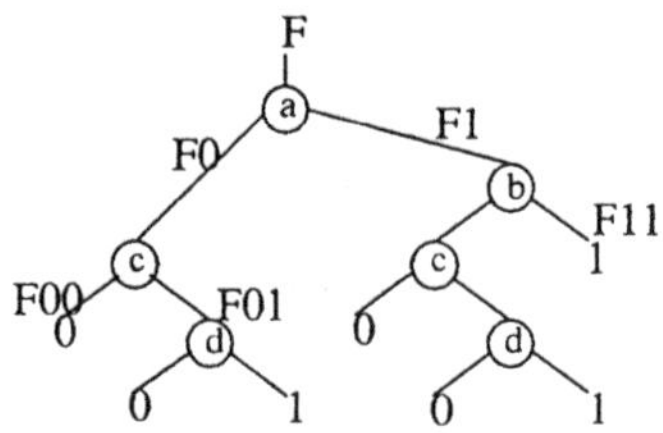


**Figure 1: BDD for F=ab+cd**

The maximally reduced graph is referred as Reduced Ordered BDD (ROBDD). For a given Boolean function and a given ordering, this graph is unique and therefore constitutes a canonical representation. This property presents crucial advantages for functional equivalence testing. Figure 2 shows the ROBDD of F= a **xor** b **xor** c.

ROBDD with negative edge is a more compact representation (Madre 88). In this representation, not only common subtrees corresponding to the classical notion of subfunctions (SF) are identified, but the complemented subtrees ($\overline{SF}$) are also detected. The subtree and its complement will be represented only once and a negative edge can be considered as an inverter. Figure 3 shows the ROBDD with negative edges of F= a **xor** b **xor** c. In this paper the term BDD denotes ROBDD with negative edges.

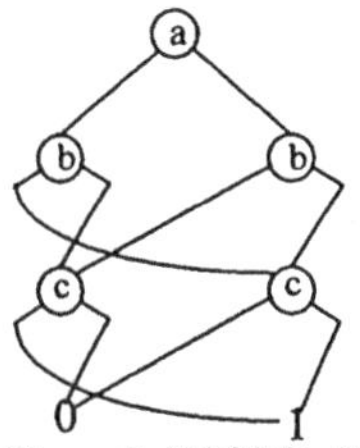


**Figure 2: ROBDD of F=a xor b xor c**

**Figure 3: ROBDD with negative edges of F**

## 2.2 Dynamic Improvement of the Ordering

After each garbage collection, before starting operations, the variable ordering is dynamically improved with the algorithm proposed in (Rudell 93). This step reduces the current node count, and makes the future operations faster to perform. It also allows to build BDDs which could not be built with any fixed ordering, as any given ordering fails during the construction. This swapping is performed on adjacent variables $v_i$ and $v_{i+1}$. This allows to only change the part of the DAG containing $v_i$ and $v_{i+1}$, instead of re-building the whole graph. Figure 8 gives the general case of the method when $F_0$ and $F_1$ are printed more than once. Note that new built nodes have to be made canonical and be looked-up in the node hash table before physically creating them. The interesting property of an adjacent swapping is that the top and bottom parts of the DAG, respectively built from $v_1$ to $v_{i-1}$ and from $v_{i+2}$ to $v_n$ are unchanged before and after the transformation. Indeed, referring to figure 4, considering F, $F_0$, $F_1$ as inputs, top part of the DAG denotes the same BDDs with the same input ordering, before and after the swapping. Thus, the canonicity property of BDDs involves that the DAG obtained after swapping $v_i$ and $v_{i+1}$ is homomorphic to the initial one. Now considering $F_{ij}$, $(i,j) \in \{0,1\}^2$, as outputs, the bottom part of the DAG denotes the same BDDs with the same ordering before and after swapping.

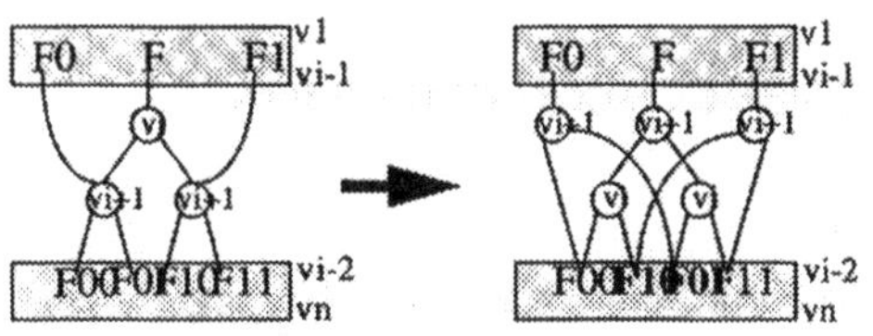


**Figure 4: Adjacent swapping**

## 2.3. Cube Set Manipulation and Implicit minimization

Classical approaches are based on Sum-of-Products (SOPs) structures which quickly appears to be very greedy in memory usage and unsuitable for logic operations like inverting or equivalence testing. So, in conjunction with BDDs which lead to a more efficient representation of Boolean functions than logic trees, an efficient cube set representation was also necessary to perform the minimization process. Madre and Coudert (Coudert 92) have proposed a BDD based meta-product structure, which requires two variables ($o_i$, $s_i$) to denote the occurrence and the sign of each input $x_i$. In a such BDD, each 1-path starting from the root node denotes a cube. The paths rooted by the Then-Edge of $o_i$ contains an occurrence of $x_i$. Then, $s_i$ expressed if $x_i$ occurs in its direct (Then-Edge) or complemented form (Else-Edge). This structure has the advantage to be compact and canonical for prime sets, and its size is not related to the number of cubes it denotes. In (Coudert 92) a breakthrough method was given to compute primes and essential primes of so large functions that they have never been handle before, and (Coudert 93b) provided a new logic minimization algorithm resolving cyclic core of all the hard Espresso problem. Nevertheless, for sparse networks like some ISCAS benchmarks the meta-products structure could blow up, even if the BDDs of these networks can be built. Minato proposed in (Minato 93a) a new cube set structure called Zero-Suppressed BDDs, or ZBDD (for Cube Set), based on new reduction rules on BDDs. This structure is definitively more suitable for cube set manipulation than meta-products, especially for sparse networks. In this structure, one variable is used to denote in which cube an input $x_i$ appears, and an other variable denote in which cube $\overline{x_i}$ appears. The major improvement consists on a new reduction rule which eliminates from the graph the variables which does not appear in the cubes to whose path they belong, namely the variables whose Then-Edge points to the "0" terminal node. Figure 7 gives the algorithm described with meta-products in (Coudert 93a), adapted here with ZBDDs using the principle of Morreale's algorithm (Morreale 70).

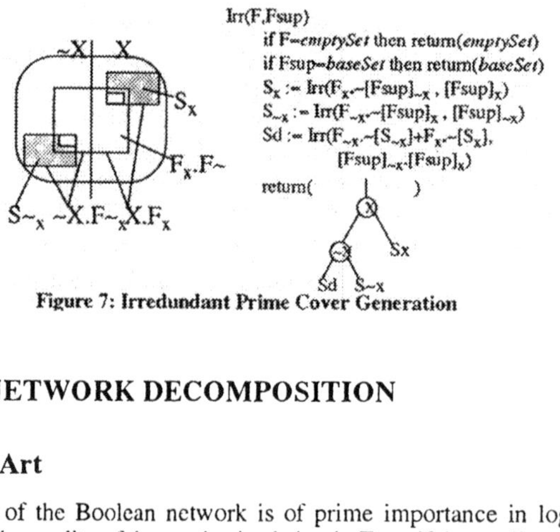


**Figure 7: Irredundant Prime Cover Generation**

# 3. BOOLEAN NETWORK DECOMPOSITION

## 3.1. State of the Art

The decomposition of the Boolean network is of prime importance in logic synthesis as it directly impacts on the quality of the synthesized circuit. Even if the quality criteria is measured in terms of area, speed, power or testability, it has been proven that the literal count is a good predictor. Algebraic methods to optimize combinational circuitry have been already proposed in the early 1950's. These methods were only suitable for small networks not exceeding 10 variables. They were nicely enhanced in the 1980's by more sophisticated techniques (Brayton 84).

Nevertheless, all these research studies were mainly driven by initial works targeted on PLA implementation. Clearly, the sum-of-products representation was very attractive to deal with Boolean networks as the programming data structure was compact, and the algorithms may use the efficiency of classical programming languages (mask on operators, and, or, xor between bytes representing the covers). Therefore, most designs of that period were handled in an efficient way as long as PLAs was an interesting issue, and the design complexity relatively small.

But the introduction and the generalization of hardware languages such as Verilog, VHDL, and the improvements in the technologies, significantly increases the size of Boolean networks. Then, industrial applications became more and more impractical to process with standard SOPs approaches. They usually consume huge memory and take large CPU time which is not compatible with the time to market criteria becoming a key of success for CAD companies. Of course, numerous heuristics solve these weaknesses but they seriously impact on the quality of the final circuitry.

The introduction of implicit techniques was a complete breakthrough in this field and creates new motivations to reconsider conventional techniques with these powerful structures. They represent the next generation in synthesis. Minimization was already successfully addressed by (Coudert 92) and (Minato 92), but the decomposition still remained a bottleneck in the flow.

Minato (Minato 93b), introduced algebraic techniques but they were very limited. In fact, Brayton's kernel extraction (Brayton 87) method was rewritten with implicit techniques and was based on the most occurring literals. Also, only algebraic division was applied, and the algorithm only considered one single function avoiding the sharing of identical logic within several functions.

Recently, Stanion (Stanion 94) proposed a more sophisticated method by producing a more complete set of divisors. Unfortunately, the weighting of these candidates was based on BDD's 1 path which is neither prime nor irredundant. So, they could not correctly link the candidate extraction to their final goal. Moreover, they provided a more sophisticated division than Minato with Boolean consideration by using cofactor techniques, but these methods could not provide the same quality than a fully Boolean division. Finally, as in the previous method, they only deal with single functions.

The method that we propose tries to cover the previous weaknesses by taking advantage of the efficiency of our new BDD and ZBDD package presented before. In fact, our goal is to develop implicit techniques allowing to consider a complete Boolean network as input of our BDD optimizer. Then, it may run sophisticated decomposition techniques (divisor extraction, exact weighting, Boolean division) on industrial networks with fast execution time.

## 3.2. Decomposition flow

A typical decomposition flow is represented in the figure 8. The first phase consists of computing a set of candidates which can be used to divide one or more functions in the Boolean network and obviously, if no candidate is found the decomposition is ended. Afterwards, on or more candidates have to be selected as divisor and it is important to define an accurate selecting criteria which correctly represents the goal of the optimization. In the technology independent optimization the more suitable criteria is the number of literals in a Boolean function. Then, for each candidate we compute a weight which represents exactly the number of literals saved in the Boolean network when this candidate is used as divisor. Moreover, in order to have a more powerful decomposition, we develop and use in our flow a method to determine the best compatible set of candidates. This allows to divide the Boolean network by several candidates in the same iteration instead of using only one. To divide the Boolean network we have develop both Boolean and algebraic divisions based on ZBDD structure. The power of these two new divisions allows us to run both of them for each candidates and keep the best result without a large cpu time penalty. All algorithm used in our flow are described more precisely in the following paragraphs.

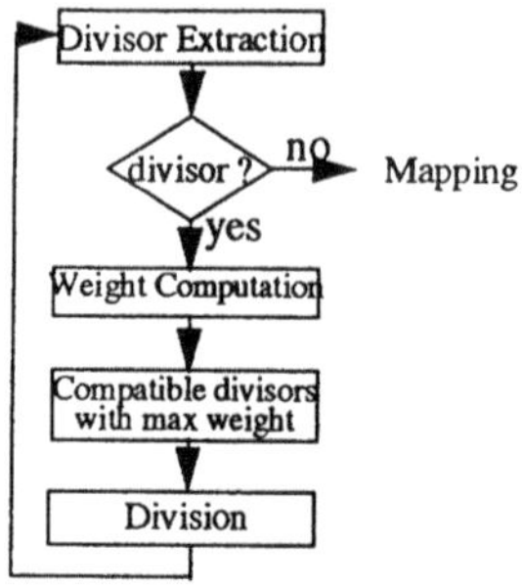


**Figure 8: Decomposition flow**

## 3.3. Divisor extraction

The most popular method for the identification of common sub-expressions in the Boolean network has been proposed by Brayton and Mcmullen at ISCAS'82. Here again, the problem is similar to the creation of optimal code for a compiler and the SOPs appear a very suitable representation for a limited complexity. In this paper, the notion of kernels of a logic expression was introduced to provide an efficient means of finding common sub-expressions. A kernel K of a function F is defined as the cube free quotient of F by a cube C.

In our flow, we have developed a new algorithm taking full advantage of the implicit structure to get all level-0 kernels of a function. One main advantage of this procedure versus classical SOP ones is that during the examination of the implicit structure graph only effective inputs are considered while SOPs require the evaluation of all inputs which allows to generate the set of level-0 kernel in a little cpu time. The pseudo code of the level-0 kernel extraction procedure is given in figure 9.

```
kernel0(zbdd, lit)
{   if nbLiterals(zbdd) < 2 then return
    while (lit:=nextInput(zbdd,lit))
    {   if not literalIsSeveralTimes(zbdd, lit) then continue;
        zbddVar := getZBDDVar(lit);
        zbdd1 := subSetWhenVarIs1 (zbdd, zbddVar);
        cubeFactor := largestCubeFactor(zbdd1, zbddVar);
        if cubeFactor=emptySet then continue;
        if cubeFactor<>baseSet
        then zbdd1 := divideAlgZBDD(zbdd1, cubeFactor);
        updateCandidatList(zbdd1);
        kernel0(zbdd1, lit);
    }
}
```

**Figure 9: kernel 0 extraction procedure**

Using only level-0 kernel candidates is not enough to have an efficient decomposition because some other candidates can be very good divisors like intersecting level-0 kernels or common cubes. Intersecting level-0 kernels are computed from the level-0 kernel set generated with the procedure previously described. For instance, if the two following level-0 kernels k1=a+b+c and k2=a+b+d are found, the *intersectKernel* procedure described below will find the candidate a+b which will be put in the candidate list. In the same way, if in the Boolean network there are two functions f and g such as f=a.b.c.d + e.f.g and g=a.b.c.e + f.g.h, we can easily see that the cubes a.b.c and f.g are shared by f and g. The procedure *commonCube* extracting common cubes in the Boolean network is described below.

Therefore, our candidate computation phase based on ZBDD generates a list of candidates containing level-0 kernels, intersecting kernels and common cubes which allows us to chose the best candidate among a large set of candidates.

```
kernelIntersect()
{   for i:=1 to (nbKernel0-1)
    {   for j:=(i+1) to nbKernel0
        {   newKernel := intersection(kerneli, kernelj);
            if newKernel<>emptySet
            then updateIntersectList(newKernel);
        }
    }
}
```

**Figure 10: kernel intersecting procedure**

## 3.4. Divisor selection

After getting the set of candidates for decomposition, we compute the weight of each of them. The candidate's weight is the number of literals saved in the Boolean network when the network is divided by this candidate. During the weight computing phase, we divide all functions in the Boolean network by the candidate and its complement by using both algebraic and Boolean division. Obviously, the weight of a candidate is the sum of the best weights got for each divided function..

```
commonCube(f, g, cube, lit)
{   if (f=baseSet OR g=baseSet) AND nbLit(cube)>1
    then updateCubeList(cube);
    while (lit:=nextInput(f, lit))
    {   zbddLit := getZBDDVar(lit);
        if nbLit(g)=1 then
            if g=zbddLit
            then zbddVarG:=baseSet
            else zbddVarG:=emptySet
        else zbddVarG:=subSetWhenVarIs1(g,zbddLit);
        if zbddVarG=emptySet then continue;
        if nbLit(f)=1 then
            if f=zbddLit
            then zbddVarF:=baseSet
            else zbddVarF:=emptySet
        else zbddVarF:=subSetWhenVarIs1(F,zbddLit);
        commonCube(zbddVarF, zbddVarG,
                   zbddMultiply(cube,zbddLit), lit);
    }
}
commonCubesExtract()
{   for all funci and funcj
        commonCube(funci,funcj, baseSet, 0);
}
```

**Figure 11: common cube extraction procedure**

Afterwards, we have to select the best candidate and divide the Boolean network by it. In fact, in order to have a powerful decomposition in terms of quality and cpu time, we select a set of candidates instead of using only one candidate at each iteration (figure 8). To select a set of candidates by keeping the right knowledge on the literal saving, we introduce the term of *compatibility* between candidates. Then, we say that 2 candidates c1 and c2 are compatible if and only if the entire Boolean network can be divided in the same way by {c1, c2} and {c2,c1}. In other work, c1 and c2 are compatible if the parts of Boolean functions affected are disjoint which means that the weight of c2 (c1) computed before and after dividing the Boolean network by c1 (c2) is the same. Then, after computing all compatibilities we build a compatibility graph and use an algorithm to find the best weighted clique which represents a set of compatible candidates.

## 3.5. Boolean division

The Boolean division problem is a complex problem to solve because it requires the computation of Boolean operations which may have a high complexity under some data structures. It is mainly because classical methods such MIS (Brayton 87) solve this problem by expressing the divisor as the *don't care* of its expression by its variable. Moreover, this method has the disadvantage to compute the complement of a divisor which is of an exponential complexity with sum of product structures instead of being immediate with BDDs. Therefore, methods using SOP should include a bunch of heuristics to handle large designs which badly impact the quality of the results. Additionally, most industrial tools make only use of algebraic techniques since they consider Boolean ones too costly. Even with the BDD structure this problem is not easy to solve since Boolean division creates a large number of nodes in the BDD structure which may easily blows up the system if no memory control is done.

In our method, we consider the Boolean division with a novel approach (figure 12). This method uses the principle of the implicit minimizer and computes the best bounds of the remainder and the dividend. First the smallest remainder is computed with smallest lower bound (Rmin) and largest upper bounds (Rmax) to increase the Boolean space. Afterwards, bounds for the dividend are computed by taking into account the result of the remainder. The figure 12 shows the principle of our Boolean division and the figure 13 shows an example of division.

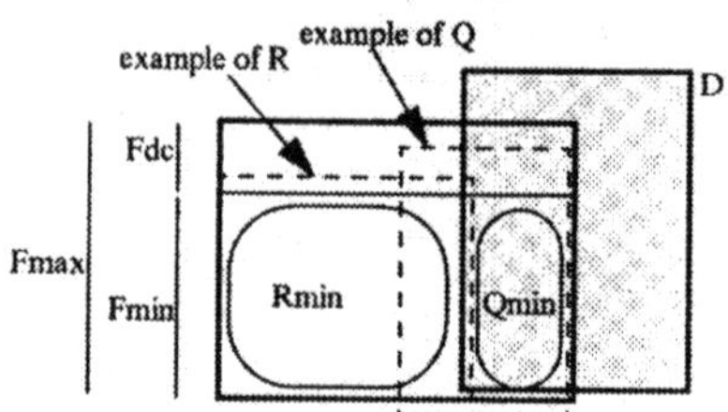


<u>F = D.Q + R</u>

0 < Q < ~D + Fmax; Fmin.~(Q.D) < R < Fmax

Rmin = Fmin.~D; Rmax = Fmax ⟶ R

Qmin = Fmin.~R; Qmax = Fmax + ~D ⟶ Q

**Figure 12: Boolean Decomposition**

Note that during execution of these operations, dynamic ordering and garbage collector control the memory used in order to avoid over size due to the creation of internal BDD nodes.

**F = a + b.c**
**D = a + c**

Rmin = Fmin.~D = (a + b.c).~a.~c = 0
Rmax = Fmax = a + b.c
→ R = 0

Qmin = Qmin.~R = a + b.c
Qmax = Fmax + ~D = a + b.c + ~a.~c = a + b + ~c
→ Q = a + b

**F = (a + c).(a + b) + 0**

**Figure 13: Example**

## 4. UNIFIED APPROACH WITH VERIFICATION

Since time to market is of prime interest for EDA companies, it is mandatory to detect as soon as possible any errors in the final implementation. For that purpose, formal verification techniques has been widely developed and adopted (Madre 90, Touati 91).

These errors can be either introduced by the system itself or by the designer who may have modified the schematic to respect a specific constraint. So, both aspect should be considered to provide an convenient system which guarantees the correctness of the design without too much CPU overhead.

The novelty of our system is to use the same data structure with its canonicity property to optimize the network and verify if the final implementation is correct regarding the initial functional description. So, there is no overhead in the computation time.

In case of errors, a report clearly tells which signal is incorrect and the designer can easily backannotate the functional description. Also, our system can be used as a prover. In that case, the designer specifies properties (for instance a traffic light is not green and red at the same time). These properties are converted in our data representation, and compared with in the specification. In that case, it provides the designer a powerful way of checking that his HDL description respects his specification.

## 5. EXPERIMENTAL RESULTS

This new Boolean optimization algorithm was implemented as part of the ASIC Synthesizer of Compass Design Automation. This program was written on Mainsail (MAINSAIL) on SPARC 1030. It was tested on standard MCNC benchmarks (MCNC 89) and compared with the factoring results of MIS (Brayton 87) from Berkeley, and CATAMOUNT from the University of Washington (Stanion 94). The first one uses classical sum of product representations, and the latest the new Boolean techniques applied on BDDs as previously described. As we are preliminary concerned by the number of literals, we use this criteria to compare all methods. It has also the interest of being independent of the mapping process and the target library, and so fair comparisons can be drawn.

Results are shown in table 1. The first column gives the name of the benchmarks provided by MCNC. The 2 others give information on the I/O connection numbers. The MIS column provides results given in (Brayton 87), and the CAT (for CATAMOUNT) were given in (Stanion 94). COM gives the results for Compass. The ratio for MIS and CAT is computed as the difference of literals with COM divided by MIS and CAT respectively. It shows an average improvement of 16.7% in terms of literals compared with MIS, and 10.5% compared with CAT. Run times to proceed these benchmarks are in a range of 10% compared to Catamount which is considered as similar for a user point of view.

| Benchmark | MIS | CAT | COM | COM vs MIS (%) | COM vs CAT (%) |
|---|---|---|---|---|---|
| rd53 | 70 | 68 | 60 | -14.3 | -11.8 |
| misex1 | 86 | 88 | 75 | -12.8 | -14.8 |
| misex2 | 164 | 163 | 117 | -28.7 | -28.2 |
| f51m | 168 | 174 | 167 | -0.6 | -4.0 |
| 5xpl | 164 | 159 | 148 | -9.8 | -6.9 |
| z4ml | 69 | 67 | 56 | -18.8 | -16.4 |
| sao2 | 183 | 174 | 201 | 9.8 | 15.5 |
| 9sym | 258 | 116 | 139 | -46.1 | 19.8 |
| vg2 | 246 | 254 | 133 | -45.9 | -47.6 |
| rd73 | 192 | 173 | 178 | -7.3 | 2.9 |
| bw | 299 | 306 | 223 | -25.4 | -27.1 |
| 9symml | 277 | 278 | 127 | -54.2 | -54.3 |
| alupla | 205 | 182 | 176 | -14.1 | -3.3 |
| duke2 | 772 | [illegible] | 528 | -31.6 | -28.3 |
| misex3 | 1053 | 1010 | 1342 | 27.4 | 32.9 |
| rd84 | 381 | 319 | 149 | -60.9 | -53.3 |
| Sum average | 4587 | 4267 | 3819 | -16.7 | -10.5 |

**Table 1: Results**

We have also evaluated this method on industrial designs, which was in fact the main evaluation criteria of this approach for a CAD company. These designs have complexity ranging between 1,000 to 50,000 gates which cover in fact the standard design complexity that actual Logic Synthesis tools should handle. It appears that 10% of these designs could not be handle with standard SOP techniques when BDD completed in few minutes. Others have been drastically speed-up (up to 100 times), and the results were ranging from -6% upto 17%. In addition, this technique allows to easily compare the complexity of the function towards its complemented form, and thus keep the better one, still allowing further improvements.

## 6. CONCLUSION AND FUTURE WORK

We have proposed new multilevel synthesis techniques based on Binary Decision Diagrams and implicit cube manipulation. These techniques have been successfully applied on industrial designs due to the efficiency of our complete dynamic approach to build and apply our BDD core algorithm. This BDD core always controls the memory usage to prevent any explosion and also to speed up the complete process which is a novel and powerful approach allowing to synthesize and optimize larger designs. We have also discussed how this BDD core is linked to each synthesis process.

Both cpu time and memory usage can still be improved by using several BDDs with different input orderings instead of using only one. Then, in the future we will improve our candidate extraction and division algorithms in order to take advantages of partioning approaches. Even the very encouraging results we got with our BDD optimizer we think that we can still greatly improve it.

Next development will focus on the extraction of Boolean divisors directly from the topology of the BDD. Also, even if literals are good estimator, they cannot reflect well the complexity of all library cells provided by CAD vendors. For instance, the integration of Low Power issues privilege the development of new AOIs (and or inverter) cells for which the complexity is not related to the number of literals. So, we plan to interface our Optimizer with a Boolean Matching System to allow a more precise weighting and selection of candidate divisors. We believe that this development would considerably improve the quality of the designs by considering earlier the design goals (area, speed, or power).

## 7. REFERENCES

Akers (1978), "Functional Testing with Binary Decision Diagrams", *Proc. of the IEEE Conf. on Fault-Tolerant Comput., pp 75-82.*

Besson, Bouzouzou, Crastes, Floricica and Saucier (1992), "Synthesis on Multiplexer-based F.P.G.A. using Binary Decision Diagrams", *Proc. of ICCD 92.*

Brayton and al. (1984), "Logic Minimization Algorithms for VLSI Synthesis", *Kluwer Academic Publishers*, Boston.

Brayton and al. (Nov. 1987), "MIS: a Multi-Level LogicOptimization System", *IEEE Transaction on CAD*, pp.1062-1081.

Bryant (1986), "Graph-Based Algorithms for Boolean Function Manipulation", *IEEE Trans. on Computers*, Vol C-35, no.8, pp. 677-691.

Coudert and Madre (1992), Implicit and Incremental Computation of Prime and Essential Prime Implicants of Boolean Functions", *29th ACM/IEEE Design Automation Conference.*

Coudert and Madre(1993a), "Toward a Symbolic Logic Minimization Algorithm", *Proc. VLSI Design'93*, Bombay, India.

Coudert, Madre and Fraisse (1993b), "A New Viewpoint on Two-Level Minimization", *30th ACM/IEEE Design Automation* Conference.

jeong et al (1991)., "Variable Ordering For FSM Traversal", *Proc. of the International Workshop on Logic Synthesis*, MCNC.

Madre and Billon (1988), "Proving Circuit Correctness using Formal Comparison Between Expected and Extracted Behaviour", *25th ACM/IEEE Design Automation Conference* , pp. 205-210.

Madre (1990), "PRIAM : un outil de verification formelle de circuits integres digitaux", *These de l'ENS des* Telecommunications, Paris.

Murgai, Brayton, Sangiovanni-Vincentelli (1992), "An Improved Synthesis Algorithm for Multiplexor-based PGAs", *29th ACM/IEEE Design Automation Conference* , Anaheim, CA, June 1992, pp. 380-386.

MCNC (1989), Introduction to Synthesis Benchmarks, *second International Workshop on Logic Synthesis*, North Carolina USA, May 1989

MAINSAIL Language manual, XIDAK inc.

Minato (1992), 'Fast Generation of Irredundant Sum-of-Products Forms from Binary Decision Diagrams", *Proc of SASIMI '92*, pp. 64-73.

Minato (1993a), "Zero-Suppressed BDDs for Set Manipulation in Combinational Problems", *30th ACM/IEEE Design Automation* Conference.

Minato (1993b), "Fast weak division method for implicit cube representation", *proceeding SASIMI"93.*

Morreale (1970), "Recursive Operators for Prime Implicant and Irredundant Normal Form Determination", *IEEE Trans. on Computers,* 1970.

Murgai, Brayton , and Sangiovanni-Vincentelli (1992), "An Improved Synthesis Algorithm for Multiplexor-based PGAs", *29th* ACM/IEEE Design Automation Conference, Anaheim, CA, pp. 380-386.

Rudell (1993), "Dynamic Variable Ordering for Ordered Binary Decision Diagrams", *Proc. of International Workshop in Logic Synthesis,* CA.

Stanion and Sechen (1994), "Boolean division and factorization using BDDs", *IEEE Transactions on CAD* vol 13, N° 9, sept. 94, pp. 1179-1184.

Touati, Brayton and Kurshan (1991), "Testing Language Containment for w-automata using BDDs", *Formal Methods in VLSI* Design, Miami, ACM, New-York.

## 8. BIOGRAPHY

Franck Poirot has got his PhD thesis in 1990 from the INPG in France under the responsability of Prof. Saucier. Then, he joined VLSI Technology as Software Developer and Compass Design Automation as Logic Synthesis project leader. Afterwards, he was appointed Program Manager to follow up the technical management of European projects. He has published more than 20 papers in high rated conferences such as DAC, ICCAD, EDAC or Euro-ASIC, and held two patents issued by the Patent and Trade-Mark Office in USA, and two others are still in the review process.
His main interests focus in several CAD aspects such as Logic Synthesis, Test, Formal Verification, Simulation, and Low Power designs.

Gerard Tarroux received the Ph.D. degree in micro-electronic from the University of Montpellier (France) in 1991. Then, he joined Compass Design Automation as Software Developer in the Logic Synthesis team. His reasearch interests include many aspects of CAD for VLSI with special emphasis on logic synthesis, FSM synthesis, formal verification, testing and low power designs.

2

# Multiple-Level Logic Optimization with Boolean Relations

*Bernd Wurth**
*Institute of Electronic Design Automation*
*Technical University of Munich, 80290 Munich, Germany*
*Tel.: +49-89-2105-3646. Fax: +49-89-2105-3696.*
*email:* wurth@regent.e-technik.tu-muenchen.de

*Norbert Wehn*
*Siemens AG, Semiconductor Division*
*81549 Munich, Germany*
*Tel.: +49-89-4144-4619. Fax: +49-89-4144-4888.*
*email:* wehn@vincent.hl.siemens.de

**Abstract**

Exact and heuristic techniques are presented to calculate the Boolean relation for an arbitrary subcircuit in a multiple-level logic circuit with an external don't care set. We are not restricted to process subcircuits which are only driven by primary inputs. The new techniques keep BDD sizes small and therefore allow the calculation of Boolean relations for many circuits of nontrivial size which could not be dealt with before. The efficiency of the techniques is demonstrated on various benchmark circuits. The developed techniques are applied to multiple-level logic optimization.

**Keywords**

Boolean Relation, Observability Relation, Don't Cares, Multiple-Level Logic Optimization, Binary Decision Diagrams

---

* The first author was supported by an Ernst von Siemens-grant given by Siemens AG.

# 1 INTRODUCTION

The ability to calculate and exploit *don't care* conditions has long been recognized an important technique in multiple-level logic synthesis. A variety of methods has been presented (Bartlett, 1988; Muroga, 1989; Damiani, 1993). Whereas a *don't care* set is calculated for a single node and specifies all the flexibility for implementing the node's function, *Boolean relations* describe all the flexibility for implementing an arbitrary multiple-output subcircuit. It has been pointed out before (Brayton, 1990) that Boolean relations are superior to don't care sets as don't care sets cannot capture the complete flexibility for implementing a multiple-output subcircuit. The reason is that for the calculation of a node's maximal don't care set, the functions of all the other nodes are not allowed to change. Boolean relations, however, allow the simultaneous modification of all nodes of a subcircuit. The example of Figure 1 illustrates the advantage of using Boolean relations.

Figure 1 shows a small section of a multiple-level logic circuit. The primary inputs and outputs of the circuit are omitted. We consider the subcircuit consisting of gate $G1$ and $G2$. The table on the right side specifies the output vectors which may be computed by the subcircuit if the respective input minterm is applied. This table represents a Boolean relation. For an input minterm with several output vectors, the values of the circuit's primary outputs do not depend on the output vector actually computed by the subcircuit. This flexibility, expressed by the Boolean relation, is due to the embedding of the subcircuit in a larger circuit. Here, the only degree of freedom exists for the input minterm $(1, 1)$, as the subcircuit may compute either the vector $(0, 1)$ or $(1, 0)$. The actual implementation of the subcircuit computes the vector $(0, 1)$. If the subcircuit is modified to compute the vector $(1, 0)$ instead, gate $G1$ is replaced by an XNOR gate, and gate $G2$ by constant 0. Gate $G3$ can be replaced by a wire from $G4$, thus yielding a total gain of two gates. It is important to note that the subcircuit cannot be optimized by don't care methods because two node functions have to be modified at the same time to exploit the degree of freedom. Methods for the minimization of Boolean relations have been developed which find good two-level implementations (Watanabe, 1991). We, in contrast, are concerned with the calculation of the Boolean relation for a subcircuit.

Techniques have been presented (Chen, 1992) which calculate Boolean relations for multiple-output subcircuits in terms of primary input variables. Other methods (Cerny, 1977; Savoj, 1993) calculate the Boolean relation of a subcircuit in terms of internal

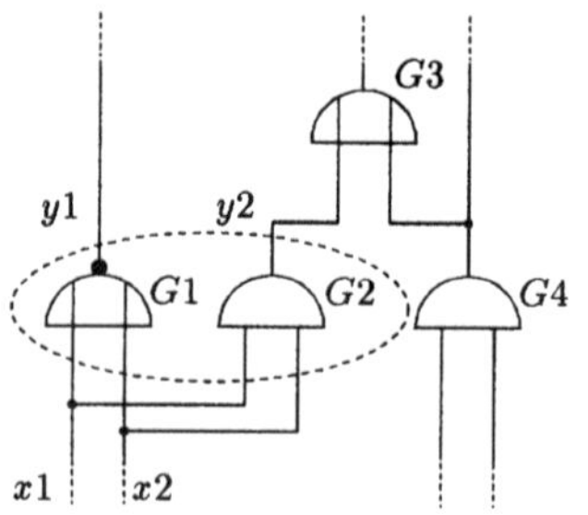


| input minterm $(x1, x2)$ | allowed output vector $(y1, y2)$ |
|---|---|
| $(0, 0)$ | $(1, 0)$ |
| $(0, 1)$ | $(0, 0)$ |
| $(1, 0)$ | $(0, 0)$ |
| $(1, 1)$ | $(0, 1), (1, 0)$ |

**Figure 1** Boolean relation for subcircuit $(G1, G2)$.

network variables. With regard to the specification of a subcircuit, the latter methods are therefore more general. Cerny and Marin laid the foundation for a theory of Boolean relations (Cerny, 1977). They present formulas to compute maximal Boolean relations for the subcircuits of a serially or parallelly two-partitioned circuit. These techniques can be used to calculate the Boolean relation for an arbitrary subcircuit. Recently, Savoj and Brayton expanded on these ideas and presented algorithms to compute maximal and sub-optimal ("compatible") Boolean relations (Savoj, 1993).

In this paper we focus on the problem of calculating maximal Boolean relations for subcircuits in a multiple-level logic circuit to use them for optimization. External don't care sets are exploited. A Boolean relation is expressed in terms of the subcircuit's outputs and inputs. New exact and heuristic techniques are presented for the calculation of Boolean relations. These methods are applicable to nontrivial circuits. We present results for the calculation and exploitation of Boolean relations.

The article is organized as follows. In Section 2 the basic terminology used throughout the paper is explained. Techniques to compute Boolean relations are discussed in Section 3. The application to multiple-level logic optimization is described in Section 4. We give conclusion and directions for future work in Section 5.

# 2 TERMINOLOGY

This section introduces some terminology for BDD (Binary Decision Diagrams) operations and Boolean relations. For the definition of a Boolean network, which is used to model a combinational circuit, we refer to standard literature (Brayton, 1990).

## 2.1 Boolean functions

We are dealing with Boolean functions $f : B^n \rightarrow B$, where $B = \{0,1\}$. Let $\mathbf{x} = (x_1, \ldots, x_n)$ denote a vector of Boolean variables and $\mathbf{x}^k \in B^n$ the $k^{th}$ minterm, i.e., $\mathbf{x}^0 = (0, \ldots, 0)$, $\mathbf{x}^1 = (0, \ldots, 0, 1)$ etc. The number of elements of a vector is denoted by $|\mathbf{x}|$, here $|\mathbf{x}| = n$. A vector of Boolean functions is given by $\mathbf{f} = (f_1, \ldots, f_m)$. We will denote the Boolean function obtained from $f(\mathbf{x}, \mathbf{y})$ by assigning the minterm $\mathbf{x}^k$ as $h^k(\mathbf{y})$, i.e., $h^k(\mathbf{y}) = f(\mathbf{x}^k, \mathbf{y})$.

The *consensus operator* of a Boolean function $f(\mathbf{x}, \mathbf{y})$ with respect to the vector $\mathbf{x}$ is given by $\mathcal{C}_{\mathbf{x}} f = h^0(\mathbf{y}) \cdot h^1(\mathbf{y}) \cdot \ldots \cdot h^{2^n-1}(\mathbf{y}) = \prod_{k=0}^{2^n-1} h^k(\mathbf{y})$.

The *smoothing operator* of a Boolean function $f(\mathbf{x}, \mathbf{y})$ with respect to the vector $\mathbf{x}$ is given by $\mathcal{S}_{\mathbf{x}} f = h^0(\mathbf{y}) + h^1(\mathbf{y}) + \ldots + h^{2^n-1}(\mathbf{y}) = \sum_{k=0}^{2^n-1} h^k(\mathbf{y})$. The consensus and smoothing operations can be computed efficiently if applied to BDD representations of Boolean functions.

## 2.2 Boolean relations

A *Boolean relation* $R$ is a binary relation from $B^n$ to $B^m : R \subseteq B^n \times B^m$. We say that an output vector $\mathbf{y}^l \in B^m$ is *allowed* for an input minterm $\mathbf{x}^k \in B^n$ iff $(\mathbf{x}^k, \mathbf{y}^l) \in R$, e.g., two output vectors are allowed for the input minterm $(1,1)$ of the Boolean relation in Figure 1. For each Boolean relation $R$ there is a *characteristic function* $\Phi : B^n \times B^m \rightarrow B$ such that $\Phi(\mathbf{x}^k, \mathbf{y}^l) = 1$ iff $(\mathbf{x}^k, \mathbf{y}^l) \in R$. As an example, the characteristic function of the

Boolean relation in Figure 1 is $\Phi(\mathbf{x},\mathbf{y}) = \overline{x_1}\cdot\overline{x_2}\cdot y_1\cdot\overline{y_2} + \overline{x_1}\cdot x_2\cdot\overline{y_1}\cdot\overline{y_2} + x_1\cdot\overline{x_2}\cdot\overline{y_1}\cdot\overline{y_2} + x_1\cdot x_2\cdot\overline{y_1}\cdot y_2 + x_1\cdot x_2\cdot y_1\cdot\overline{y_2}$.

We are dealing with characteristic functions of Boolean relations and represent them by BDDs. Because of the close correspondence between Boolean relations and characteristic functions, we sometimes use the term Boolean relation for a characteristic function. If not stated otherwise, the terms Boolean relation and characteristic function refer to a *maximal* Boolean relation, i.e., no input minterm/output vector-pair could be added to the Boolean relation without violating the circuit's specification.

Let a network $N$ implement a Boolean function $\mathbf{z} = \mathbf{f}(\mathbf{x})$. The characteristic function which is the relational representation of the multiple-output function $\mathbf{z}$ is

$$\Phi_N^I(\mathbf{x},\mathbf{z}) = \prod_{i=1}^{|\mathbf{Z}|} (z_i \overline{\oplus} f_{z_i}(\mathbf{x}))\,. \tag{1}$$

A Boolean function $\mathbf{f}: B^n \to B^m$ is *compatible* with $R$ iff $\forall \mathbf{x}^k \in B^n$, $(\mathbf{x}^k, \mathbf{f}(\mathbf{x}^k)) \in R$. The goal of Boolean relation minimization (Watanabe, 1991) is to find a function with minimum cost which is compatible with $R$.

## 3 NEW TECHNIQUES TO COMPUTE BOOLEAN RELATIONS

In this section, new techniques to calculate the Boolean relation for a subcircuit are derived. These techniques exploit the fact that in many practical cases a combinational circuit's specification is not an arbitrary relation, but a multiple-output function plus an *external don't care* set. An *external don't care* set denotes all input vectors that are never produced by the environment at the circuit's inputs. For example, during the synthesis of the combinational logic of a finite state machine, the set of unreachable states can be used as an external don't care set. Existing methods (Cerny, 1977; Savoj, 1993) handle arbitrary relational specifications. This, however, implies unnecessarily expensive computations in the above-mentioned case of a specification with an external don't care set.

Our new exact technique allows the computation of maximal Boolean relations in much larger circuits than existing methods if the circuit is specified with or without an external don't care set. This is achieved by the application of a new formula for the calculation of the *satisfiability relation*, yielding smaller BDD representations, and the use of efficient image computation techniques. Experimental results are given to compare the new techniques with existing approaches.

### 3.1 Satisfiability relation

In (Cerny, 1977) formulas are given to calculate the characteristic functions for the subcircuits in a serially or parallelly two-partitioned circuit. Two formulas needed in the sequel are restated here without explanation. A network $N$, serially decomposed into the subnetworks $L$ and $U$ (lower and upper part) is given, see Figure 2(a). The characteristic functions for $L$, $U$ and $N$ are $\Phi_L(\mathbf{x},\mathbf{y})$, $\Phi_U(\mathbf{y},\mathbf{z})$ and $\Phi_N(\mathbf{x},\mathbf{z})$.

If $\Phi_N(\mathbf{x},\mathbf{z})$ and $\Phi_U(\mathbf{y},\mathbf{z})$ are given and $\Phi_U = \Phi_U^I$ according to equation (1), the maximal Boolean relation $\Phi_{L_{max}}(\mathbf{x},\mathbf{y})$ is

$$\Phi_{L_{max}}(\mathbf{x},\mathbf{y}) = \mathcal{S}_{\mathbf{z}}\left(\Phi_N(\mathbf{x},\mathbf{z}) \cdot \Phi_U^I(\mathbf{y},\mathbf{z})\right). \tag{2}$$

As this Boolean relation follows from the restricted observability of the outputs of $L$ at the primary outputs, it is called *observability relation.* The observability relation determines for each minterm $\mathbf{x}$ a set of allowed vectors $\mathbf{y}$, for which the upper part $U$ computes allowed values at the primary outputs $\mathbf{z}$.

If $\Phi_N(\mathbf{x},\mathbf{z})$ and $\Phi_L(\mathbf{x},\mathbf{y})$ are given, the maximal Boolean relation $\Phi_{U_{max}}(\mathbf{y},\mathbf{z})$ is calculated as

$$\Phi_{U_{max}}(\mathbf{y},\mathbf{z}) = \mathcal{C}_{\mathbf{x}}\left(\Phi_N(\mathbf{x},\mathbf{z}) + \overline{\Phi_L(\mathbf{x},\mathbf{y})}\right). \tag{3}$$

As this Boolean relation follows from the restricted satisfiability of the inputs of $U$ from the primary inputs, it is called *satisfiability relation.*

Whereas equation (2) exploits the fact that $\Phi_U = \Phi_U^I$, formula (3) holds for general Boolean relations $\Phi_L$. We will use a formula for the satisfiability relation which exploits $\Phi_L = \Phi_L^I$.

**Lemma 1** *If $\Phi_U(\mathbf{y},\mathbf{z})$ and $\Phi_L^I(\mathbf{x},\mathbf{y})$ are given, and $\Phi_N(\mathbf{x},\mathbf{z}) = \mathcal{S}_{\mathbf{y}}(\Phi_L^I(\mathbf{x},\mathbf{y}) \cdot \Phi_U(\mathbf{y},\mathbf{z}))$, the maximal Boolean relation $\Phi_{U_{max}}(\mathbf{y},\mathbf{z})$ is calculated as*

$$\Phi_{U_{max}}(\mathbf{y},\mathbf{z}) = \Phi_U(\mathbf{y},\mathbf{z}) + \overline{\mathcal{S}_{\mathbf{x}}\left(\Phi_L^I(\mathbf{x},\mathbf{y})\right)}. \tag{4}$$

*Proof.* We restate the lemma in the following way: Iff $\mathbf{z}^l$ is an output vector allowed for the input minterm $\mathbf{y}^k$, then $\Phi_{U_{max}}(\mathbf{y}^k,\mathbf{z}^l) = 1$. There are two cases: First, the input minterm $\mathbf{y}^k$ is computed by $L$ for some primary input vector, i.e., $\mathbf{y}^k$ is satisfiable. Then no additional degree of freedom is added to the Boolean relation of $U$ by the presence of subnetwork $L$. We have $\Phi_{U_{max}}(\mathbf{y}^k,\mathbf{z}^l) = 1$ iff $\Phi_U(\mathbf{y}^k,\mathbf{z}^l) = 1$ because the second term in equation (4) evaluates to 0. Second, the input minterm $\mathbf{y}^k$ cannot be computed by $L$, i.e., $\mathbf{y}^k$ is not satisfiable. Then any vector $\mathbf{z}^l$ is an allowed output of $U$. It holds that $\Phi_{U_{max}}(\mathbf{y}^k,\mathbf{z}^l) = 1$ because the second term in equation (4) evaluates to 1. □

This formula, which can also be applied if an external don't care set is present, is relevant for multiple-level logic optimization because during resynthesis of $U$ some network $L$ is given. Consequently, $\Phi_L = \Phi_L^I$, which is exploited by equation (4). Using equation (4) instead of equation (3) has significant practical advantages as will be shown below.

## 3.2 Boolean relation of a subcircuit

Our goal is to calculate the Boolean relation for an arbitrary subcircuit or *block* of a Boolean network $N'$. Through structural analysis of network $N'$, a subnetwork $N$ is obtained which contains all nodes functionally related to the considered block. The network $N$ can be further divided into three subnetworks, i.e., the block $B$, the lower part $L$ and the upper part $U$. The subnetwork $U$ consists of all the nodes which may influence the

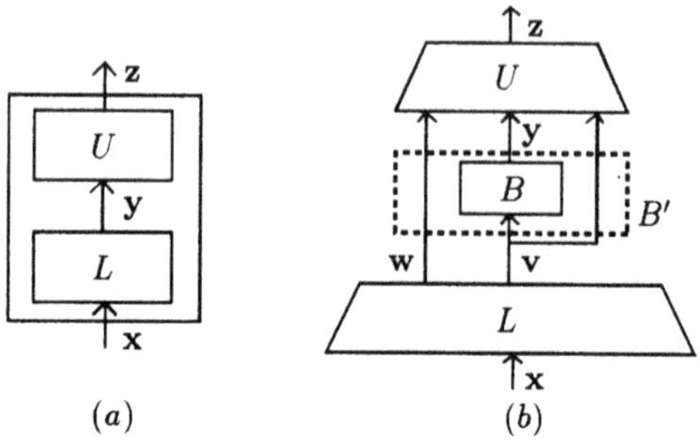


**Figure 2** Decomposed network $N$.

*observability* of the block's outputs at the primary outputs. Likewise, the subnetwork $L$ holds all the nodes which may influence the *satisfiability* of the block inputs or the inputs of $U$. Figure 2(b) shows the network $N$ with its three parts. As the figure illustrates, block inputs can directly go into $U$, and some outputs of $L$ may directly feed $U$.

For notational convenience, the same symbol, e.g. $v_i$, is used to denote a node and its variable in the Boolean network. We will denote the Boolean function of a node $w_i$ in terms of the primary inputs $\mathbf{x}$ by $f_{w_i}(\mathbf{x})$. The Boolean functions of other nodes are denoted analogously. For each primary output $z_i$ two Boolean functions, $f^a_{z_i}$ and $f^b_{z_i}$, are needed in the sequel. The function $f^a_{z_i}$ is the Boolean function of the primary output node $z_i$ in terms of the variables $\mathbf{v}$ and $\mathbf{w}$, $f^a_{z_i}(\mathbf{v},\mathbf{w})$. The function $f^b_{z_i}$ is the function of $z_i$ in terms of the variables $\mathbf{v},\mathbf{w}$ and $\mathbf{y}$, $f^b_{z_i}(\mathbf{v},\mathbf{w},\mathbf{y})$.

The new technique for the calculation of a Boolean relation is based on the following theorem.

**Theorem 1** *Given a block $B$ in a network $N$, which may have an external don't care set $exdc(\mathbf{x})$, the characteristic function of the maximal Boolean relation of $B$ is given by*

$$\Phi_{B_{max}}(\mathbf{v},\mathbf{y}) = \mathcal{C}_{\mathbf{W}}\left(\Phi_{obs}(\mathbf{v},\mathbf{w},\mathbf{y}) + \overline{\Phi_{sat}(\mathbf{v},\mathbf{w})}\right), \tag{5}$$

*with the observability relation* $\Phi_{obs}(\mathbf{v},\mathbf{w},\mathbf{y}) = \prod_{i=1}^{|\mathbf{Z}|}\left(f^a_{z_i}(\mathbf{v},\mathbf{w})\overline{\oplus}f^b_{z_i}(\mathbf{v},\mathbf{w},\mathbf{y})\right)$, *and the satisfiability function* $\Phi_{sat}(\mathbf{v},\mathbf{w}) = \mathcal{S}_{\mathbf{X}}\left(\overline{exdc(\mathbf{x})}\cdot\prod_{i=1}^{|\mathbf{V}|}(v_i\overline{\oplus}f_{v_i}(\mathbf{x}))\cdot\prod_{i=1}^{|\mathbf{W}|}(w_i\overline{\oplus}f_{w_i}(\mathbf{x}))\right)$.

*Proof.* Combination of equation (2) and equation (4) yields the Boolean relation for $B'$ (see Figure 2(b)):

$$\Phi_{B'_{max}}(\mathbf{v},\mathbf{w},\mathbf{y}) = \mathcal{S}_{\mathbf{Z}}\left(\Phi_{BU}(\mathbf{v},\mathbf{w},\mathbf{z})\cdot\Phi_U(\mathbf{v},\mathbf{w},\mathbf{y},\mathbf{z})\right) + \overline{\mathcal{S}_{\mathbf{X}}\left(\Phi_L(\mathbf{x},\mathbf{v},\mathbf{w})\right)}. \tag{6}$$

Using $\Phi_{BU} = \Phi^I_{BU}$ of the cascade of $B$ and $U$, $\Phi_{BU}(\mathbf{v},\mathbf{w},\mathbf{z}) = \prod_{i=1}^{|\mathbf{Z}|}\left(z_i\overline{\oplus}f^a_{z_i}(\mathbf{v},\mathbf{w})\right)$, and similarly $\Phi_U = \Phi^I_U$, $\Phi_U(\mathbf{v},\mathbf{w},\mathbf{y},\mathbf{z}) = \prod_{i=1}^{|\mathbf{Z}|}\left(z_i\overline{\oplus}f^b_{z_i}(\mathbf{v},\mathbf{w},\mathbf{y})\right)$, we obtain for the first term on the right side of equation (6) the *observability relation*:

$$\begin{aligned}\Phi_{obs}(\mathbf{v},\mathbf{w},\mathbf{y}) &= \mathcal{S}_{\mathbf{Z}}\left(\Phi_{BU}(\mathbf{v},\mathbf{w},\mathbf{z})\cdot\Phi_U(\mathbf{v},\mathbf{w},\mathbf{y},\mathbf{z})\right)\\ &= \mathcal{S}_{\mathbf{Z}}\left(\prod_{i=1}^{|\mathbf{Z}|}\left(z_i\overline{\oplus}f^a_{z_i}(\mathbf{v},\mathbf{w})\right)\cdot\prod_{i=1}^{|\mathbf{Z}|}\left(z_i\overline{\oplus}f^b_{z_i}(\mathbf{v},\mathbf{w},\mathbf{y})\right)\right)\\ &= \prod_{i=1}^{|\mathbf{Z}|}\left(f^a_{z_i}(\mathbf{v},\mathbf{w})\overline{\oplus}f^b_{z_i}(\mathbf{v},\mathbf{w},\mathbf{y})\right).\end{aligned}$$

For the (not complemented) second term in equation (6), use of $\Phi_L = \Phi_L^I$ according to equation (1) and consideration of external don't cares yields the *satisfiability function* $\Phi_{sat}$. To obtain $\Phi_{B_{max}}$ from $\Phi_{B'_{max}}$, formula (8) in (Cerny, 1977) is applied: $\Phi_{B_{max}}(\mathbf{v},\mathbf{y}) = \mathcal{C}_{\mathbf{W}}(\Phi_{B'_{max}}(\mathbf{v},\mathbf{w},\mathbf{y}))$. □

Note that the onset of $\Phi_{sat}$ is the image of the possible primary input minterms, which are in the offset of $exdc(\mathbf{x})$, by the functions $\mathbf{f}_w, \mathbf{f}_v$. For comparison, equation (7) shows how the Boolean relation for block $B$ is computed in (Cerny, 1977) and similarly in (Savoj, 1993),

$$\Phi_{B_{max}}(\mathbf{v},\mathbf{y}) = \mathcal{C}_{\mathbf{W},\mathbf{X}}\left(\Phi_{LB}(\mathbf{x},\mathbf{v},\mathbf{w},\mathbf{y}) + \overline{\Phi_L^I(\mathbf{x},\mathbf{v},\mathbf{w})}\right), \tag{7}$$

where the characteristic function for the cascade of $B$ and $L$ is denoted by $\Phi_{LB}$. The derivation of equation (7) is given in the appendix. Note that equation (7) has also been used in (Kukimoto, 1992) in the context of design rectification for FPGAs.

Despite their structural similarity, the two formulas (5) and (7) differ in several important aspects: First, the observability relation $\Phi_{obs}$ in formula (5) depends on the internal variables $\mathbf{v}, \mathbf{w}, \mathbf{y}$. The corresponding function $\Phi_{LB}$ in formula (7) additionally depends on the primary input variables $\mathbf{x}$. Furthermore, the satisfiability function $\Phi_{sat}$ in formula (5) depends on the internal variables $\mathbf{v}, \mathbf{w}$ only, whereas the function $\Phi_L$ in formula (7) also depends on the variables $\mathbf{x}$. In the worst case, if the block is located at the primary outputs, $\Phi_L$ and $\Phi_{LB}$ represent the functionality of almost the complete network $N$. As a consequence, the BDD representation of the argument of the consensus operation is much larger for formula (7) than for formula (5). Finally, to calculate the satisfiability function $\Phi_{sat}$ in formula (5), the characteristic function $\Phi_L$ need not be calculated explicitly. This is achieved by the use of image computation techniques, which have originally been developed for automata verification (Coudert, 1990). The satisfiability function is computed using image computation techniques as shown in (Savoj, 1991). Note that $\Phi_L$ has to be calculated explicitly in formula (7).

These differences are essential, because during the calculation of the block's characteristic function according to formula (5) much smaller BDDs are created. This enables us to handle larger circuits. Due to smaller BDD sizes, the calculation according to the new technique also proves to be significantly faster. However, since our technique is based on the exploitation of a special kind of specification (multiple-output function plus external don't care set), we cannot deal with arbitrary relational specifications.

We have implemented both methods according to the new formula (5) and formula (7)

**Table 1** Comparison of Boolean relation computation techniques

| *Circuit* | *In* | *Out* | *Literals* | *BDD nodes* | | | *CPU time in sec* | | |
|---|---|---|---|---|---|---|---|---|---|
| | | | | *previous* | *new* | *Ratio* | *previous* | *new* | *Ratio* |
| apex7 | 49 | 37 | 351 | 52776 | 12387 | 0.23 | 413.4 | 37.6 | 0.09 |
| b9 | 41 | 21 | 256 | NF | 65846 | - | - | 27.1 | - |
| c8 | 28 | 18 | 301 | 2450 | 826 | 0.34 | 3.6 | 1.4 | 0.39 |
| cordic | 23 | 2 | 194 | 32327 | 1987 | 0.06 | 163.7 | 5.5 | 0.03 |
| count | 35 | 16 | 174 | 1392 | 524 | 0.38 | 3.3 | 1.6 | 0.48 |
| example2 | 85 | 66 | 432 | 124280 | 17346 | 0.14 | 64.2 | 32.0 | 0.50 |
| f51m | 8 | 8 | 319 | 70 | 70 | 1.0 | 0.1 | 0.1 | 1.0 |
| i4 | 192 | 6 | 340 | NF | 1158 | - | - | 4.4 | - |
| k2 | 45 | 45 | 3047 | 131533 | 20961 | 0.16 | 868.6 | 248.0 | 0.29 |
| lal | 26 | 19 | 257 | 1101 | 799 | 0.73 | 2.7 | 2.4 | 0.89 |
| pcler8 | 27 | 17 | 102 | 102583 | 28187 | 0.27 | 47.5 | 8.3 | 0.17 |
| term1 | 34 | 10 | 977 | NF | 149592 | - | - | 139.7 | - |
| ttt2 | 24 | 21 | 592 | 2307 | 2034 | 0.88 | 5.9 | 3.8 | 0.64 |
| s298 | 17 | 20 | 244 | 143862 | 6227 | 0.04 | 200.9 | 16.8 | 0.08 |
| s344 | 24 | 26 | 269 | NF | 47622 | - | - | 42.2 | - |
| s444 | 24 | 27 | 352 | NF | 35460 | - | - | 89.7 | - |
| s526 | 24 | 27 | 445 | NF | 57609 | - | - | 284.0 | - |
| s641 | 54 | 42 | 539 | NF | 63027 | - | - | 1751.2 | - |

using the same shared ROBDD package, which is based on (Brace, 1990). For the ordering of the BDD variables, a simple depth first search heuristic was used.

Table 1 shows results for the computation of Boolean relations for some MCNC benchmark circuits. The first column gives the name of the circuit. The next two columns give the number of primary inputs and outputs. For sequential circuits, the number of flip-flops is added to the number of inputs and outputs. The fourth column shows the circuit size in literals. The next two columns show the maximum number of BDD nodes during the calculation of a block's Boolean relation, where **previous** denotes the method based on formula (7), **new** the new technique based on formula (5). The following column **Ratio** shows the ratio between the BDD sizes obtained with the new and the previous method. The last three columns show the CPU time in seconds for the two different methods and again the ratio. An entry NF signifies that the shared ROBDD grew larger than 150000 nodes. Calculation was aborted at that point. The circuits have been partitioned into two-output blocks using a heuristic described in the next section. All experiments have been performed on a DECstation 5000/200.

There are some circuits, e.g. f51m, where both methods yield identical maximal BDD sizes. This happens when the characteristic functions $\Phi_{B'}$ of the blocks and not the intermediate functions have the largest BDD representation.

The results emphasize two aspects. First, Boolean relations can generally be calculated with significantly smaller BDDs and also much faster using the new technique. Many circuits, e.g. almost all sequential circuits in Table 1, cannot be handled by the previously known method. Secondly, the BDDs for large circuits are usually still large. Even the new method fails for the large sequential circuits not given in Table 1. This is mainly due to the large number of outputs $w_i$ of subnetwork $L$ which directly feed $U$.

To tackle this problem, we propose a heuristic. The functions $f_{w_i}$ involved in the image computation strongly influence BDD sizes and computation time. Therefore, the outputs $w_i$ of the lower part $L$ are partitioned into two sets $W'$ and $W''$, where $W'$ contains a prespecified number of nodes which are structurally "close" to the block. By calculating the image only for the functions of the nodes in $V$ and $W'$, it is assumed that the functions of the nodes in $W''$ may produce any vector. This simplification decreases the result quality. The Boolean relations are not maximal any more, but their computation is in general sped up and becomes possible for large circuits. Note that Table 1 gives only results for the calculation of maximal Boolean relations.

# 4 APPLICATION TO MULTIPLE-LEVEL LOGIC OPTIMIZATION

We apply the new techniques to multiple-level logic optimization. Our algorithm is as follows. A network is first partitioned into blocks. The nodes in a block have to be "related" to one another in order to successfully exploit Boolean relations. We cluster nodes such that the nodes of a block have a large number of common successors and predecessors. For each block, the maximal Boolean relation according to formula (5) is calculated and exploited.

To exploit the Boolean relation, the optimal function compatible with the Boolean relation has to be found. Our cost function is the number of literals. As we do not have a Boolean relation minimizer at hand, we find a good implementation by exhaustively optimizing compatible functions with ESPRESSO (Brayton, 1994). First, however, we check if the Boolean relation allows to set a node function to constant 0 or 1, or to replace one block output function by another one. These situations usually yield high gain.

The algorithm was applied to a number of sequential benchmark circuits. First, for each circuit the set of unreachable states was calculated. These unreachable states served as external don't care set during optimization. The circuits were preprocessed by the first part of the SIS script *script.rugged*, which basically performs cube and kernel extraction. The circuits after preprocessing are the initial circuits.

We simplified the initial circuits with our algorithm. For comparison, the rest of the SIS script *script.rugged* was applied. Whereas our algorithm calculates and exploits Boolean relations, the command *full_simplify*, which is called in the script *script.rugged*, employs node simplification by don't care methods. Table 2 shows the obtained results. The column **Init** lists the number of literals of the initial circuits. The column **SIS 1.1** shows the results obtained by *script.rugged*, the column **BooRel** the number of literals obtained by our algorithm.

The size of the initial circuits could be reduced significantly using Boolean relations. BooRel performs better than SIS in terms of literals. However, CPU times are larger for BooRel because the computation of Boolean relations is more expensive than the computation of don't care sets. We believe that even better results can be obtained if the following problems are resolved.

First of all, the problem of partitioning a circuit has to be examined more thoroughly. Experiments showed that the optimization quality is very sensitive to the partitioning. Furthermore, a Boolean relation minimizer is necessary.

**Table 2** Optimization results (in literals)

| *Circuit* | *In* | *Out* | *Init* | *SIS 1.1* | *BooRel* |
|---|---|---|---|---|---|
| s298 | 17 | 20 | 126 | 97 | 89 |
| s344 | 24 | 26 | 154 | 142 | 137 |
| s382 | 24 | 27 | 166 | 150 | 137 |
| s386 | 13 | 13 | 136 | 120 | 111 |
| s400 | 24 | 27 | 160 | 143 | 136 |
| s444 | 24 | 27 | 158 | 140 | 138 |
| s526 | 24 | 27 | 222 | 154 | 135 |
| s641 | 54 | 42 | 193 | 185 | 153 |

## 5 CONCLUSION

We have developed techniques to calculate the Boolean relation for a multiple-output subcircuit with arbitrary inputs. We exploit the fact that in many cases the circuit is specified by a multiple-output function and an external don't care set, which is a special relational specification. Experimental results show the superiority of our technique in terms of computation time and space to previous approaches. With the new technique, the BDDs remain much smaller during the calculation of characteristic functions. This enables us to apply the technique to many circuits of non-trivial size which could not be dealt with before. We applied the computation of Boolean relations to multiple-level logic optimization. Other possible applications are design rectification for FPGAs (Kukimoto, 1992) and technology mapping. Ongoing research concerns the partitioning of circuits into subcircuits, which has shown a significant influence on the quality of optimization results.

## 6 APPENDIX

With equation (2), the characteristic function $\Phi_N$ for the complete network, and the characteristic function $\Phi_U^I$ according to equation (1) for the upper part, we get the characteristic function for the subnetwork consisting of $B$ and $L$:

$$\Phi_{LB}(\mathbf{x},\mathbf{v},\mathbf{w},\mathbf{y}) = \mathcal{S}_{\mathbf{z}}\left(\Phi_N(\mathbf{x},\mathbf{z})\cdot\Phi_U^I(\mathbf{v},\mathbf{w},\mathbf{y},\mathbf{z})\right).$$

With formula (3), formula (8) in (Cerny, 1977), and the characteristic function according to equation (1) for the lower part, $\Phi_L^I$, formula (7) is obtained:

$$\Phi_{B_{max}}(\mathbf{v},\mathbf{y}) = \mathcal{C}_{\mathbf{w},\mathbf{x}}\left(\Phi_{LB}(\mathbf{x},\mathbf{v},\mathbf{w},\mathbf{y}) + \overline{\Phi_L^I(\mathbf{x},\mathbf{v},\mathbf{w})}\right).$$

## 7 REFERENCES

Bartlett, K. A.; Brayton, R. K.; Hachtel, G. D.; Jacoby, R. M.; Morrison, C. R.; Rudell, R. L.; Sangiovanni-Vincentelli, A. and Wang, A. R. (1988) Multilevel logic simplification using implicit don't cares. *IEEE Transactions on Computer-Aided Design*, **7**, 723-40.

Brace, K. S.; Rudell, R. L. and Bryant, R. E. (1990) Efficient implementation of a BDD package. *27th ACM/IEEE Design Automation Conference DAC*, 40-5.

Brayton, R. K.; Hachtel, G. D.; McMullen, C. T. and Sangiovanni-Vincentelli, A. L. (1984) *Logic minimization algorithms for VLSI synthesis*. Kluwer Academic Publishers, Boston, MA.

Brayton, R. K.; Hachtel, G. D. and Sangiovanni-Vincentelli, A. L. (1990) Multilevel logic synthesis. *Proceedings of the IEEE*, **78**, 264-300.

Cerny, E. and Marin, M. A. (1977) An approach to unified methodology of combinational switching circuits. *IEEE Transactions on Computers*, **26**, 745-56.

Chen, K.-C. and Fujita, M. (1992) Efficient sum-to-one subsets algorithm for logic optimization. *29th ACM/IEEE Design Automation Conference, DAC* 443-8.

Coudert, O. and Madre, J. C. (1990) A unified framework for the formal verification of sequential circuits. *IEEE/ACM International Conference on Computer-Aided Design ICCAD*, 126-9.

Damiani, M. and Micheli, G. D. (1993) Don't care set specification in combinational and synchronous logic circuits. *IEEE Transactions on Computer-Aided Design*, **12**, 365-88.

Kukimoto, Y. and Fujita, M. (1992) Rectification method for lookup-table type FPGAs. *IEEE/ACM International Conference on Computer-Aided Design ICCAD*, 54-61.

Muroga, S.; Kambayashi, Y.; Lai, H. C. and Culliney, J. N. (1989) The transduction method - design of logic networks based on permissible functions. *IEEE Transactions on Computers*, **38**, 1404-24.

Savoj, H.; Brayton, R. K. and Touati, H. J. (1991) Extracting local don't cares for network optimization. *IEEE International Conference on Computer-Aided Design ICCAD*, 514-7.

Savoj, H. and Brayton, R. K. (1993) Observability relations for multi-output nodes. *Proceedings International Workshop on Logic Synthesis*

Watanabe, Y. and Brayton, R. K. (1991) Heuristic minimization of multiple-valued relations. *IEEE/ACM International Conference on Computer-Aided Design ICCAD*, 129-32.

## 8 BIOGRAPHY

Bernd Wurth, Ernst von Siemens scholar, is working towards his Ph.D. degree at the Department of Electrical Engineering of the Technical University of Munich, Germany. He studied Electrical Engineering at the University of Illinois at Urbana-Champaign and the Technical University of Munich, Germany, where he received a Dipl.-Ing. degree in 1992. His research interests are in logic synthesis and optimization and in test methods.

Norbert Wehn received the engineering degree and the Ph.D. degree in electrical engineering from the Darmstadt University of Technology, Germany, in 1984 and 1989, respectively. Since 1984 he has been working in the VLSI group at the Darmstadt University of Technology at CAD for physical design problems, defect-tolerant circuits and architectural synthesis. In 1991, he joined Siemens and is now working in the Semiconductor Division on multimedia. Dr. Wehn has authored or coauthored more than 20 papers on international conferences and journals in the cited fields.

# BONSAI : A pragmatic approach to Logic Synthesis and Formal Verification

*H.N. Nguyen and L. Ducousso*

*Hardware Development Paris-Angers*
***BULL S.A.** 78340 Les Clayes-sous-Bois, FRANCE*
*Tel : +33.1.30.80.60.19, email : H.N.Nguyen@frcl.bull.fr*

## Abstract

This paper describes the large scale application of logic synthesis and formal verification to the design of the CPU and caches of the high-end series of the Bull DPS7000 mainframe family. The logic CAD suite used for supporting the design of this system proved its efficiency on very complex integrated circuits. The key feature of this logic design environment is the methodology that integrates a set of logic synthesis and formal verification techniques to build an effective logic-design system to support the design of high-performance circuits.

## Keywords

Computer-Aided Design, Logic Synthesis, Formal Verification.

## 1. INTRODUCTION AND MOTIVATION

This paper describes the original computer-aided logic design approach used in developing the high range version of the Bull DPS7000 mainframe family. The task at hand was the design of a family of new high performance mainframes, based on CISC architecture, using a submicron CMOS technology in a semi-custom, very large scale integrated chip technology. Aggressive performance goals and shorter design time than prior projects of similar complexity were our main objectives.

Design tools development at Bull S.A., during the previous several years, demonstrated that a comprehensive system-scale verification process was feasible, and that our goal should be a first-pass design capable of running the proprietary GCOS7 operating system.

At the start-up phase of the development project and after considering a number of competitive alternatives, it appeared that the design objectives could be achieved by a combination of adopting a well-established structured semi-custom chip design methodology, developing and using logic synthesis and formal verification to improve the design time and provide reliability, and by establishing a closer communication between the designer and the CAD developer.

The purpose of this paper is to present the BONSAI logic design system and its associated methodology developed to support the synthesis and formal verification of high performance circuits (sections 2 and 3). Our efforts were mainly dedicated to the timing improvement aspects which appear both in the ability to insert customized gates in the standard cell synthesis process and in the links with layout as discussed in section 4. New formal comparison techniques were also developed to handle more efficiently large and complex logic.

The development of the Logic Synthesis and Formal Verification tools was supported in part by the Jessi-AC8 Project.

## 2. A STRUCTURED CUSTOM CHIP DESIGN METHODOLOGY

The overall system design flow is illustrated in Figure 1.a.

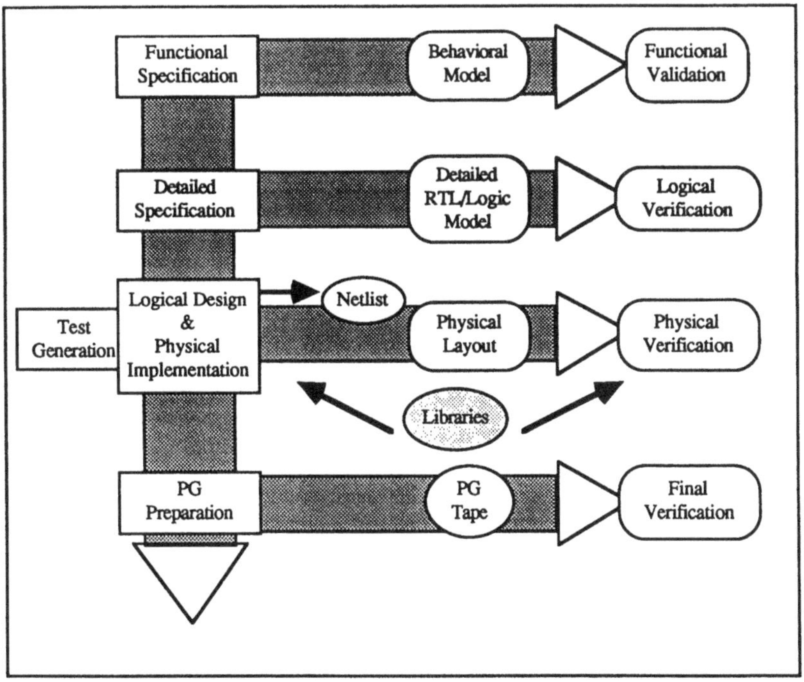


Figure 1.a : System Design Flow

## 2.1 Hardware Modelling and Validation

In many ways, the design process for the earlier VLSI DPS7000 mainframe computer set the tone for all subsequent VLSI designs at Bull. The process was characterized by massive simulation of a full system model, running in all modes of operation. Two types of simulation models were used (cf. Figures 1.a and 1.b) : the first type was designed as a high-level software breadboard used to develop and check out the microcode before the chip hardware was available. The second type was developed as a relatively detailed register transfer level model of the actual physical partitions and design concepts of the chip themselves. This model was dedicated to pilot the synthesis tool suite or used directly by the logic designers to develop the circuit-level representations of parts of the design. The models were described in a proprietary hardware description language oriented toward synchronous design (called LDS), where object (signal, latch) names follow an imposed naming convention that guarantees unique identification and improves readability. Due to the large amount of data, a Zycad SDE hardware accelerator was used for non-regression simulation.

## 2.2 Hardware Design Methodology

The hardware design methodology is targeted for a mixed custom and semi-custom design environment (Figure 1.b). In the adopted top-down approach, the design starts at the highest level and progresses toward lower levels of the hierarchy and the upper level defines the requirements of the design for the lower levels. Performance tuning is achieved during the planning and design phase, thus providing the ability to perform a one-pass design (indeed, two passes are necessary to meet all the performance constraints). At each level of the abstraction, validation is performed by simulation of the whole model of the CPU.

## 2.3 Logic Synthesis Requirements

The CPU and caches of the DPS7000 system are composed of VLSI circuits which are decomposed into macro-blocks during the early floor planning phase. While structured logic blocks (e.g. operators of the data path, ROMS) are either designed manually or using general purpose [Nguyen 90] or specific module generators, logic synthesis is applied to the design of the control parts (control logic blocks and micro-sequencers) .

The logic style of the synthesized blocks is a mixed regular and random logic associated with a multiclocking and multiphase scheme. The usage of multi-input latches, precharged and tristate signals represents another characteristics of the synthesized logic. In addition to these considerations, the main design constraints for these blocks are drastic timing constraints and narrow area margins.

As a result of these considerations, synthesis is a highly iterative process and our main concern has been to ensure a rapid and smooth convergence of this process toward the desired result. This implies the ability of the logic synthesis system to take into account hand tuning needed to improve logic speed in problem areas (e.g. the use of non-standard gates as illustrated in Figure 2). In addition, logic design was synchronized with the semi-custom physical design process for complete design verification.

## 2.4 Formal Verification Requirements

Checking plays a significant role in our methodology. Among the many different checking functions, formal techniques are used to perform logical-to-physical checking and model verification. Logical-to-physical checking is intended to verify that the physical implementation of the blocks matches their logical descriptions while model verification is used to prove the equivalence of models (e.g. different representations or versions of a block) and to check some basic properties (e.g. scan path connectivity, exclusive control signals at a multiplexing node, ...etc.). The main requirements for the usage of formal verification is its ability to handle large and complex logic (e.g. manipulation of Boolean expression with hundreds of variables in its support).

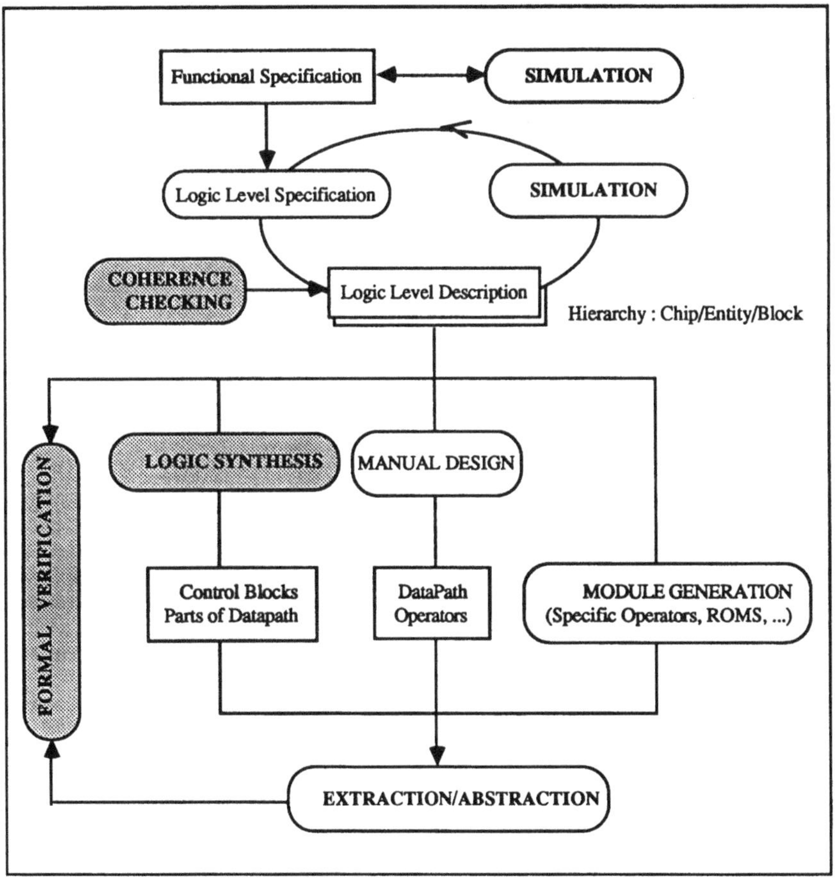


Figure 1.b Logic Desin Methodology

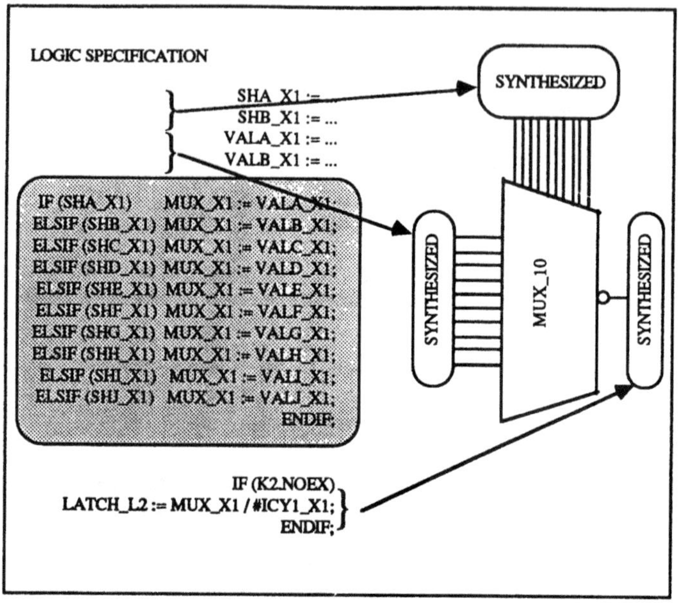


Figure 2 : Introduction of custom gate to improve timing

## 3. FEATURES OF THE BONSAI SYSTEM

We consider the problem of logic synthesis which is that part of the design path that starts with a RT/Logic specification of hardware that is to be mapped on a library of predefined primitives and ends with a netlist of gates which, after the physical implementation phase, satisfies both the timing and area constraints. The availability of a set of tools covering the whole logic synthesis path and addressing verification issues is essential for a fast design of VLSI circuits.

The following sections describe some of the major components and features of the BONSAI system.

### 3.1 Methodology of Usage

Logic synthesis involves the transformation of RT/Logic level specifications to a technology-dependent gate-level description. In order to cope with the timing and area constraints, the imposition of parts of the design as structural constraints is allowed during the synthesis process as shown in Figure 2. The structural constraints are mainly used for special purposes such as :

- Introduction of customized gates (e.g. large switch multiplexors, combination of latch and tristate gate) to improve the timing performance on critical paths,
- Hardware duplication or introduction of additional stages (logical layers) to reduce the overall circuit delay.

In presence of structural constraints, the system automatically checks for the coherence of the two overlapping views of the design and synthesizes the remaining part of the functional specification while taking into account the already designed part. Thus the convergence of the design process is ensured after a limited number of (re-)synthesis steps. This represents the pragmatic aspect of our approach in solving difficult timing problems : capture of the designer's knowledge instead of tuning multiple parameters of the algorithm until obtaining the desired results. In addition, the structural constraints are input as complement to the logic specifications, thus avoiding to rewrite the initial specification which is an error prone task (validation of the model is performed by simulation of the whole CPU model).
As a consequence, Formal Verification is built-in the system : it is used both inside the synthesis procedure to check for the coherence of the specification and the structural constraints, and independently to prove that the input models used for massive simulation are equivalent to the low-level descriptions generated by the synthesis system or abstracted from the layouts (Figure 3).

## 3.2 Overview of the BONSAI System

### 3.2.1 Features

The originality of our logic design system lies in the strategy that combines synthesis and verification techniques. The main feature of the system is built on the concept of a Boolean Network as detailed in [Hachtel 87][De Geus 87]. However, it is necessary to extend those definitions in order to type the nodes and to handle sequential components although only the combinational parts interconnecting these components are considered in our works.

Definition 1
A Boolean Network N is a 3-uple (F,PO,M) where **F** = {**Fj** / j=1,...,m} is an incompletely specified function. With each **Fj** is associated a logic variable vj in the set V = {vj / j=1,...,m} called the intermediate variable set. The specified Primary Output set PO {1,...,m} identifies the subset {vj / j $\in$ PO} of the observable outputs of the **Fj** while the memory set M$\subset${1,...,m} defines the subset {vj / j $\in$ M} of the memory variables. In our case, these memory variables are used to model latches. The behaviour of a latch with input d, output q and control c is defined by the following function :

if c(t+1) then q(t+1) := d(t+1) else q(t+1) := d(t) endif;

Definition 2
A representation of a Boolean Network is a 4-uple (F,DC,PO,M) such that {Fj / j=1,...,m} (resp. {DCj / j=1,...,m}) is a set of m given representations of the on-sets (resp. DC don't care sets) of Fj.

Let us note the support set of Fj by SUPP(Fj), then we will have t = / $\cup$j SUPP(Fj) / > m. The last t - m logic variables vj, j > m have no associated Fj and are identified as the Primary Input set PI. In addition, we can state without lost of generality that : for all j=1,...,m SUPP (DCj)$\subset$ SUPP(Fj).
Note that the don't care set plays an important role in the verification process : For an operator of a data path, the set of control signals is most of the time generated by a control logic block and therefore it is not complete. Besides, the DC set is also necessary to handle tristate signals.

Notation : In the following sections, we will note O = PO $\cup$ M and I = PI $\cup$ M.

Definition 3

For each variable vj, j=1,...,m+n we define the Fan In set FIj as follows :

if j > m then FIj = $\phi$ else FIj = {k / vk $\in$ SUPP(FIj)}

Similarly, The Fan Out set FOj is defined as : FOj = { k / vj $\in$ FIk }The transitive Fan In and Fan Out sets associated to a variable vj are defined in the same way as follows :

TFIj = { k / $\exists$ a sequence [k1,...,kr] such that k1 = k, kr = j

and for q = 1, ...,r-1, q $\in$ FIq+1, and for q, 1<q<r, q$\notin$ M }

TFOj = {k / j $\in$ TFIk}

### 3.2.2 Functions

As diagrammed in Figure 3, the main modules of the BONSAI system are the following :

• Assistance to specification : The system verifies that the Boolean Network is well-defined, i.e. satisfying the following criterions :

i/ $\forall$ j =1,...,m, j $\notin$ TFIj

ii/ $\forall$ j $\in$ I, $\exists$ i $\in$ O such that i $\in$ TFOj.

iii/ The control signals at all nodes of M are mutually exclusive

These rules express the facts that (i) there is no functional loop and (ii) every intermediate variables or primary inputs are effectively used, so checking these criterions allow to detect errors in writing the model.

In addition, it provides consistency checking and symbolic simulation to detect incoherence in the specification such as unused or unassigned variables, ill-defined conditional assignment or multiple assignments exclusivity of the control signals at a multiplexing node.

• (Re-)Synthesis with structural constraints : The synthesis procedure is classically implemented in two steps : Logic minimization and technology mapping. While logic minimization adopts a symbolic method to transform the initial Network into a primary and non redundant Boolean Network [Hachtel 87], the technology mapping is based on a mixed algorithmic and rule-based approach. The rule-based mapping procedure is used improve the netlist produced by the algorithmic step and to solve local problems such as fitting fan in/Fan out, synchronization of latch controls, etc ...

• Formal Proof : The role of the Formal Prover is to check for the equivalence of two Boolean Networks defined as follows :

Boolean Networks N1=(F1,PO,M) and N2=(F2,PO,M) are said to be equivalent if : $\forall$ j $\in$ O (= PO $\cup$ M), z1(j) = z2(j) on the definition domain of j, where z denotes the IO map a Boolean Network (as defined in [Hachtel 87]) with the following extension z : I --> O instead of z : PI --> PO).

The definition domain of a node is defined as the complement of the set ( $\cup$ DC(k) / k $\in$ TFIj).

Practically, the adopted approach is based on a canonical representation of Boolean functions called Compacted Decision Diagram [Nguyen 92] which is an outgrowth of the Binary Decision Diagram [Bryant 86]. Used in conjunction with a swap strategy, this representation allows to handle efficiently large designs by reducing the memory occupation and improving the computation time.

• Netlist Optimization : The improvement of the quality of the synthesized netlist taking into account the interconnect delays extracted from the layout is described in details in the next section.

• Schematic Generation : Most of the time, large schematics are not readable. To tackle this problem, the schematic generator provides local views (eg. path between two nodes, logical cone whose output belongs to O and inputs are elements of I, and introduces timing barriers (represented by memory nodes or I/O ports) to decompose the network into successive logical slices .

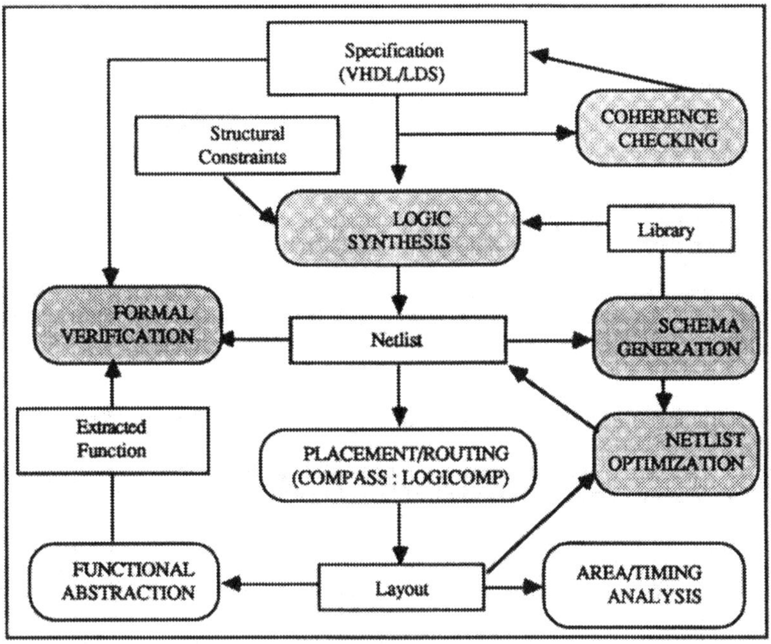


Figure 3.a : Features of the BONSAI System

## 3.3 Implementation

The BONSAI system is implemented in Le_Lisp [Chailloux 90] which is a Lisp dialect enhanced with object-oriented facilities, and runs on a large range of workstations (BULL/DPX5000, BULL/DPX20, IBM/RS6000, SUN3/4).

# 4. LOGIC SYNTHESIS LINKS WITH LAYOUT

Because interconnect delays become very critical as feature sizes shrink and tends to become predominant (especially in the design of submicron ASICs), some of the benefits provided by synthesis at the front end design cycle are lost during the layout phase : There is a growing disparity between the estimates done in synthesis and the real, post-route delays. As a consequence, a lot of time is spent in fixing timing problems after the layout is done.

## 4.1 Analysis of the Interconnect Delays

Measures of the influence of routing on the timing are represented in Figures 4.a, 4.b and 4.c : Figure 4.a shows the distribution of the wire length as a function of the number of pins of the nets while in Figure 4.b we have represented the distribution of the ratio wire capacitance / Fanin . Finally, Figure 4.c represents the variation of the mean value of this ratio versus the size of the block for a set of small blocks. We also note that those measures vary in a random manner when the ratio of the block (height / width) diverges from 1. From these measures, we can deduce that it is almost impossible to a-priori predict the interconnect delays as a function of the size of the block and of the number of interconnected pins during the synthesis procedure. As a consequence estimation performed at the synthesis step is not accurate especially when the underlying technology is submicronic.

## 4.2 Links Between Synthesis and Layout and Related Works

There are two different approaches commonly adopted by logic synthesis vendors (e.g. SYNOPSYS Design Compiler 3.0 and COMPASS ChipPlanner) :

• Top-down approach : Constraints (net priorities, path delays, net capacitance and timing) are passed from synthesis to a timing-driven layout tool. This method appears to be of limited success due to the limitations of the layout tools : Only a "reasonable" (< 10) number of constraints can be handled efficiently (e.g. in our application, only net priorities are accepted by the COMPASS/Logicomp tool as constraints and the results become rapidly unpredictable when the number of constraints is too large).

• Feedback Approach : Resynthesis with back-annotated delay values face the difficult problem of convergence : Obviously with the change of the logic structure, the back-annotated values become wrong and the placement/routing heuristics will produce others timing discrepancy.

## 4.3 In-Place Optimization

Based on the previous analysis, we have finally adopted the pragmatic approach commonly called "in-place optimization" : when timing constraints are not met due to interconnect delay, the netlist optimizer (Figure 3) automatically replace cells by more powerful ones without changing their footprints, thus leaving unchanged the netlist and the placement/routing of the block [Nguyen 93]

Our implementation of this procedure has the following characteristics :

• Timing Model : Delay = Gate's Inherent Delay + F * (Σ Input Load + Wiring Delay). While the wiring delay is estimated as a function of the interconnected pins during the synthesis step, it is extracted from the layout in the netlist optimization procedure.

• Relaxation of the Fan in / Fan out constraints to allow convergence of the procedure. Effectively, a more powerful gate yields more input capacitance.
• Each logic gate can be mapped on a family of cells (having the same width), the most powerful cell of which is reserved exclusively to the in-place optimization procedure (Figure 5).

# 5. RESULTS

The BONSAI system has been successfully used in the Auriga2 development project for the design of control logic blocks of the circuits composing the CPU and the shared cache. One chip of the shared cache has been entirely synthesized (i.e.. including the datapath) . The results are summarized in Table 1. In addition, we have performed the following statistical measures on the improvement of timing performance :
• The structural constraints represent about 5 % of the total number of synthesized gates,
• The timing improvement due to the in-place optimization module is estimated about 10%.

| SYNTHESIS | | |
|---|---|---|
| CPU | Shared Cache | |
| 3 Processors + Private Cache | SHA chip | SHD chip |
| Synthesis of all Control Blocks (Total # 200 K Transistors) | Synthesis of Control Blocks (Total # 50K Transistors) | Entirely Synthesized (Total # 100K Transistors) |
| FORMAL VERIFICATION | | |
| Verification of all the Blocks (control logic blocks, datapath operators, ROMS) (# 4 M Transistors)<br>Verification of the Scan Paths of all entities | | |

Table 1 : Achieved Results

Other applications of BONSAI include the design of a processor which allows communication between cluster of the Escala machine (designed at Bull) through serial links. The hardware part of this project is a chip containing about 1.2 M of transistors, designed in a CMOS 0.35 micron technology and entirely synthesized and formally verified with BONSAI.

# 6. CONCLUSIONS

We have presented in this paper an overview of the logic synthesis and verification system BONSAI developed to support the design of the Auriga2 CPU and caches of the DPS7000 mainframe. The originality of our work lies in the methodology that integrates a set of state-of-the-art design tools to make an effective logic-design system. A major strength of this methodology is the manner in which it allows rapid convergence of the design process in presence of drastic timing constraints and its ability to handle large and complex logic. This methodology is targeted for the semi-custom design environment, where a high degree of flexibility is required in the design process.

Current and future works include the development of tighter links with floorplanning and the evolution toward higher level of abstraction.

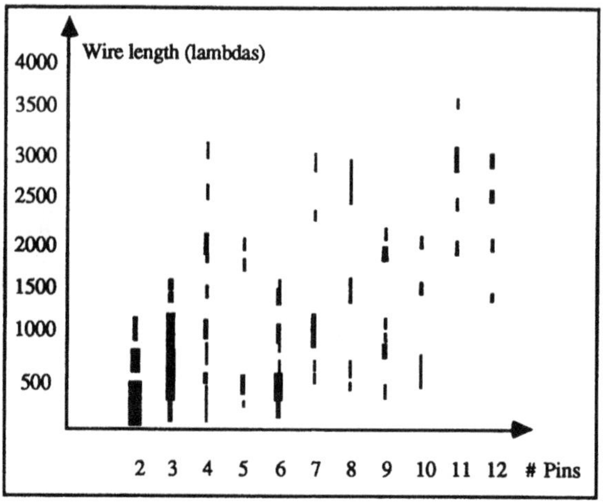


Figure 4.a : Distribution of the wire length

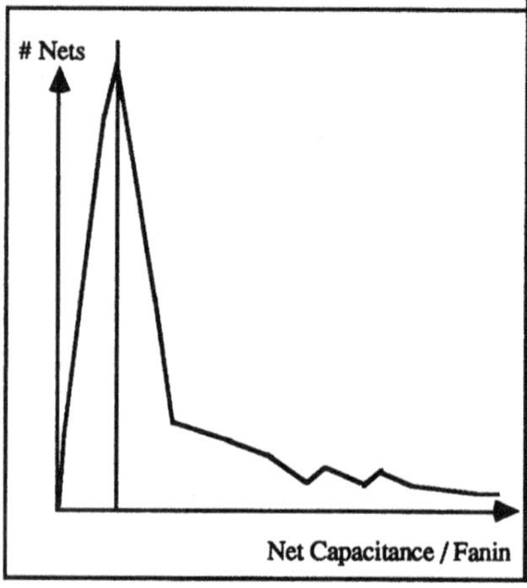


Figure 4.b : Distribution of net capacitance / fanin

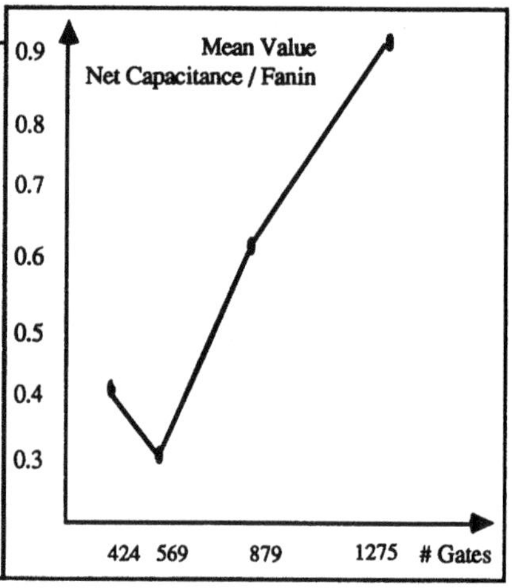


Figure 4.c : Mean value vs the size of the block

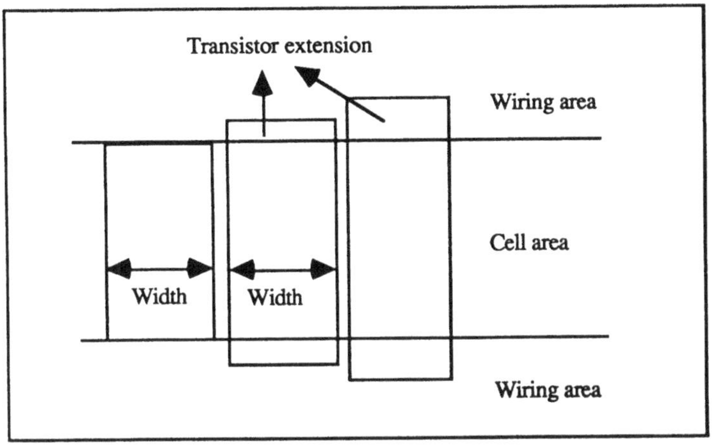


Figure 5 : Usage of cell for in-place optimization

## References

[Brayton 90] R.K. Brayton, GD. Hachtel and A.L. Sangiovanni-Vincentelli : "Multilevel Logic Synthesis", Proceedings of the IEEE, Vol. 78, N° 2, pp. 264-300, Feb 1990.

[Bryant 86] R.E. Bryant : "Graph-based Algorithms For Boolean Function Manipulation", IEEE Transactions on Computers, 35(8), pp. 677-691, 1986.

[Chailloux 90] J. Chailloux, M. Devin et als, "Le_Lisp Version V15.23 : Manuel de Référence", I.N.R.I.A. , July 1990.

[De Geus 87] A.J. De Geus and D.G. Gregory : "The SOCRATES Synthesis and Optimization System", in Design Systems for VLSI Circuits, Martinus Nijhoff Publishers, pp 473-498, 1987.

[Hachtel 87] G.D. HACHTEL and R.M. JACOBY, "Verification Algorithms for VLSI Synthesis", in Design Systems for VLSI Circuits, Martinus Nijhoff Publishers, pp 249-300, 1987.

[Nguyen 90] H.N. Nguyen and L. Ducousso, "PLAYL : A General-Purpose Data Path Assembler", Proc. EUROASIC'90, Paris May 29-June 1, 1990.

[Nguyen 92] H.N. Nguyen, L. Ducousso, M. Thill and P. Vallet, "A Pragmatic Approach to the Automation of the Logic Design Process", Proceedings of the EDAC'92, Brussels, February 1992.

[Nguyen 93] H.N. Nguyen, L. Ducousso and M. Thill, "Optimal Use of Cell to Improve Timing Performance in Logic Synthesis", PATMOS'93, La Grande-Motte, October 1993.

PART TWO

# Cell Assignment Based on BDD

4

# Combined approach of ROBDDs and structural analysis in the mapping and matching of logic functions

*Trullemans A.-M., Jacobi R. and Zhang Q.*
*UCL-Laboratoire de Microélectronique*
*Place du Levant, n° 3, B-1348 Louvain-la-Neuve, Belgium,*
*Phone: (+32) 10 47 25 65 - Fax: (+32) 10 47 86 67*
*e-mail: atrullemans@dice.ucl.ac.be*

**Abstract**

The technology mapping - final step of the logic synthesis - maps the decomposed Boolean function on physical cells. We address here the decomposition and the matching steps. We present two different ROBDD-based techniques to handle the decomposition problem, and compare them. For handling the matching step, we analyse a heuristics based on symmetry and develop a new structural approach, based on controlling value analysis and observation function deduction. This last appears to be efficient regarding the CPU time for checking the match with basic cells, mostly when don't cares are present, and should be particularly interesting to handle the complex cells of FPGAs. Benchmarks are presented which validate the various heuristics.

**Keywords**

ROBDD, technology mapping, decomposition, matching

## 1 INTRODUCTION

Technology mapping is the last and decisive step in logic synthesis. It is usually performed after a technology independent minimization, which generates an optimized multi-level logic network, and a decomposition step, which splits that network into simpler subfunctions, easier to manipulate. The mapping itself consists in covering the network by implementing each node (or set of nodes) with one cell (or set of cells) selected from the target library, while trying to meet the design constraints.

Gate matching is the process of selecting the library cells to implement the Boolean functions associated to the network's nodes. Network covering is the transformation of the whole Boolean network into an acceptable design by selecting which subsets of the network nodes shall be collapsed and mapped to a single cell. During the covering step, the gate matching is called several times to evaluate alternative solutions, and the mapping cost is thus heavily dependent on the matching cost.

We will concentrate on the two basic steps of decomposition and matching, and analyse innovative techniques based on ROBDD manipulation and on structural analysis of logic circuits.

Starting from a minimized description, the decomposition step will select a set of subfunctions to produce a Boolean network. We have analysed two decomposition methods based on ROBDD representations (Bryant, 1986): the direct decomposition and the path oriented decomposition. The direct decomposition is a very fast approach that extracts shared subgraphs and replaces them by new variables in the original ROBDD. The path oriented decomposition is a particular algebraic decomposition performed on ROBDDs. These two methods were tested inside the LOGOS synthesis system (Jacobi, 1993-2), on a set of benchmarks. These experiments show that, in most cases, the direct decomposition performs better than the path oriented decomposition. The results were compared to SIS (Brayton, 1987) and we find that the direct decomposition gives most of the time better results in terms of complexity of the decomposed networks.

The matching step consists in selecting in the given library the particular cell that implements a node of the Boolean network (the node function). The phase of the function, and the ordering and phase of the input variables are unknown, and this correspondence has to be established before the node function and the cell can be compared. The number of checks is directly related to the number of inputs, which greatly complicates the problem when handling technologies with complex logic cells, like field programmable gate arrays (FPGAs). The first approach proposed follows the traditional ones, which solve the problem by tree and graph matching techniques (Darringer, 1984), (Keutzer, 1987), (Aho, 1983). We reduce the complexity of the check by classifying the library cells and the functions to be mapped considering their symmetry properties. This classification method is based on the flexibility of the ROBDD structures.

However, some library cells cannot be represented as tree circuits, and this limits the applicability of the tree-based approaches. Recently, research works have focused on deriving a characteristic signature for each input variable and/or phase (Schlichtmann, 1992), (Cheng, 1993), (Mohnke, 1993). The signature method is efficient but suffers from the aliasing problem: two different input variables or input phases may produce the same signature, which leaves many variants to be checked by ROBDD comparison.

We propose here an alternative method, based on fault propagation analysis, and which is applicable to functions with don't cares: the library cell is treated as an incorrect implementation of the node function, and its structure is analysed based on the observation functions that are derived from the node function (Zhang, 1994). Using our method, incorrect input variable permutations can be easily detected, and for a correct input variable permutation, the input phase assignment can be directly determined. This reduces considerably the different ROBDDs to be checked.

## 2 DEFINITIONS

We define hereafter some terms used in the rest of this paper:

A *variable* is a symbol representing a single coordinate in the Boolean space.

A *literal* is a variable or its complement.

A *Boolean network* (BN) is a directed acyclic graph where each node $n_i$ is associated to a Boolean function $f_i$ and a literal $y_i$. An *edge* connecting node $n_i$ to node $n_j$ indicates that the variable $y_i$ is used explicitely by the node function $f_i$.

The *support* of $f_i$ is the set of variables used. A *path* in the BN is a set of successive edges that connects two nodes. The number of edges in a path $p$ is the *length* of $p$. The set of nodes connected to node $n_i$ by paths with length = 1 is the *fanin* set of $n_i$. The *fanout* set of $n_i$ is the set of nodes reached from $n_i$ by paths with length = 1.

The ROBDDs are defined as proposed by (Bryant, 1986). Each variable has an index that identifies its level in the ROBDD. A *level* $l_i$ of a ROBDD is the set of nodes associated with the input variable $x_i$.

A *1-path* in a ROBDD is the set of arcs that connect the root to the 1 terminal. The set of 1-paths in a ROBDD forms a disjoint cover of the Boolean function it represents.

A *minterm* is a vertex in the *n*-dimensional Boolean space.

The *cofactor* of $f$ with respect to the positive (negative) phase of an input variable $x_i$, denoted as $f_{xi=1}(x)$ ($f_{xi=0}(x)$), is the function obtained by setting $x_i$ to 1 (0) in the function $f$.

The *Boolean difference* of a function $f$ with respect to an input variable $x_i$, is the function defined as: $\partial f / \partial x_i (x) = f_{xi=1}(x) \oplus f_{xi=0}(x)$, where $\oplus$ is the exclusive-or operator. This function is also called the observation function of the input variable $x_i$, because it indicates that changes at the input $x_i$ can be observed at the output of the function $f$.

An input of a logic gate is said to have a *controlling value* if its value prevents other inputs from affecting the output of the gate. It is well known that the controlling value for an AND gate is 0, the controlling value for an OR gate is 1, and no controlling value exists for inverters and EXOR gates.

## 3 DECOMPOSITION PROBLEM

### 3.1 Direct decomposition

Normally, ROBDDs are not related to the structure of a logic circuit: they are only a graph representation of a Boolean function. However, we could extract structural information from the ROBDD: each node in the diagram is the root of a subdiagram that stands for a subfunction. The *direct decomposition* consists in extracting these subfunctions from the ROBDD to find a *convenient* Boolean network representation: a decomposition adequate to the target technology (simple functions for the usual standard cells or gate arrays, functions with limited number of inputs - 4 or 5 - for lookup table FPGAs).

The *extraction* of a ROBDD subfunction starts with the selection of an interesting subdiagram, according to some objectives. This subdiagram corresponds to a node $r$ in the ROBDD, the root of the subdiagram, and the corresponding subfunction will create a new intermediate variable $y_i$ in the Boolean network. A new ROBDD node $r_i$ depending on this variable is introduced in the graph, replacing the root node $r$. All the fathers of node $r$ now point to $r_i$. The node $r_i$ represents the whole subfunction: its low and high sons are connected to the terminals {0,1}. In consequence, all the nodes and edges that were used only in the extracted subdiagram may be removed from the ROBDD. An example of this decomposition is shown in Figure 1.

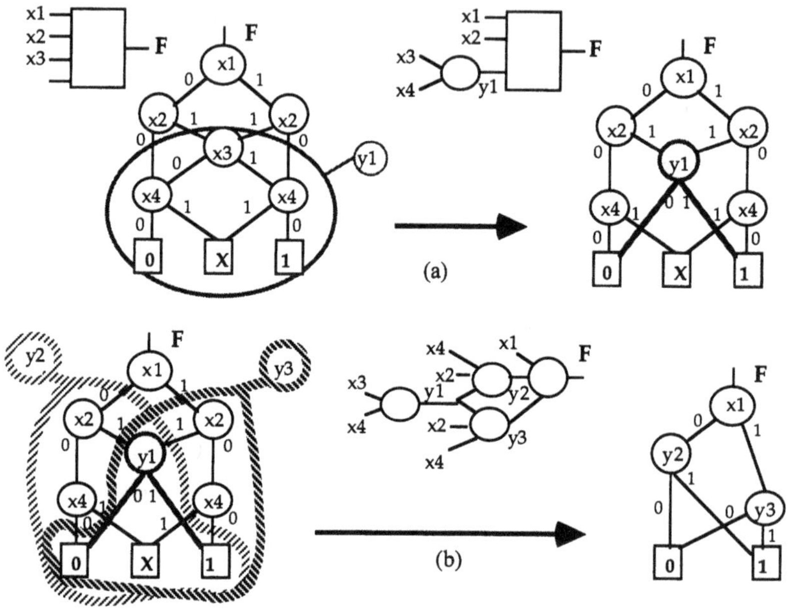


**Figure 1** Example of direct decomposition of an ROBDD.

Three criteria were used to extract a subdiagram from the ROBDD:

1. Select a node with at least *NF* fathers.
2. Select a node with a subfunction depending on at least *NI* inputs.
3. Select a node whose subdiagram has a size equal or greater than *NN*.

Those values should be specified by the designer. The first criterion is related to the amount of sharing of a subfunction in the ROBDD: the number of fathers of a node *n* is the number of subfunctions using or depending on the subfunction associated to that node. It is a direct measure of the fanout of *n*. Each father of *n* in the ROBDD corresponds to a node that will have *n* in its fanin. Thus the fanin/fanout constraints may be analysed directly in the graph. The second criterion is an approximation of the function complexity measured by the number of variables it depends on. It is directly related to the complexity of the cells in the library. The third criterion provides an alternative where the size of the ROBDD of the subfunctions is used as a cost function.

## 3.2 Path oriented decomposition

The *path oriented decomposition* method is based on the fact that each path in a ROBDD corresponds to a cube of a disjoint cover of the function. Thus covering all the 1-paths of the ROBDD is a way of covering the function.

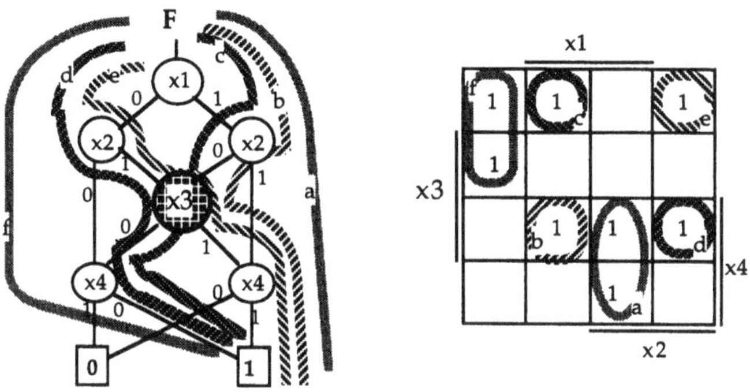


**Figure 2** Path cover of an ROBDD.

To decompose the ROBDD into a set of functions, one has to select subsets of the ROBDD paths and to cover them with the extracted subfunctions. This is done by exploiting the graphical nature of the ROBDD, in order to get a faster algorithm:

1. Select a node in *M*, the global ROBDD, according to some cost function: *n* is called the *splitting node*. Let *S* be the set of paths through *n*.
2. Build *QD*, a new ROBDD with *S*. All arcs of *QD* that are not defined point to the 0 terminal.
3. Let *S1* be the set of the remaining paths of *M*: build *R*, a new ROBDD with *S1*. All the undefined arcs point to the 0 terminal.
4. The sum of the ROBDDs *QD* and *R* is a cover of *M*: $M = QD + R$.

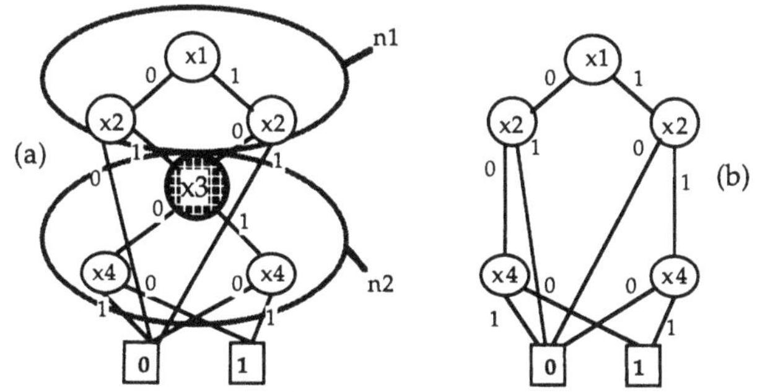


**Figure 3** Path oriented decomposition: ROBDD corresponding to *QD* (a) and *R* (b).

A good criterion to select the splitting node is hard to define. The heuristic used here is to select, among the nodes that split the ROBDD in subfunctions with support of similar size, the node that covers more 1-paths. The support of the functions are thus reduced at each decomposition step, leading to simpler functions. A more detailed analysis of the splitting nodes can be found in (Jacobi, 1993-2).

## 3.3 Experimental results

We checked both algorithms on some experiments: the results are presented in Table 1. For each circuit, the ROBDD size is first reduced using a swapping technique (Jacobi, 1991), (Calazans, 1992), then we apply a decomposition algorithm. Direct decomposition was tested with three parameter settings : (*NF*=2, *NI*=2, 3, 4). The quality of the result is measured via a set of three numbers: (*Nodes-ROBDD-Arcs*). *Nodes* is the number of generated subfunctions (nodes of the Boolean network - corresponds to the size of the network). *ROBDD* is the total number of nodes of the ROBDD of each subfunction (complexity of each subfunction), and *Arcs* is the number of interconnections in the Boolean network (interconnection complexity). These parameters give an estimate of the complexity of the multi-level representation.

**Table 1** Decomposition benchmarks - (Nodes-ROBDD-Arcs)

| | *Direct Decomposition* | | | *Path Decomposition* | *Sis* |
|---|---|---|---|---|---|
| | *NF=2 NI=2* | *NF=2 NI=3* | *NF=2 NI=4* | | |
| 5xp1 | 65-276-162 | 52-237-143 | **38-187-119** | 87-427-236 | 42-244-120 |
| alu3 | 55-231-135 | 35-168-106 | **28-142-95** | 49-244-138 | **28**-192-145 |
| bw | 102-420-279 | 90-382-266 | **63-276**-227 | 228-1040-568 | 65-400-**216** |
| dekoder | 17-61-39 | 15-55-37 | 12-**42**-34 | **7**-47-**25** | 10-55-30 |
| dk27 | 22-74-49 | 15-51-40 | 13-**43**-35 | 16-73-41 | **11-55-30** |
| f51m | 62-275-163 | 57-260-153 | **37-197-117** | 116-518-278 | 42-243-119 |
| fo2 | 15-60-36 | 11-48-31 | 8-35-24 | **4-30-16** | 12-56-32 |
| misex1 | 34-148-92 | 32-139-90 | **18-91**-68 | 48-229-130 | **18**-118-**67** |
| misex2 | 86-322-184 | 54-223-153 | 42-**181**-136 | 44-276-185 | **30**-190-**113** |
| r53 | 21-95-55 | 19-89-53 | **14-69-41** | 19-101-53 | **14**-85-**41** |
| r73 | 41-193-113 | 39-187-111 | 31-**159**-96 | 46-286-153 | **28**-193-**93** |
| r84 | 57-265-155 | 54-256-152 | 44-223-137 | **26-219-102** | 37-337-140 |
| z4 | 24-128-60 | 12-73-36 | **8-49-26** | 12-80-40 | 12-64-30 |

We give in the last column the results obtained with SIS (Brayton, 1987). The best results are indicated in boldface character. Direct Decomposition with *NI*=4 gives the best result in most of the cases. Sometimes Path Oriented gives the best one, but it may also produce very bad solutions (bw and f51m). These algorithms appear to be fast (time range from a few seconds to at most a few minutes, implemented in LISP on a personal computer), so we propose to apply both and select the best solution. The results are slightly better than SIS in terms of the complexity of a Boolean network where each node function is represented by a ROBDD. We have supposed that all subfunction's ROBDDs have the same variable ordering.

The time complexity of the direct decomposition method (Jacobi, 1993-2) is linear with respect to the ROBDD size. The path oriented decomposition method is more complex: every node of the subfunctions is checked for redundancy, which is done by checking for implication between its son's functions. This gives a global time complexity bound by $O(n^2)$, where $n$ is the ROBDD size. This is to be compared to the Boolean methods involving logic minimization, which is a NP-complete problem.

The ROBDD extensions proposed in the literature, namely the strong canonical form and the negative edges, would be usefully used in these direct and path oriented decomposition methods. The negative edges allows a fast verification of the existence and use of the complement of the extracted subfunctions in both methods (this is implemented here by an equivalence checking algorithm). The strong canonical form will present interest in the path oriented decomposition, for it keeps the global ROBDD always updated.

# 4 MATCHING PROBLEM

## 4.1 Defining the matching problem

Let $f(x_1, \ldots, x_n)$ be the function of a node in the Boolean network to be mapped, and $g(y_1, \ldots, y_n)$ the function of a cell in the library. For simplicity, we assume that $f$ is a completely specified function with one output.

The matching problem between the two functions $f$ and $g$ can be stated as follows: Does there exist an input variable permutation $\pi$, an input phase assignment $\phi$ and an output phase assignment $\theta$, such that $f(\mathbf{x}) = \theta\{ g[\pi\phi(\mathbf{y})]\}$. The number of variants to be considered are: $n! * 2^n * 2$.

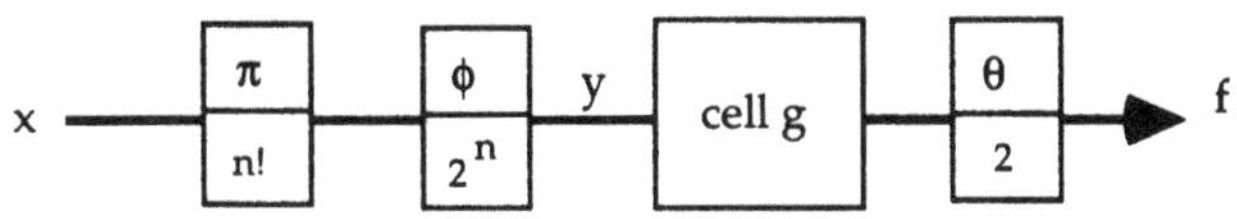


**Figure 4** The matching problem: input and output transformations.

## 4.2 Exploiting symmetry

### 4.2.1 Symmetry classification

The BDD structure, i.e., the BDD graph without node values and with unlabelled arcs, represents a family of functions. This is exemplified in Figure 5.

Considering the ROBDD representation, the matching of a node function with a cell can be realized by first finding two isomorphic ROBDD structures and then extracting from them the variables and output phases. The isomorphism between the graphs can be discovered by classifying the ROBDDs according to their symmetry classes. If *BDDf* is the ROBDD of the node function to be matched, the ROBDD matching steps are:

1. Identify the symmetry classes (SymCs) of the *BDDf*.

2. Sort the classes in the descending order of cardinality.
3. Match the node function ROBDD with the cells that have the same number of SymCs and the same SymCs cardinality.

This algorithm is based on the fact that two functions $f$ and $g$ can match only if they have the same number of symmetry classes and if for each symmetry class of $f$ there is a corresponding symmetry class in $g$ with the same cardinality.

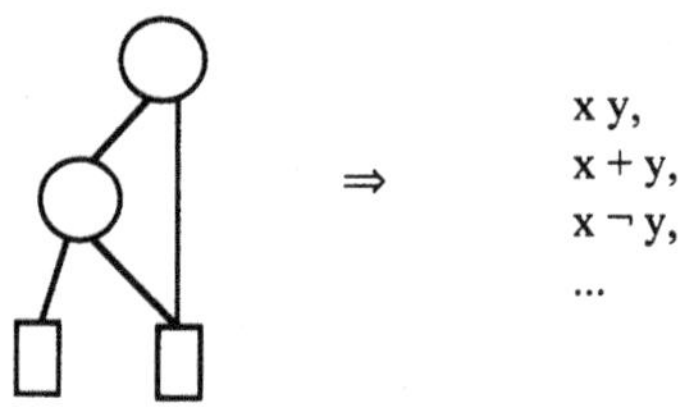


**Figure 5** BDD structure and its family of functions.

To find the symmetry classes of a function, we just have to check the ROBDD of the function: two variables are symmetric in a ROBDD if they can be exchanged without modifying the topology of the graph. If we already have one ROBDD of the function, we could simply apply incremental techniques in order to avoid the cost of building a new ROBDD. If we exchange two adjacent variables we can easily check if the ROBDD will be modified or not. The next theorem allows to check this without exchanging the variables (Jacobi, 1993-2):

**Theorem 1** Let $S$ be the set of subgraphs associated to two adjacent variables in the ROBDD ordering ($x_i$, $x_{i+1}$) which have at least 2 nodes. Each subgraph has one root and can lead to 2, 3 or 4 different nodes in the ROBDD. $x_i$ and $x_{i+1}$ are *phase free* symmetric either if all successor in $S$ lead to only 2 nodes in the ROBDD or if they lead to three nodes: two nodes are associated to paths $\neg x_i \neg x_j$ and $x_i x_j$ in the graph and the third one is associated to paths $\neg x_i x_j$ and $\neg x_j x_i$.

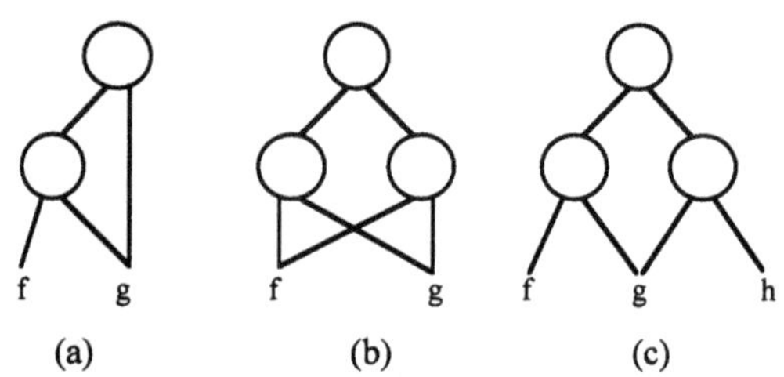


**Figure 6** Examples of adjacent variables subgraphs for check of phase free symmetry.

This idea is illustrated in Figure 6. Subgraphs (a) and (b) are phase free symmetric with only two different nodes, while (c) is phase free symmetric with three different ones.

The algorithm to generate the phase free symmetry classes uses the incremental techniques proposed in (Jacobi, 1991) to swap adjacent variables in the ROBDD ordering. The number of steps is upper bounded by $n(n\text{-}1)/2$. Usually there are less comparisons due to the transitive property of symmetric variables: if one variable belongs to a SymCs it does not need to be further compared. After all variables are pairwise tested, the ROBDD is classified by putting the SymCs with larger cardinality closer to the root.

### 4.2.2 Matching algorithm

```
function MatchBDD (mbd, order): matchlist;
begin
        (mbd,order) = ClassBDD (mbd, order);
        vector = GetSCsVector(mbd);  /* Symmetry class vector */
        G = GetCells(vector);
        foreach cell in G
           if (match = MatchCell(mbd, G.mbd, order))
           then put match in matchlist; /* match = cell and inverters needed */
        return (matchlist);
end;
function MatchCcell (bddf, bddg, order): match;
begin
        initialize PhaseVector;
        if (RecMatchCell (bddf, bddg, PhaseVector))
           then return (BuildMatch ( order, PhaseVector))
           else return (NULL);
end;
function RecMatchCcell (bddf, bddg, phv)
begin
        case (bddf,bddg) of
           (they have different indices): return(false);
           (both are terminals):
              if terminals are distinct then invert the output;
              return(true);
           (only one is a terminal): return (false);
           ((RecMatchCell (lowv (bddf), lowv (bddg), phv) and
              RecMatchCcell (highv (bddf), highv (bddg) phv)):
              return(true);
           ((RecMatchCell (lowv (bddf), highv (bddg), phv) and
              RecMatchCcell (highv (bddf), lowv (bddg) phv)):
              phv[index(bddf)] = NEGATED;
              return(true);
end;
```

**Figure 7** Matching algorithm based on symmetry classes.

Figure 7 presents the matching algorithm. The function *MatchBDD* takes a ROBDD (mbd) with its variable order, and returns the set of matches that can implement the function. Each match is formed by the cell and the set of inputs/output inverters generated by the matching algorithm. The function is first classified according to its symmetry classes. Then the SymCs cardinality vector is used to address the set of cells in the library that are candidates to match the function. The function *MatchCell* takes the ROBDD of the function (bddf) and of the cell (bddg), and returns the input assignment with the respective input and output phases, or NULL if they are not matchable. Function *RecMatchCell* checks the matching by traversing both ROBDDs with direct or complemented match at each node.

## 4.3 Structural analysis

The controlling value analysis and observation function deduction method was first proposed for the location of multiple design errors in an incorrect gate-level circuit (Zhang, 1994). It can also be used for checking the match between a node function $f$ and a library cell $g$. In this section, we first give a brief review of this method, and then show its application to the matching problem.

### 4.3.1 Controlling value analysis and observation function deduction

Consider the required function $f(\mathbf{x})$, and its proposed - and maybe incorrect - implementation by the circuit $C$, which realizes the function $g(\mathbf{x})$. For simplicity, we assume that the circuit $C$ is exclusively composed of inverters and AND gates. To locate the design errors within the circuit $C$, its primary inputs are first initialized with the corresponding observation functions derived from the function $f(\mathbf{x})$. This is based on the following theorem (Zhang, 1994).

**Theorem 2** The function $g(\mathbf{x})$ is said to be equivalent to the function $f(\mathbf{x})$, or equivalent to its complement if, and only if $\partial g / \partial x_i (\mathbf{x}) = \partial f / \partial x_i (\mathbf{x})$ for any input $x_i$.

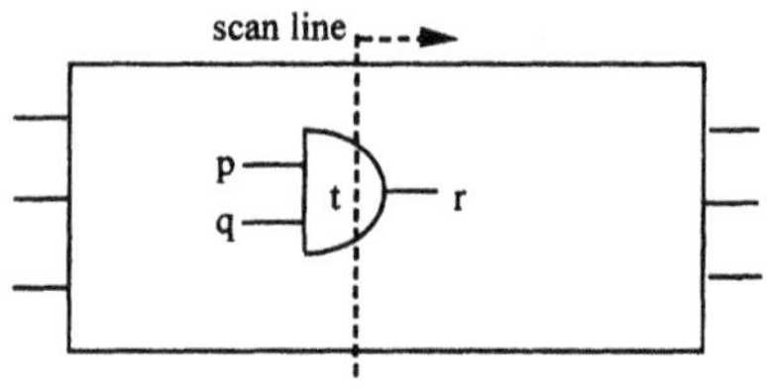


**Figure 8** A logic gate $t$ in the incorrect circuit $C$.

The detection of the possible design errors associated with the logic gate $t$ in the incorrect circuit $C$, can be derived from the controlling value analysis. We assume that the observation functions of the gate inputs $p$ and $q$ with respect to the required function $f(\mathbf{x})$ are known. We present here two theorems adapted from (Zhang, 1994).

**Theorem 3** The gate inputs $p$ and $q$ are the correct inputs of the logic gate $t$ if, and only if

$$\partial f / \partial p (\mathbf{x}) \cdot q(\mathbf{x}) = 0 \text{ or } \partial f / \partial p (\mathbf{x}) \cdot \neg q (\mathbf{x}) = 0,$$

$$\text{and } \partial f / \partial q (\mathbf{x}) \cdot p(\mathbf{x}) = 0 \text{ or } \partial f / \partial q (\mathbf{x}) \cdot \neg p(\mathbf{x}) = 0.$$

**Theorem 4** If $\partial f / \partial p\,(\mathbf{x}) \cdot q(\mathbf{x}) = 0$, the input $q$ has the controlling value 1 for the gate $t$, otherwise it has the controlling value 0.

If the observation functions of the inputs of a logic gate $t$ cannot satisfy Theorem 3, an interconnection design error is detected. Furthermore, if a gate input is proved to have the controlling value 1 for an AND gate, a missing inverted error at the signal line is detected.

If, for a given gate $t$ with inputs $p$, $q$ and output $r$, there is no design error or all design errors have been corrected (for example by adding inverters), the observation function of the gate output $r$ can be deduced from that of the gate inputs $p$, $q$ according to the following theorem.

**Theorem 5** $\partial f / \partial r\,(\mathbf{x}) = \partial f / \partial p\,(\mathbf{x}) + \partial f / \partial q\,(\mathbf{x})$.

### 4.3.2 Application to the matching problem

In the problem of matching a node function *f* with a library cell g, most of the input variables, at least for many practical functions (and if the function has no don't cares), can be identified by characteristic signatures (Schlichtmann, 1992), (Cheng, 1993), (Mohnke, 1993) and symmetric variables (Keutzer, 1987), (Mailhot, 1993), (Jacobi, 1993-1). Thus, the total number of possible input variable permutations and the total number of possible input phase assignments have already been reduced, but there may remain ambiguity for some variables (aliases). We will explain the strategy of the structural analysis on the example of Figure 9, where we have to match the node function *f* (represented here for simplicity by its Karnaugh map) with some cell g, which will be analysed as an equivalent netlist of elementary gates.

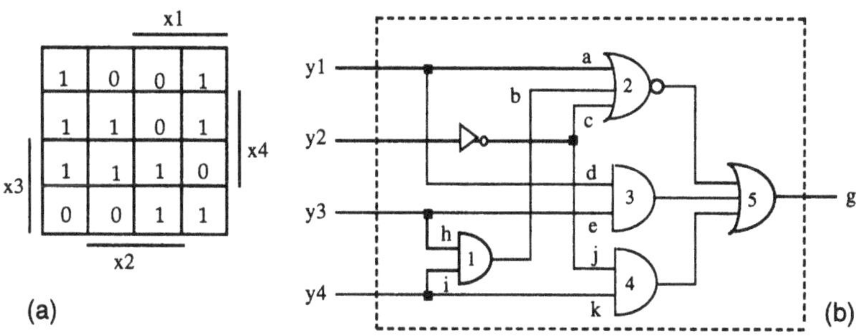


**Figure 9** Example of match: (a) Karnaugh map of function f. (b) netlist equivalent of library cell g.

**Table 2** The possible input variable permutations and phases

| | *Input variable permutation* $\pi$ | | | | *Input phase assignment* $\phi$ | | | |
|---|---|---|---|---|---|---|---|---|
| *No.* | *y1* | *y2* | *y3* | *y4* | *y1* | *y2* | *y3* | *y4* |
| 1 | x1 | x3 | x2 | x4 | ? | ? | 1 | 0 |
| 2 | x1 | x3 | x4 | x2 | ? | ? | 0 | 1 |
| 3 | x3 | x1 | x2 | x4 | ? | ? | 1 | 0 |
| 4 | x3 | x1 | x4 | x2 | ? | ? | 0 | 1 |

In this particular example, after checking signatures and symmetry, it remains only 4 cases to be checked (Table 2).

For each of these input variable permutations, the inputs of the library cell $g$ are initialized with the observation functions derived from the node function $f$. The design errors considered here for the library cell $g$ are only wire exchanges at the inputs and missing inverters at the inputs and/or at the output. Based on these observation functions, the current input variable permutation $\pi$ is first verified according to Theorem 3, then the input phase assignment $\phi$ is verified according to Theorem 4.

With respect to the input variables of the node function $f$, the observation functions can be computed as:

$$\partial f / \partial x_1 (\mathbf{x}) = x_3 \neg x_4 + x_2 \neg x_3 x_4 + \neg x_2 x_3. \quad (1)$$

$$\partial f / \partial x_2 (\mathbf{x}) = \neg x_3 \neg x_4 + x_1 x_4. \quad (2)$$

$$\partial f / \partial x_3 (\mathbf{x}) = x_1 x_4 + x_1 x_2 + \neg x_1 \neg x_2 \neg x_4. \quad (3)$$

$$\partial f / \partial x_4 (\mathbf{x}) = \neg x_1 x_2 + \neg x_2 x_3. \quad (4)$$

Note that the observation function $\partial f / \partial x_i(\mathbf{x})$ contains the input combinations on which the input $x_i$ can be observed, or in other words, the stuck-at-faults at the input $x_i$ that can be propagated. These observation functions will be used to check the input variable permutation, and also to determine the input phase assignment.

For each of the possible input variable permutations, the corresponding inputs of the library cell $g$ are initialized with these observation functions.

Let us consider the variable permutation $\pi(y_1, y_2, y_3, y_4) = (x_1, x_3, x_4, x_2)$ as shown in Figure 10. For this variable permutation, the input phase assignment is:

$$\phi(y_1, y_2, y_3, y_4) = (?, ?, 0, 1). \quad (5)$$

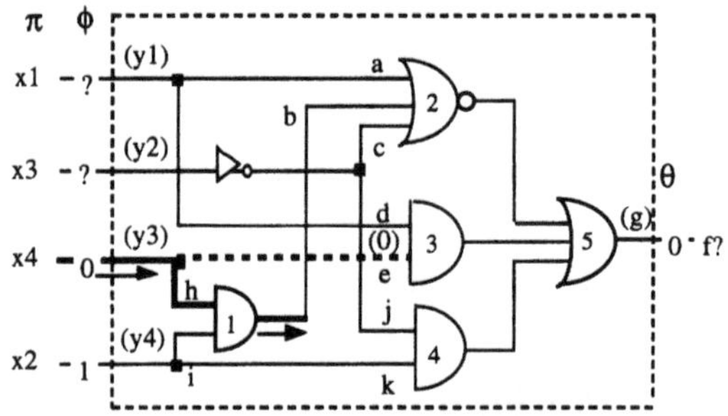


**Figure 10** Checking the input variable permutation $\pi(y_1, y_2, y_3, y_4) = (x_1, x_3, x_4, x_2)$.

According to the structure of the library cell $g$, the function $[\partial f / \partial x_4 (\mathbf{x}) \cdot \neg d (\mathbf{x})]$ can be used as partial observation function of the signal line $h$. If $\phi(y_1) = 0$, the observation function of the signal line $h$ is: $\partial f / \partial h (\mathbf{x}) = \neg x_1 x_2 + \neg x_1 \neg x_2 x_3$. It indicates that the signal line $h$

can be observed in spite of the logic values at the input $x_2$, which is contradictory to the fact that the gate 1 is an AND gate. If $\phi(y_1) = 1$, we have $\partial f / \partial h\,(\mathbf{x}) = x_1 \neg x_2 x_3$. This function does satisfy Theorem 3. Thus, the input phase of the variable $y_1$, $\phi(y_1) = 1$, is determined according to Theorem 4. Similarly, for the signal line $i$, we have $\partial f / \partial i\,(\mathbf{x}) = x_1 x_3 x_4$ if $\phi(y_2) = 0$. This observation function also satisfies Theorem 3, and the input phase of the variable $y_2$, $\phi(y_2) = 0$, is determined.

According to Theorem 5, the observation function of the signal line $b$ can be deduced as: $\partial f / \partial b\,(\mathbf{x}) = \partial f / \partial i\,(\mathbf{x}) + \partial f / \partial h\,(\mathbf{x})$. The same analysis process continues for the remaining logic gates. In fact, not all logic gates in the library cell need to be analysed because the possible design errors occur only at the primary inputs. In our method, only the logic gates whose inputs contain the primary inputs are considered. If no conflicts have been found, the ROBDDs for the subcircuit $f$ and for the library cell $g$ are then constructed for equivalence checking.

In our example, for the variable permutation $\pi(y_1, y_2, y_3, y_4) = (x_1, x_3, x_4, x_2)$, the input phase assignment $\phi(y_1, y_2, y_3, y_4) = (1, 0, 0, 1)$ and the output phase assignment $\theta = 0$, the ROBDDs of the two functions $f$ and $g$ are isomorphic. Thus, the library cell $g$ is a match of the node function $f$.

### 4.3.3 Experimental results

**Table 3** Check of matchable cells

| *Library cell* | | *complete ROBDD match* | | *structural analysis* | |
|---|---|---|---|---|---|
| *Cell name* | *nb inputs* | *BCT* | *CPU (sec)* | *BCT* | *CPU (sec)* |
| AO04D1 | 3 | 36 | 0.23 | 1 | 0.00 |
| FN02D1 | 3 | 1 | 0.02 | 1 | 0.00 |
| OA04D1 | 3 | 36 | 0.25 | 1 | 0.02 |
| AO01D1 | 4 | 12 | 0.03 | 1 | 0.03 |
| AO5D1 | 4 | 236 | 1.60 | 1 | 0.00 |
| AO08D1 | 4 | 347 | 2.38 | 1 | 0.00 |
| FN03D1 | 4 | 49 | 0.37 | 1 | 0.02 |
| FN04D1 | 4 | 49 | 0.35 | 1 | 0.00 |
| OA01D1 | 4 | 58 | 0.42 | 1 | 0.02 |
| OA05D1 | 4 | 236 | 1.65 | 1 | 0.00 |
| OA08D1 | 4 | 340 | 2.35 | 1 | 0.03 |
| AO03D1 | 5 | 3084 | 23.28 | 1 | 0.00 |
| AO07D1 | 5 | 2908 | 21.75 | 1 | 0.05 |
| OA03D1 | 5 | 3457 | 25.82 | 1 | 0.03 |
| OA07D1 | 5 | 2641 | 19.23 | 1 | 0.03 |
| AO02D1 | 6 | 3892 | 30.77 | 1 | 0.00 |
| AO06D1 | 6 | 4441 | 3.52 | 1 | 0.03 |
| OA02D1 | 6 | 4234 | 22.92 | 1 | 0.03 |
| OA06D1 | 6 | 3892 | 31.25 | 1 | 0.03 |
| ex-A07 | 7 | 556852 | 6019.85 | 1 | 0.02 |

We implemented a simple ROBDD check and the structural check for basic cells in a prototype written in C++. We checked the match with standard cells of an industrial library (library VSC450L offered in the COMPASS CAD tools), composed mostly of simple cells.

The node function to be mapped is defined as the library function, with the addition of 5% random don't cares. In these tests, of course, signature can not be used, for it does not apply if don't cares are present. In Table 3, we compared the number of ROBDD checks (BCT) to be handled to confirm the correspondence between the function and the cell, and the CPU time (in seconds) on a SUN Sparc20 workstation. The last one (ex-A07) is a test on some more complex cell artificially constructed, to verify how the algorithm handles the complexity.

**Table 4** Check of non matchable cells

| *Library cell* | | *complete ROBDD match* | | *structural analysis* | |
|---|---|---|---|---|---|
| *Cell name* | *nb inputs* | *BCT* | *CPU (sec)* | *BCT* | *CPU (sec)* |
| AO04D1 | 3 | 48 | 0.28 | 0 | 0.00 |
| FN02D1 | 3 | 48 | 0.35 | 6 | 0.03 |
| OA04D1 | 3 | 48 | 0.32 | 0 | 0.02 |
| AO01D1 | 4 | 384 | 2.58 | 0 | 0.02 |
| AO5D1 | 4 | 384 | 2.55 | 0 | 0.02 |
| AO08D1 | 4 | 384 | 2.67 | 0 | 0.02 |
| FN03D1 | 4 | 384 | 2.57 | 0 | 0.02 |
| FN04D1 | 4 | 384 | 2.68 | 0 | 0.00 |
| OA01D1 | 4 | 384 | 2.57 | 0 | 0.02 |
| OA05D1 | 4 | 384 | 2.67 | 0 | 0.02 |
| OA08D1 | 4 | 384 | 2.63 | 0 | 0.02 |
| AO03D1 | 5 | 3840 | 28.93 | 0 | 0.03 |
| AO07D1 | 5 | 3840 | 28.02 | 0 | 0.08 |
| OA03D1 | 5 | 3840 | 28.60 | 0 | 0.03 |
| OA07D1 | 5 | 3840 | 27.95 | 0 | 0.05 |
| AO02D1 | 6 | 46080 | 370.60 | 0 | 0.15 |
| AO06D1 | 6 | 46080 | 369.23 | 0 | 0.03 |
| OA02D1 | 6 | 46080 | 365.30 | 0 | 0.25 |
| OA06D1 | 6 | 46080 | 369.88 | 0 | 0.10 |
| ex-A07 | 7 | 645120 | 5763.65 | 0 | 0.37 |

In Table 4, we introduced a modification in the function, so that the cell could not be a match, and of course the ROBDD checks are to be considered for all the possible permutations and phases.

In every case, the structural analysis appears to be very rapid. It handles very well the complexity.

## 5 CONCLUSION

The two decomposition methods presented here produce multi-level networks from a ROBDD description. The direct decomposition is very fast and produces good results. Of course, to be effective it needs a previous reordering to reduce the ROBDD size. It is easy to tune the complexity of the node functions (support size of the function). Fine granularity decomposition may be obtained by restricting the support size to two variables, for example. This produces a larger network, but with simpler node functions. Reducing the amount of sharing of a subfunction will also reduce the network size, by identifying subfunctions that are used several times in the multi-level representation. It should be used concurrently with the support size, as an alternative cost function. The path oriented decomposition has not yet produced interesting results : it has to be improved by a deeper analysis of the splitting node selection. In conclusion, the direct decomposition seems particularly interesting for the synthesis on multiplexor-based cells, for instance. Current research includes the application of this algorithm in the logic synthesis of MUX-based FPGAs, and the development of new Boolean decomposition techniques based on partial symmetry detection in ROBDDs.

We have analysed various methods to solve the matching problem. Characteristic signatures and symmetric variables are really efficient to reduce the search space of the matching problem. But it may leave ambiguities between some input variables and for some phases. Moreover, these techniques are not applicable in the presence of don't cares. Using the new structural analysis proposed here, incorrect variable permutations can be easily checked, and the input phase assignment can be directly determined. Compared with the ROBDD based techniques (Mailhot, 1993), the time needed to compute the ROBDDs for the library cell with all the possible input variable permutations and input phase assignments, can be saved.

This structural matching technique will be implemented in a complete technology mapper, along with the proposed decomposition method, and tested on realistic examples, mapping on standard cells and on FPGAs.

## 6 REFERENCES


Aho, A., Hopcroft, J. and Ullman, J. (1983) *Data Structures and Algorithms*. Addison-Wesley, Reading, Massachusetts.

Brayton, R. K., Rudell, R., Sangiovanni-Vincentelli, A. and Wang, A. (1987) MIS: A multiple-Level Logic Optimization System. *IEEE Trans. on Computer-Aided Design*, **CAD-6**, no 6.

Bryant, R.E. (1986) Graph-Based Algorithm for Boolean Function Manipulation. *IEEE Trans. on Computers*, **C-35**, no 8.

Burgun, L. (1993) Multi-level Synthesis Based on Binary Decision Diagrams. *VIII SBMicro*, Campinas.

Calazans, N., Zhang, Q., Jacobi, R., Yernaux, B. and Trullemans, A-M. (1992) Advanced Ordering and Manipulation Techniques for Binary Decision Diagrams. *Proceedings EDAC*, Bruxelles.

Cheng, D. I. et al (1993) Verifying equivalence of functions with unknown input correspondence. *EDAC/EURO-ASIC*, Paris.

Darringer, J. A. et al. (1984) LSS: A System for Production Logic Synthesis. *IBM Journal of Research and Development*, **28**, no 5.

Jacobi, R., Calazans, N. and Trullemans, C. (1991) Incremental Reduction of Binary Decision Diagrams. *Proceedings of International Symposium on Circuits and Systems*, Singapore.

Jacobi, R. et al (1993-1) Boolean mapping with binary decision diagrams. *IFIP Workshop on Logic and Architecture Synthesis*, Grenoble.

Jacobi, R. (1993-2) *A Study of the Application of Binary Decision Diagrams on Multi-level Logic Synthesis*. Doctoral Thesis, Université Catholique de Louvain, Belgium.

Keutzer, K. (1987). DAGON: Technology Binding and Local Optimization by DAG Matching. *24th ACM/IEEE Design Automation Conference*.

Mailhot, F. et al (1993) Algorithms for technology mapping based on binary decision diagrams and on Boolean operations. *IEEE Trans. on Computer-Aided Design*, **CAD-12**, no 5.

Mohnke, J. et al (1993) Permutation and phase independent Boolean comparison. *EDAC/EURO-ASIC*, Paris.

Schlichtmann, U. et al (1992) Characterization of Boolean functions for rapid matching in EPGA technology mapping. *29th DAC*.

Zhang, Q. (1994) *Logic verification and design error diagnosis for combinational circuits*. Ph.D. Dissertation, Université Catholique de Louvain, Belgium.

## 7 BIOGRAPHY

**Anne-Marie Trullemans-Anckaert** was born in Brussels, Belgium, in 1944. She received the engineering degree and the Doctorat en Sciences Appliquées from the Université Catholique de Louvain (UCL), Belgium, in 1967 and 1974, respectively. From 1967 she undertook research, at the Laboratoire de Microélectronique of UCL, in circuit simulators, symbolic layout editors and digital CAD tools, and was involved in the teaching of microelectronics. She is now Associated Professor at UCL. Her interest is in the development of CAD tools for the simulation and the design of MOS VLSI. Recently, she concentrate her research on digital circuits synthesis tools from HDL.

**Ricardo Pezzuol Jacobi** received his B.S. degree in electrical engineering and his M.S. degree in Computer Science at the Universidade Federal do Rio Grande do Sul (UFRGS), Brazil, in Dec. 82 and Aug. 86, respectively. He received his Doctoral degree from Université Catholique de Louvain, Belgium, in Dec. 93. His research interest includes logic synthesis for FPGAs, layout oriented technology mapping and software and hardware codesign. He is an adjoin professor of the Informatics Institute at the UFRGS since Dec. 1989.

**Qinhai Zhang** received the B.Sc. degree in physics and the M.Sc. degree in electrical engineering, both from Fudan University, Shanghai, China. In 1995, he received the Ph.D. degree from the Université Catholique de Louvain, Belgium. Currently he is a research assistant at the Université Catholique de Louvain, Belgium. He was an assistant from 1985 to 1987, and a lecturer from 1988 to 1990, in the E.E. department at Fudan University, Shanghai, China. In 1990, he joined the Electronics Department of Université Catholique de Louvain. His research interests include logic synthesis, formal verification, design error diagnosis and circuit layout.

# 5

# Efficient ROBDD based computation of common decomposition functions of multi-output boolean functions *

*Christoph Scholl* [†], *Paul Molitor* [‡]
[†] *Universität des Saarlandes*
*Department of Computer Science, D 66041 Saarbrücken, FRG.*
[‡] *Martin-Luther Universität Halle*
*Department of Computer Science, D 06099 Halle (Saale), FRG.*

**Abstract**

One of the crucial problems multi-level logic synthesis techniques for multi-output boolean functions $f = (f_1, \ldots, f_m) : \{0,1\}^n \rightarrow \{0,1\}^m$ have to deal with is finding sublogic which can be shared by different outputs, i.e., finding boolean functions $\alpha = (\alpha_1, \ldots, \alpha_h) : \{0,1\}^p \rightarrow \{0,1\}^h$ which can be used as common sublogic of good realizations of $f_1, \ldots, f_m$.

In this paper we present an efficient ROBDD based implementation of this COMMON DECOMPOSITION FUNCTIONS PROBLEM (CDF).

Formally, CDF is defined as follows: Given $m$ boolean functions $f_1, \ldots, f_m : \{0,1\}^n \rightarrow \{0,1\}$, and two natural numbers $p$ and $h$, find $h$ boolean functions $\alpha_1, \ldots, \alpha_h : \{0,1\}^p \rightarrow \{0,1\}$ such that $\forall 1 \leq k \leq m$ there is a decomposition of $f_k$ of the form

$$f_k(x_1, \ldots, x_n) = g^{(k)}(\alpha_1(x_1, \ldots, x_p), \ldots, \alpha_h(x_1, \ldots, x_p), \alpha_{h+1}^{(k)}(x_1, \ldots, x_p), \ldots, \alpha_{r_k}^{(k)}(x_1, \ldots, x_p), x_{p+1}, \ldots, x_n)$$

using a minimal number $r_k$ of single-output boolean decomposition functions.

## 1 INTRODUCTION

The long term goal for logic synthesis is the automatic transformation from a behavioral description of a boolean function to near-optimal netlists, whether the goal is minimum delay, minimum area, or some combination. Most of the approaches attacking the multi-level logic synthesis problem use gate count as optimization criterion. A survey can be found in Brayton (1990). Alternatively, some recent papers (e.g. Hwang (1992), Lai (1993), Schlichtmann (1993)) propose an approach different from the one addressed above. This approach to multi-level logic synthesis which originates from Ashenhurst (1959), Curtis (1961), and Karp (1963) is based on minimizing communication complexity. The methods

*This work was supported in part by DFG grant SFB 124 and the Graduiertenkolleg of the Universität des Saarlandes

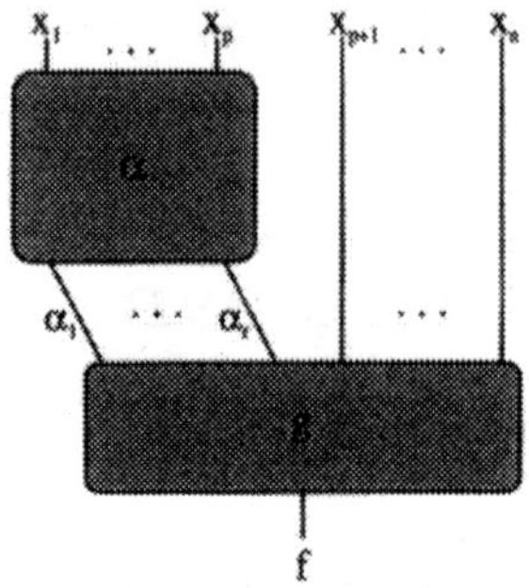

**Figure 1** Decomposition of a boolean function $f : \{0,1\}^n \to \{0,1\}$

used to reduce communication complexity employ functional decomposition, i.e., given a boolean function $f : \{0,1\}^n \to \{0,1\}$ they are looking for boolean functions $\alpha$ and $g$, such that $f(x_1,\ldots,x_n) = g(\alpha(x_1,\ldots,x_p), x_{p+1},\ldots,x_n)$ holds for all $(x_1,\ldots,x_n) \in \{0,1\}^n$ (see Figure 1). $\alpha = (\alpha_1,\ldots,\alpha_r):\{0,1\}^p \to \{0,1\}^r$ is a multi-output boolean function. The goal is to find decompositions where the number $r$ of decomposition functions (i.e. the number of wires between block $\alpha$ and block $g$) is minimal. (If $r < p$, then the decomposition is called non-trivial.)

A fundamental step in logic synthesis is the identification of common sublogic. Methods to identify common sublogic by (algebraic or boolean) division were developed by Brayton et al. and included in the SIS package (Sentovich (1992)). In this paper we present a method to identify common sublogic for the case that boolean functions are realized by decomposition. This method comes into play when we have to process multi-output boolean functions $f = (f_1,\ldots,f_m) :\{0,1\}^n \to \{0,1\}^m$. Note that even if the original function $f$ has only 1 output ($m = 1$), in most cases we need a generalization to multi-output boolean functions when we apply functional decomposition recursively to $\alpha$ and $g$.

All $f_i$ are decomposed as single-output functions, but in order to identify common sublogic we make use of our freedom in the choice of the decomposition functions to compute such decomposition functions which can be used in the decomposition of as many $f_i$ as possible. Unlike Lai (1994) we avoid to compute the huge set of all possible decomposition functions for all $f_i$ to choose common decomposition functions of the functions $f_i$ from these sets. We present an algorithm which directly computes a maximum number of common decomposition functions of $f_1,\ldots,f_m$.

In contrast to Molitor/Scholl (1994), which was based on function tables and decomposition charts, we efficiently make use of REDUCED ORDERED BINARY DECISION DIAGRAMS (ROBDD) during the computation of common decomposition functions. ROBDDs (Bryant (1986)) are compact representations for many of the boolean functions encountered in typical applications. In this paper we show that it is possible to carry out all necessary steps based on ROBDDs. This increases the efficiency of the approach in a high degree. In particular, we show that the computation of *common decomposition functions* for the decomposition of several single-output functions can be performed efficiently based on ROBDD's.

Benchmarking results show the new method to be efficient with respect to layout size, signal delay and running time.

| $f_1$ | $x_4$ 00 … 1<br>$x_5$ 00 … 1<br>$x_n$ 01 … 1 |
|---|---|
| $x_1x_2x_3$ | |
| 0 0 0 | row pattern 1 |
| 0 0 1 | row pattern 1 |
| 0 1 0 | row pattern 2 |
| 0 1 1 | row pattern 2 |
| 1 0 0 | row pattern 3 |
| 1 0 1 | row pattern 3 |
| 1 1 0 | row pattern 3 |
| 1 1 1 | row pattern 2 |

| $f_2$ | $x_4$ 00 … 1<br>$x_5$ 00 … 1<br>$x_n$ 01 … 1 |
|---|---|
| $x_1x_2x_3$ | |
| 0 0 0 | row pattern 1 |
| 0 0 1 | row pattern 2 |
| 0 1 0 | row pattern 3 |
| 0 1 1 | row pattern 4 |
| 1 0 0 | row pattern 2 |
| 1 0 1 | row pattern 2 |
| 1 1 0 | row pattern 1 |
| 1 1 1 | row pattern 4 |

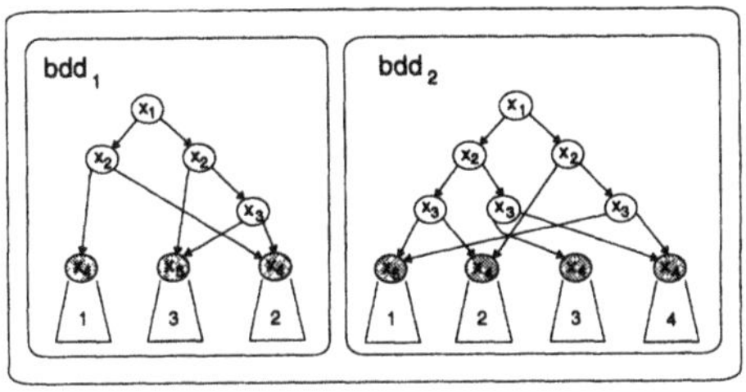


**Figure 2** Charts and ROBDDs of the single-output boolean functions $f_1$ and $f_2$ which are decomposed with respect to $(\{x_1, x_2, x_3\}, \{x_4, \ldots, x_n\})$. Each chart obviously consists of 8 rows. A row pattern is associated to each row. There are three (four) different row patterns in $M(f_1)$ $(M(f_2))$ denoted by the numbers 1, 2, and 3 (1 to 4). Thus, $r_1 = r_2 = 2$ holds.

## 2 BASIC DEFINITIONS

**Definition 1** *A* decomposition of a multi-output boolean function $f : \{0,1\}^n \rightarrow \{0,1\}^m$ *with respect to the input partition* $\{X_1, X_2\}$ $(X_1 = \{x_1, \ldots, x_p\}, X_2 = \{x_{p+1}, \ldots, x_{p+q}\}$, $p+q=n)$ *is a representation of $f$ of the form*
$f_k(x_1, \ldots, x_n) = g^{(k)}(\alpha_1^{(k)}(X_1), \ldots, \alpha_{r'_k}^{(k)}(X_1), X_2)$ $(\forall k \in \{1, \ldots, m\})$,
*with functions* $\alpha_i^{(k)} : \{0,1\}^p \rightarrow \{0,1\}$, *and* $g^{(k)} : \{0,1\}^{r'_k+q} \rightarrow \{0,1\}$. $\alpha_i^{(k)}$ *are called* decomposition functions of $f_k$. $g^{(k)}$ *is called* composition function of $f_k$.

With respect to a given input partition $\{X_1, X_2\}$, a single-output function $f_k$ can be represented as a $2^p \times 2^q$ matrix $M(f_k)$, the *decomposition matrix of $f_k$* or the *chart of $f_k$* with respect to $\{X_1, X_2\}$. (For illustration see Figure 2.) Each row and column of $M(f_k)$ is associated with a distinct assignment of values to the inputs in $X_1$ and $X_2$, respectively, such that $f_k(X_1, X_2) = M(f_k)[X_1, X_2]$ where $M(f_k)[X_1, X_2]$ represents the element of $M(f_k)$ which lies in the row associated with $X_1$ and the column associated with $X_2$.

Note that $(\alpha_1^{(k)}, \ldots, \alpha_{r'_k}^{(k)})$ of definition 1 encodes the rows of chart $M(f_k)$. Of course, the following property has to hold.

**Encoding Property:** If the row pattern of row $(v_1, \ldots, v_p) \in \{0,1\}^p$ differs from the row pattern of row $(v'_1, \ldots, v'_p) \in \{0,1\}^p$, then $(\alpha_1^{(k)}, \ldots, \alpha_{r'_k}^{(k)})$ has to assign different codes to $(v_1, \ldots, v_p)$ and $(v'_1, \ldots, v'_p)$.

The minimum number of communication wires required between the subcircuit which encodes the rows of $M(f_k)$ and the composition function $g^{(k)}$ is $\lceil \log p_1^{(k)} \rceil$ where $p_1^{(k)}$ is the number of distinct row patterns in $M(f_k)$. $r_k$ will denote value $\lceil \log p_1^{(k)} \rceil$ in the following. In the following we will always consider decompositions with minimal number $r_k = \lceil \log p_1^{(k)} \rceil$ of decomposition functions.

If $f_k$ is given by a ROBDD $bdd_k$, the minimal number of decomposition functions can be determined in an easy way too: For all $(v_1, \ldots, v_p) \in \{0,1\}^p$ the row pattern belonging to row $(v_1, \ldots, v_p)$ of $M(f_k)$ equals the function table of the cofactor $(f_k)_{x_1^{v_1} \cdot \ldots \cdot x_p^{v_p}}$ (with $x_i^0 = \overline{x_i}$ and $x_i^1 = x_i$). Thus the problem of determining the number $p_1^{(k)}$ of different row patterns of $M(f_k)$ is equivalent to the problem of computing the number of different cofactors $(f_k)_{x_1^{v_1} \cdot \ldots \cdot x_p^{v_p}}$. The ROBDD of the cofactor $(f_k)_{x_1^{v_1} \cdot \ldots \cdot x_p^{v_p}}$ is given by the sub-bdd of $bdd_k$ whose root is reached by starting at the root of $bdd_k$ and then following the path corresponding to $(v_1, \ldots, v_p)$. The roots of these cofactors are called *linking nodes* (the shaded nodes in Figure 2). Since $f_k$ is given by a ROBDD, the number of different linking nodes of $bdd_k$ obviously equals the number of different cofactors. The computational complexity of determining the number of different linking nodes is at most linear in the size of $bdd_k$ since it can be determined by traversing $bdd_k$ in a depth first search manner.

The rows of chart $M(f_k)$ induce a partition of $\{0,1\}^p$ into equivalence classes $K_1^{(k)}, \ldots, K_{p_1^{(k)}}^{(k)}$ such that $v, v' \in \{0,1\}^p$ belong to the same class $K_j^{(k)}$ if and only if the two corresponding row patterns of $M(f_k)$ are identical. We denote the corresponding equivalence relation by $\equiv_k$ and the set of the equivalence classes $\{K_1^{(k)}, \ldots, K_{p_1^{(k)}}^{(k)}\}$ by $\{0,1\}^p/\equiv_k$.

Since every equivalence class $K_j^{(k)}$ is associated to exactly one linking node $n_j^{(k)}$ and vice versa, we are able to compute $K_j^{(k)}$ from the ROBDD $bdd_k$ for $f_k$. We receive a BDD for the characteristic function of $K_j^{(k)}$ if we replace $n_j^{(k)}$ by the constant 1 and all other linking nodes by constant 0.

# 3 CDF

To compute decomposition functions (with domain $X_1$) of a multi-output function $f$ which are used by different single-output functions $f_k$, we have to consider the following problem which will be denoted by CDF.

**Given:** Let $f = (f_1, \ldots, f_m) : \{0,1\}^n \to \{0,1\}^m$ be a multi-output boolean function, $A = \{X_1, X_2\}$ with $X_1 = \{x_1, \ldots, x_p\}$ and $X_2 = \{x_{p+1}, \ldots, x_n\}$ be an input partition, and $h$ be a natural number with $h \leq r_k$ $(= \lceil \log p_1^{(k)} \rceil)$ $(\forall k)$.

**Find:** $h$ single-output boolean functions $\alpha_1, \ldots, \alpha_h$, which can be used as decomposition functions of every single-output function $f_k$ for $k = 1, \ldots, m$ such that there is a decomposition of $f_k$ with minimal number $r_k$ decomposition functions of the form
$f_k(x_1, \ldots, x_n) = g^{(k)}(\alpha_1(X_1), \ldots, \alpha_h(X_1), \alpha_{h+1}^{(k)}(X_1), \ldots, \alpha_{r_k}^{(k)}(X_1), X_2)$.
Of course, such $h$ boolean functions need not to exist. We have proven the problem whether such functions $\alpha_1, \ldots, \alpha_h$ exist to be NP-complete. Nevertheless, we have to solve CDF†. An algorithm which is applicable from the practical point of view (as shown by the benchmarking results) is presented in this section.

## 3.1 Theoretical Background

We start with a theoretical result working towards a solution to CDF. It gives a condition necessary and sufficient that $h$ single-output functions $\alpha_1, \ldots, \alpha_h : \{0,1\}^p \to \{0,1\}$ are common decomposition functions of $f_1, \ldots, f_m$. It is a generalization of a lemma shown by Karp (1963). For this, we need the following notations: Let $\theta^{(k)} : \{0,1\}^p \to \{1, \ldots, p_1^{(k)}\}$ be the function which maps $v \in \{0,1\}^p$ to the index $j$ of the class $K_j^{(k)}$ to which it belongs.

†The (maximal) value of parameter $h$ of CDF is determined by logarithmic search. After that we solve CDF for subsets of $\{f_1, \ldots, f_m\}$, but only for such subsets $\{f_{i_1}, \ldots, f_{i_l}\}$ where all pairs $f_{i_j}$ and $f_{i_k}$ have at least one common decomposition function. Note that this question is *not* NP-complete, but can be decided efficiently by dynamic programming. Also note that the algorithms for the computation of common decomposition functions given in this paper can be generalized in a canonical manner for the case that some of the decomposition functions $\alpha_i^{(k)}$ $(i > h)$ are already predetermined. More details of how the CDF algorithm is integrated in the tool can be found in Molitor/Scholl (1994).

Furthermore, for given $\alpha_{1,\dots,h}$ [‡] and all $a \in \{0,1\}^h$, let $S_a^{(k)}$ be the set $\{\theta^{(k)}(v);\ \alpha_{1,\dots,h}(v) = a\}$ of those classes which contain a row mapped to $a$ by $\alpha_{1,\dots,h}$. ($\alpha_{1,\dots,h}$ is not able to tell these rows apart (see the Encoding Property).) Note that $S_a^{(k)}$ and $S_{a'}^{(k)}$ need not to be disjoint for $a \neq a'$, and that the number $|S_a^{(k)}|$ of elements of $S_a^{(k)}$ equals the number of *distinct* row patterns of $M(f_k)$ mapped to $a$ by $\alpha_{1,\dots,h}$.

**Lemma 1** *$\alpha_1, \dots, \alpha_h$ are common decomposition functions of $f_1, \dots, f_m$ with respect to $\{X_1, X_2\}$ such that there is a decomposition of $f_k$ with minimal number of decomposition functions of the form*
$f_k(x_1, \dots, x_n) = g^{(k)}(\alpha_1(X_1), \dots, \alpha_h(X_1), \alpha_{h+1}^{(k)}(X_1), \dots, \alpha_{r_k}^{(k)}(X_1), X_2)\ \ \forall k \in \{1, \dots, m\}$
*if and only if* $\max\{|S_a^{(k)}|;\ a \in \{0,1\}^h\}$ *is* $\leq 2^{r_k - h}$ $(\forall k)$.

*Proof.* Since $(\alpha_{1,\dots,h}, \alpha_{h+1,\dots,r_k}^{(k)})$ has to assign different values to rows of chart $M(f_k)$ with different row patterns (see the Encoding Property), $\alpha_{h+1,\dots,r_k}^{(k)}$ has to assign different values to those rows which cannot be told apart by $\alpha_{1,\dots,h}$. As $\alpha_{h+1,\dots,r_k}^{(k)}$ can produce at most $2^{r_k-h}$ different values, the statement of the lemma follows. □

## 3.2 Solution

$h$ common decomposition functions $\alpha_1, \dots, \alpha_h$ can be computed (on principle) by a (simplified) branch and bound algorithm (see also Molitor/Scholl 1994). It constructs the function table of $\alpha_{1,\dots,h}$ row by row (assigning function values to all elements of $\{0,1\}^p$). Branches are pruned as soon as the condition of lemma 1 is violated for the initial part of the function table of $\alpha_{1,\dots,h}$ constructed so far.

In order to speed up the branch and bound algorithm and to receive 'simpler' decompositon functions we restrict our search for common decomposition functions to a subclass of functions, which we will call 'equivalence preserving decomposition functions'[§]:

**Definition 2** *A decomposition function $\alpha_i : \{0,1\}^p \to \{0,1\}$ of a boolean function $f_k : \{0,1\}^n \to \{0,1\}$ is said to* preserve equivalences *if $\alpha_i(v) = \alpha_i(v')$ holds for every $v, v' \in \{0,1\}^p$ with $v \equiv_k v'$.*

Common equivalence preserving decomposition functions $\alpha_1, \dots, \alpha_h$ of $f_1, \dots, f_m$ have to assign the same value to $v$ and $v' \in \{0,1\}^p$ whenever there is a $k \in \{1, \dots, m\}$ such that the rows of $M(f_k)$ corresponding to $v$ and $v'$ have identical row patterns. More formally, let
$v \sim v' \overset{\text{def}}{\iff} (\exists 1 \leq k \leq m)\ v \equiv_k v'$,
then the corresponding equivalence relation partitions the rows, i.e. $\{0,1\}^p$, into equivalence classes $E_1, \dots, E_l$ such that common equivalence preserving decomposition functions have to assign the same value to each $v \in E_i$. We will denote the set of these equivalence classes by $\{0,1\}^p/_\sim$.

Now we can modify our branch and bound algorithm, such that it makes assignments not to single elements of $\{0,1\}^p$ but to whole classes $E_i$. Since $l$ mostly is much smaller than $2^p$, this approach considerably reduces the running time (see also section 4).

---

[‡] $\alpha_{1,\dots,h}$ denotes the tuple $(\alpha_1, \dots, \alpha_h)$.

[§] In practical applications functions $f_k$ often have some desirable properties like symmetry in some variables or independence of some variables. Equivalence preserving decomposition functions 'preserve such properties'.

**Table 1** Experimental results.

| Circuit | Running time *mulop* | *mulopII* | ratio | CDF | Layout size *sis* | *mulopII* | ratio | Signal delay *sis* | *mulopII* | ratio |
|---|---|---|---|---|---|---|---|---|---|---|
| 9symml | 1.40 | 1.23 | 1.14 | 0.69% | 1194336 | 201400 | 5.93 | 27.6 | 13.6 | 2.03 |
| C17 | 0.32 | 0.15 | 2.13 | 0.01% | 28800 | 31744 | 0.91 | 4.2 | 4.2 | 1.00 |
| cm138a | 1.01 | 0.18 | 5.61 | 0.89% | 103896 | 87480 | 1.19 | 5.8 | 6.8 | 0.85 |
| cm151a | 4.16 | 1.09 | 3.82 | 0.13% | 95312 | 177712 | 0.54 | 12.6 | 16.4 | 0.77 |
| cm152a | 2.15 | 0.50 | 4.30 | 0.36% | 85536 | 106704 | 0.80 | 10.0 | 13.2 | 0.76 |
| cm162a | 350.65 | 3.32 | 105.62 | 0.26% | 131976 | 192000 | 0.69 | 12.0 | 13.2 | 0.91 |
| cm163a | 2923.31 | 2.35 | 1243.96 | 0.08% | 144008 | 164416 | 0.88 | 13.0 | 10.4 | 1.25 |
| cm82a | 0.38 | 0.21 | 1.81 | 0.01% | 74784 | 61600 | 1.21 | 7.2 | 7.0 | 1.03 |
| cm85a | 7.46 | 3.73 | 2.00 | 0.27% | 165456 | 180000 | 0.92 | 10.2 | 11.0 | 0.93 |
| cmb | 1836.13 | 2.52 | 728.62 | 0.05% | 204792 | 123496 | 1.66 | 9.4 | 6.8 | 1.38 |
| decod | 26.15 | 2.56 | 10.21 | 1.93% | 140448 | 119496 | 1.18 | 6.2 | 5.0 | 1.24 |
| f51m | 3.14 | 1.83 | 1.71 | 0.28% | 561184 | 251392 | 2.23 | 51.0 | 18.4 | 2.77 |
| majority | 0.44 | 0.08 | 5.50 | 0.01% | 42200 | 39168 | 1.00 | 7.8 | 6.6 | 1.18 |
| parity | 111.06 | 1.37 | 81.07 | 0.00% | 99408 | 96976 | 1.03 | 5.0 | 5.0 | 1.00 |
| z4ml | 0.66 | 0.76 | 0.87 | 0.16% | 156288 | 103896 | 1.50 | 16.2 | 9.8 | 1.65 |
| $\sum$ | | | | | 3228K | 1937K | 1.67 | 198.2 | 147.4 | 1.34 |

As already mentioned in section 2 ROBDDs $bdd_1^{(k)}, \ldots, bdd_{p_1^{(k)}}^{(k)}$ for the characteristic functions of the equivalence classes $K_1^{(k)}, \ldots, K_{p_1^{(k)}}^{(k)}$ with respect to $\equiv_k$ can easily be computed from the ROBDDs of the $f_k$. To compute the characteristic functions of the equivalence classes with respect to $\sim$, we implicitely construct a graph $G = (V, E)$ where the set $V$ of vertices is given by the ROBDDs $bdd_j^{(k)}$ representing the equivalence classes $K_j^{(k)}$. At the end, there is an undirected edge $\{bdd_{j_1}^{(k_1)}, bdd_{j_2}^{(k_2)}\}$ if and only if $bdd_{j_1}^{(k_1)} \wedge bdd_{j_2}^{(k_2)} \neq \emptyset$, i.e., iff $K_{j_1}^{(k_1)} \cap K_{j_2}^{(k_2)} \neq \emptyset$. Obviously, there is a one-to-one relation between the set of the connected components (in the graph-theoretical sense) of $G$ and the set of the equivalence classes $\{0,1\}^p/_\sim$. For every class $E_i$, there is a connected component $CC_i$ of $G$ such that the logical-or of the ROBDDs $bdd_j^{(k)}$ (for any fixed $k$) corresponding to vertices of $CC_i$ results in a representation of $E_i$ and vice versa.

## 4 EXPERIMENTAL RESULTS

We applied our tool, which uses the CDF algorithm described above as basis, to a number of benchmarks of the 1991 MCNC multi-level logic benchmark set. We will call the ROBDD based implementation of our tool *mulopII*. Our former implementation working on charts will be called *mulop* (Molitor/Scholl 1994).

Columns 1–3 of Table 1 show running times (in CPU seconds, measured on a SPARC-station 10/30 (64 MByte RAM)) of our ROBDD based implementation *mulopII* compared to those of our former version *mulop*. Experiments prove our ROBDD based version to be much more efficient than the former version.

The next column of Table 1 shows the fraction of running time which is used in the computation of common decomposition functions compared to the total running time of the tool *mulopII*. It shows that only a very small fraction of the total running time is used for the computation of common decomposition functions. The running time is dominated by the computation of good input partitions, *not* by the computation of common decomposition functions. This confirms our approach to compute common decomposition functions rather than to encode linking nodes in a straightforward manner.

Columns 5–7 of Table 1 show a comparison between *sis* (Sentovich (1992)) and *mulopII* with respect to layout size¶. For almost two thirds of the benchmark set, our approach dominates (or is as good as) that of *sis* with respect to layout size. Nevertheless, the signal delays of our realizations for more than two thirds of the circuits considered are better (or equal) than those of the realizations synthesized by *sis* (see columns 8–10 of Table 1).

## 5 CONCLUSION

We have presented a ROBDD based technique for computing common decomposition functions of multi-output boolean functions. This algorithm has been integrated in our multilevel synthesis tool which has been presented in Molitor/Scholl (1994) where more details of how the CDF algorithm is integrated can be found. The benchmarking results show that most of the circuits constructed by our synthesis tool are very efficient. They also prove it to be applicable in terms of running time.

## REFERENCES

R.L. Ashenhurst. The decomposition of switching functions. In *Proceedings on an International Symposium on the Theory of Switching* held at Comp. Lab. of Harvard University, pages 74–116, 1959.

R.K. Brayton, G.D. Hachtel, and A. L. Sangiovanni-Vincentelli. Multilevel logic synthesis. *Proceedings of the IEEE*, 78(2):264–300, February 1990.

R.E. Bryant. Graph-based algorithms for boolean function manipulation. *IEEE Trans. on Computers*, C-35(8):677–691, August 1986.

H.A. Curtis. A generalized tree circuit. *J. Assoc. Comput. Mach.*, 8:484–496, 1961.

T. Hwang, R.M. Owens, and M.J. Irwin. Efficient computing communication complexity for multilevel logic synthesis. *IEEE Trans. on CAD*, CAD-11(5):545–554, May 1992.

R.M. Karp. Functional decomposition and switching circuit design. *Journal of Society of Industrial Applied Mathematics*, 11(2):291–335, June 1963.

Y. Lai, M. Pedram, and S. Vrudhula. BDD based decomposition of logic functions with application to FPGA synthesis. In *IEEE/ACM Design Automation Conference DAC93*, pages 642–647, 1993.

Y. Lai, K. Pan, and M. Pedram. FPGA Synthesis using Function Decomposition. In *Proceedings of ICCD94*, pages 30–35, 1994.

P. Molitor, C. Scholl. Communication based multilevel synthesis for multioutput boolean functions. In *Proceedings of the 4th Great Lakes Symposium on VLSI, Notre Dame, Indiana*, March 1994.

U. Schlichtmann. Boolean Matching and Disjoint Decomposition for FPGA Technology Mapping. In *Proceedings of the IFIP Workshop on Logic and Architecture Synthesis*, pages 83–102, 1993.

E. Sentovich et al. SIS: a system for sequential circuit synthesis. Department of EE and CS, UC Berkeley, May 1992.

¶The technology library consists of the 2-input gates from *stdcell2_2.genlib* available in *octtools*. Placement and routing was done by *TimberWolf* integrated in *octtools*.

6

# Circuit depth optimization by BDD based function decomposition

*Alexander I. Kornilov, Tatiana Yu. Isaeva*
*Research Institute of VLSI CAD Systems,*
*8a, Mazhorov per., Moscow E-023, 105023 RUSSIA,*
*phone: 7 (095) 369 19 75 fax: 7 (095) 369 51 07*
*e-mail: korn@sapran.msk.su*

**Abstract**

We present a BDD-based algorithm for logic functions' decomposition that is aimed at speed-up of circuits constructed as straightforward mapping of BDDs. Existing algorithms (including BDD-based) aimed at circuits' size optimization do not solve this problem. We propose a novel decomposition formula; BDDs' graphical properties are explicitly used. This allows to reduce circuits' performance sufficiently. A class of BDDs occurring in arithmetic units' synthesis is described for which the algorithm can be applied iteratively.

**Keywords**

High-performance circuits, functions' decomposition, Binary-Decision Diagrams.

## 1 INTRODUCTION

This paper is the result of the research carried out in the field of high-performance combinational circuits synthesis. The proposed algorithm is based on Reduced Ordered Binary Decision Diagrams (further called BDD's for short) proposed by R.E.Bryant (Bryant,1986), that provide a convenient representation for most of the logic functions in practical use. We do not concern the problem of variable ordering since there exist several algorithms that find a good ordering if possible. This function representation is widely used in the field of logic synthesis. BDD's proved to be an effective structure for the implementation of the traditional synthesis methods such as function decomposition and factorization, ISOP form construction. Again several synthesis algorithms were developed that construct a network as the straightforward mapping of its BDD structure (e.g., in static CMOS base, multiplexor base or FPGA base). For such circuits the assumed BDD depth directly defines their performance. However, a number of familiar modern algorithms minimize BDD size only, leaving behind the problem of depth optimization.

Familiar methods of decomposition (including BDD-based decomposition) aimed at simplification of Boolean functions do not solve the problem at least for such important class of Boolean functions as carry functions in arithmetic units.

We made an attempt to take advantage of the fact that we use BDD structure both as the circuit model and as the function representation convenient for minimization. In this paper we describe a BDD based function decomposition algorithm aimed at the speed-up of the circuit constructed as the straightforward BDD implementation. The main goal of such decomposition is to assign a complicated function instead of a variable to the BDD root and hence reduce the BDD depth.

In the first section we remind the familiar Shannon's decomposition. The second section is devoted to the proposed algorithm description. In the third section we describe a class of BDD's connected with arithmetic units for which the algorithm is especially good due to the opportunity to apply it iteratively.

## 2 SHANNON'S DECOMPOSITION

Consider the Shannon's decomposition formula with respect to the first k variables.

$$(1)\quad f(x_1,...,x_n) = \bigvee_{\forall (a_1,...,a_k),\ a_i\in\{0,1\}} x_1^{a1} \,\&\, x_2^{a2} \,\&...\&\, x_k^{ak} \,\&\, f(a_1,a_2,...,a_k,x_{k+1},...,x_n).$$

We can rewrite this equivalence in the following way:

$$(2)\quad f(x_1,...,x_n) = \bigvee_{i\in I} g_i(x_1,...,x_k) \,\&\, h_i(x_{k+1},...,x_n).$$

Here $I=\{1,...,w\}$, where w is the number of different cofactors, $H = \{h_i,\ i\in\ \}$ is the set of these cofactors and $G = \{g_i,\ i\in I\}$ is the set of corresponding characteristic functions:

$$g_i(a_1,...,a_k).= \begin{cases} 1, & \text{if } h_i(x_{k+1},...,x_n) \equiv f(a_1,a_2,...,a_k,x_{k+1},...,x_n) \\ 0, & \text{otherwise.} \end{cases}$$

Consider BDD implementing the function f for a certain variable ordering. For definiteness assume that the variables are ordered in the natural way: $x_1,x_2,...,x_n$. Assume each internal BDD vertex v has an index of variable it is labeled, terminal vertices having index n+1. Let us denote this value index(v).

From formula (2) we immediately obtain a disjunctive decomposition of function f. According to this formula we can draw a crossing line that divides function f BDD into higher and lower parts, the former consisting of vertices v such that index(v)≤k, and the latter consisting of vertices v whose index(v)>k. Note that the lower part of the BDD implement the system of cofactors for a given k.

There exist several BDD patterns that at once lead to good function decomposition. E.g., if there are only two cofactors and one of them is constant, the function is implemented as disjunction (for constant 1) or conjunctions (for constant 0) of two functions. If the two cofactors are complementary functions the function is implemented as EXOR of two functions.

In general, if the number of cofactors is comparatively small we obtain a good function implementation. This case (disjunctive decomposition) was described in (Lai,1993), as well as a nondisjunctive decomposition algorithm for size optimization.

Function f may just be implemented as the independent implementation of system of characteristic functions G, system of cofactors H and formula (2). The total depth of this implementation is the sum of depths of formula (2) implementation ($\approx \log_2 w$, if it is implemented as a pyramidal structure) and maximal depth for the systems G and H (for these systems can be implemented in parallel). System H is implemented as the lower part of the initial BDD. Implementation for each of the functions $g_i$ can be obtained from the upper part of the initial BDD. Thus we finally obtain implementation of less depth, not exceeding $\log_2 w + \max(k, n-k)$.

Unfortunately if the number of different cofactors is big this implementation is impractical both for size and depth minimization.

## 3 DECOMPOSITION ALGORITHM FOR DEPTH REDUCTION

In this section we describe the advanced decomposition algorithm for depth reduction.

Let us transform the formula (2) in the following way:

$$(3) \qquad f(x_1,\ldots,x_n) = \bigvee_{i \in I_0,} g_i(x_1,\ldots,x_k) \,\&\, h_i(x_{k+1},\ldots,x_n) \vee g_0(x_1,\ldots,x_k) \,\&\, h_0(x_1,\ldots,x_n),$$

where $I_0$ is a subset of I, the function

$$g_0(x_1,\ldots,x_k) = \bigvee_{i \in I \setminus I_0,} g_i(x_1,\ldots,x_k)$$

and the function

$$h_0(x_1,\ldots,x_n) = \begin{cases} \bigvee_{i \in I \setminus I_0,} g_i(x_1,\ldots,x_k) \,\&\, h_i(x_{k+1},\ldots,x_n), & \text{if } g_0(x_1,\ldots,x_k) = 1; \\ \text{don't care}, & \text{otherwise.} \end{cases}$$

Note that unlike the previous case this formula contains only a subset of cofactors and characteristic functions; and the rest of them are gathered in the last term, that includes the special function $h_0$, depending on the whole set of input variables.

We choose the set of indexes $I_0$ so that to break only long paths in the BDD. Further we describe how we do this.

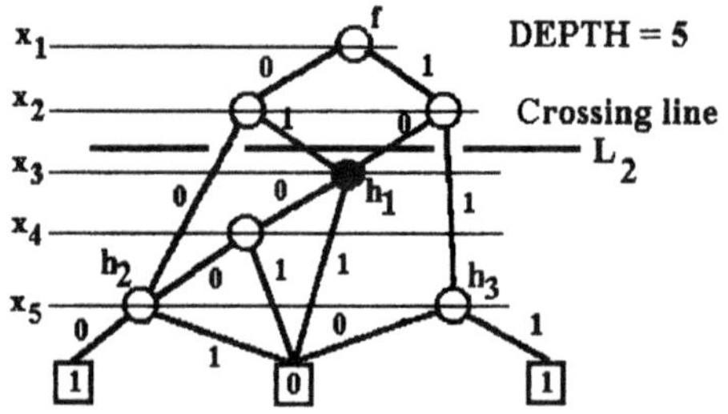


$f = x_1(x_2 x_5 \vee \bar{x}_2 \bar{x}_3 \bar{x}_4 \bar{x}_5) \vee \bar{x}_1(x_2 \bar{x}_3 \bar{x}_4 \bar{x}_5 \vee \bar{x}_2 \bar{x}_5)$

Figure 1. BDD example: $k = 2$; $l_0 = 3$; $I = \{1,2,3\}$; $I_0 = \{1\}$.

<u>Definition.</u> The edge cutset $E_k$ of the BDD F at level k is the set of edges (v1,v2) for which index(v1)<=k, and index(v2)>k.

That is, if we draw a line dividing the BDD into upper and lower parts, the former consisting of the vertices with indexes less or equal than k, and the latter -- of the vertices with the indexes greater than k, Ek is the set of all the BDD edges, crossing that line.

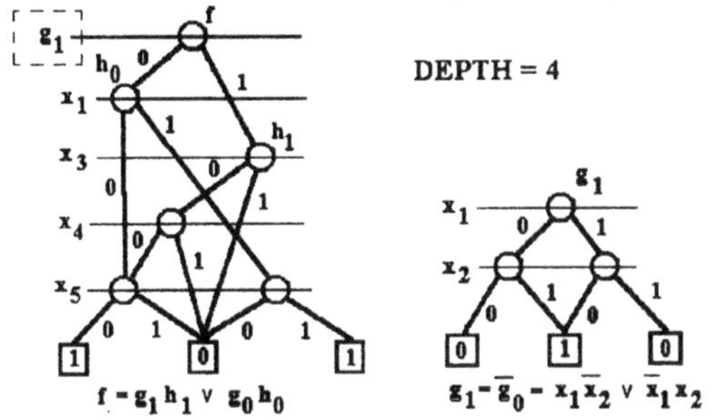


Figure. 2. Resulting implementation for the example function.

Given a set E of BDD edges denote Ends(E) the set of all the end vertices of the edges from E. Clear that the set Ends($E_k$) is the set of all the BDD vertices corresponding to the functions from H. Let us set the critical length $l_0$

<u>Definition.</u> $E_k(l_0)$ is the set of edges from $E_k$, belonging to the paths that are longer than $l_0$.

Clear that Ends($E_k(l_0)$) $\subseteq$ Ends($E_k$), and in many cases these sets are not the same. In Figure.1 an example of such a BDD is given. We choose $I_0$ as the set of indexes of the cofactors corresponding to the BDD vertices from Ends($E_k(l_0)$).

As in the previous case the system of cofactors $H_0 = \{h_i, i \in I_0\}$ is implemented by the lower part of BDD, and BDD's for the system of characteristic functions $G_0 = \{g_i, i \in I_0 \cup \{0\}\}$ can be built using the upper part of the BDD. Let us describe construction of these BDD's in more details.

Each of the characteristic functions can be considered as the reachability function (as defined by N. Ishiura in (Ishiura,1992)) between the initial BDD root and the BDD vertex representing the corresponding cofactor. We can build BDD for a function $g_i$ ($i \neq 0$) in the following way:

- consider a subgraph consisting of all the paths connecting BDD root to the vertex representing cofactor $h_i$ (preserving all the labels);
- consider the initial BDD root as the new BDD root;
- replace the vertex corresponding to the cofactor $h_i$ with the terminal vertex 1;
- add missing complementary edges directing them to the terminal vertex 0.

Function $g_0$ is implemented as the other characteristic function save that we replace by terminal vertex 1 all the vertices corresponding to the cofactors $h_i$, $i \in I \backslash I_0$.

We can see that the depths of the BDD's for the cofactors and the characteristic functions are determined by the path's' lengths in the lower and upper BDD parts, and thus we can just estimate them as the number of input variables of these functions.

Unlike these functions, the special function $h_0$ depends on the whole set of input variables. Of course, some of them can be insufficient, but we don't know that beforehand. But for this function we can explicitly build BDD with depth not exceeding the chosen critical depth.

Consider all the paths in the initial BDD connecting its root to the terminal vertices passing through the BDD vertices from the set Ends($E_k$)\Ends($E_k(l_0)$). These vertices correspond to the cofactors that are not implemented separately. These paths form a directed acyclic graph that is not a BDD for some of the complementary edges are missing, including the edges from $E_k(l_0)$. For every vertex with only one pull-down edge let us insert an edge directed to the same vertex as its complementary edge. Thus we obtain a new non reduced BDD implementing the function $h_0$ (setting its don't care values). Its depth does not exceed the critical depth because all the edges from the critical paths cutset are removed, and the inserted edges do not bring in long paths.

The total implementation depth is defined by the depth of the formula (3) implementation ($\approx \log_2|I_0|$); the depth of BDD's for the systems $G_0$ and $H_0$; the depth of function $h_0$ BDD. Thus it does not exceed $\log_2|I_0|+\max(l_0,k,n-k)$. If $k \approx n/2$ the depth is $\approx \log_2|I_0|+\max(l_0,n/2)$. Taking advantage of function $h_0$ don't care values often leads to the reduction by 1 of its BDD depth (See Figure.2).

It is important that for fixed crossing line level k and critical depth $l_0$ the time complexity of the proposed algorithm is linear on the number of BDD vertices. But of course to minimize the particular implementation depth a certain search may be carried out, e.g., a minimal cutting set may be found.

## 4 ALGORITHM APPLICATION FOR ARITHMETIC UNITS

There is a BDD type that often occurs in arithmetic units synthesis -- BDD's, that contain a single long path, the subBDD's at the ends of the complement edges along this path being of constant depth (e.g., subBDD's representing input variables and their complements). In this particular case the BDD edge cutset consists of a single edge, and the formula (3) can be rewritten as follows:

$$(4) \quad f(x_1,\ldots,x_n) = \quad g_1(x_1,\ldots,x_k) \,\&\, h_1(x_{k+1},\ldots,x_n) \vee \hat{}g_1(x_1,\ldots,x_k) \,\&\, h_0(x_1,\ldots,x_n)$$

BDD's for the functions $h_1$, $g_1$ and $h_0$ are of the same type as for the function f. This allows the iterative use of the method, thus yielding pyramidal structure instead of the initial linear structure. The structure depth is reduced from O(n) to $O(\log_2 n)$ while its size just slightly increases.

Basing on these principles we have developed a methodology to synthesize high-performance arithmetic units. We have designed a look ahead carry generator (LACG) as one of the applications of this methodology.

Let us consider a well known LACG (Motorola) based on the traditional 2-level synthesis principles for the purpose of comparison. To implement this generator in a basis of static CMOS circuits 116 transistors are required. Two 4-input NAND elements are connected sequentially in this circuit and produce the main part of the whole delay that is rather high.

Our LACG is based on the BDD technology principles (Kornilov,1988). To implement this generator in a basis of static CMOS circuits only 80 transistors are required. In this case the delay producing circuit consists of two sequentially connected passive multiplexors and two invertors with performance approximately equivalent to a performance of 3-inputs NAND gate (it also has the depth equal 3). Hence we can claim that the proposed generator is at least two times faster than first one and requires for its implementation approximately 30% less transistors.

**Table 1:** Comparison of the LACG implementations for different synthesis styles.

| Two-level Synthesis | | BDD-based Decomposition | |
|---|---|---|---|
| Output | Fall Delay/Rise Delay | Output | Fall Delay/Rise Delay |
| GE | 4.60/4.90 | AF | 3.28/2.72 |
| PR | 1.06/4.30 | MF | 1.20/2.56 |
| CO | 5.30/5.72 | CO | 4.14/3.86 |

The developed methodology enables to construct high-performance arithmetic units without using LACG.

This particular algorithm was described in (Kornilov,1990).

## 5 CONCLUSIONS

The proposed algorithm allows to reduce sufficiently the BDD representation depth without big size increase, and thus optimize performance of the circuits obtained by straightforward mapping algorithms. In some cases the structure depth may be reduced from linear to logarithmic in the number of input variables.

## REFERENCES

Bryant R. E. (1986) Graph-Based Algorithms, *IEEE Trans. Comput., Vol. C-35, no.8, Aug. 1986, 677-691.*

Lai Yung-Te , Pedram Massound (1993) BDD Based Decomposition of Logic Function with Application to FPGA Synthesis, *Proceedings of 30th ACM/IEEE Design Automation Conference, 1993, 642-647.*

Ishiura N. (1992) Synthesis of Multi-Level Logic Circuits from Binary Decision Diagrams. *Synthesis and Simulation Meeting and International Interchange, Proccedings of SASIMI-92, 1992, 74-83.*

Motorola H4C Series Design Reference Guide, 7-165.

Kornilov A. I. et al. (1988) A Carry Forming Device, *Inventor's Certificate of the USSR No. 1608648, G 06 F 7/50.*

Kornilov A. I. (1990) High Performance Arithmetic Device Construction Using a Universal Logical Base, *Communication Facilities Engineering. Ser. Microelectronic Hardware, Vol. 1-2 (12-13), 1990, 41-47 (in Russian).*

7

# Symmetry Based Variable Ordering for ROBDDs

*Dirk Möller*, Paul Molitor*, Rolf Drechsler***

**Martin-Luther-Universität Halle*

*Institut für Informatik, Fachbereich Mathematik und Informatik, Martin-Luther-Universität Halle, 06099 Halle, Germany.*

*Tel: +49-345-622522. Fax: +49-345-622514.*

*email:* {moeller,molitor}@informatik.uni-halle.de

***Johann Wolfgang Goethe-Universität Frankfurt*

*Fachbereich 20 - Informatik, J. W. Goethe-Universität, Robert-Mayer-Straße 11-15, 60054 Frankfurt am Main, Germany.*

*Tel: +49-69-79828472. Fax: +49-69-79828250.*

*email:* drechsle@kea.informatik.uni-frankfurt.de

**Abstract**

Reduced Ordered Binary Decision Diagrams (ROBDDs) are a data structure frequently used for representation and manipulation of Boolean functions. Since the size of ROBDDs is extremely sensitive to the variable order a lot of heuristics to get a good variable order have been developed. For the class of partially symmetric Boolean functions this paper presents a new general method to improve quality of ordering heuristics based on the exchange of variables. Statistical and benchmark results are given to show the efficiency of our approach.

**Keywords**

variable ordering, symmetric Boolean functions

*Supported in part by DFG grant SFB 124/B6

**Supported in part by DFG grant Be 1176/4-2

# 1 INTRODUCTION

Binary Decision Diagrams (BDDs) as a data structure for representation of Boolean functions were first introduced by Lee (1959) and further popularized by Akers (1978) and Moret (1982) . In the restricted form of ROBDDs they gained widespread application, because ROBDDs are a canonical representation and allow efficient manipulations (Bryant 1986) . Some fields of application are logic design verification, test generation, fault simulation, and logic synthesis (Malik 1988, Bryant 1992). Most of the algorithms using ROBDDs have running time polynomial in the size of the ROBDDs. The sizes themselves depend on the variable order used. Thus, there is a need to find a variable order that minimizes the number of nodes in an ROBDD.

As an example of the application of ROBDDs consider the use of Field Programmable Gate Arrays (FPGA) in the construction of Combinational Logic Circuits (CLC). A BDD has a direct correspondence to a CLC when each node of the BDD is substituted by a multiplexer. Since it is straightforward to map these multiplexer circuits on a FPGA whose logic blocks are based on multiplexers, BDDs have become a good framework for logic synthesis. The usage of BDDs also simplifies the technology mapping and the routing for FPGAs (Brown 1992). Here it is already useful to save only a few nodes using a good order. Additionally, these circuits have nice testability properties (Becker 1992).

The existing methods for finding good variable orders can be classified into three categories. The first are initial heuristics starting from a CLC (Malik 1988, Fujita 1988, Fujii 1993), the second are gradual improvement heuristics based on the exchange of variables in the ROBDD (Ishiura 1991, Fujita 1991, Rudell 1993, Felt 1993), and the third are exhaustive methods to find a mininum order (Friedman 1990). The gradual improvement heuristics play an important role in dynamic variable reordering, since these methods are able to handle large circuits that cannot be represented without dynamic methods (Rudell 1993).

In this paper we consider partially symmetric functions, i.e., functions that are invariant under the permutation of some input variables. Given its ROBDD representation the symmetries of a Boolean function can be determined efficiently (Möller 1993). Knowing a Boolean function to be symmetric allows the application of special logic synthesis tools that can improve the results of the design (Hurst 1977, Kim 1991, Drechsler 1995). Furthermore, knowing the variables of a function which are symmetric often restricts the search space of a logic design problem what may yield in a remarkable decrease of running time for that problem. Such problems are, e.g. the equivalence test of Boolean functions for that the input correspondence is not known (Cheng 1993, Lai 1992, Mohnke 1993) and technology mapping (Mailhot 1990). In this paper, we especially show that symmetry properties can be efficiently used to construct good variable orders for ROBDDs using modified gradual improvement heuristics.

There are three facts that create the basis of our ordering improvements:

1. The exchange of two symmetric variables does not change the size of the ROBDD, because the function remains the same.
2. The size of the ROBDD of any totally symmetric function $f : \{0,1\}^n \rightarrow \{0,1\}$ is $O(n^2)$ (Wegener 1984, Bryant 1986).

3. The value of a function which is symmetric in some variables $\{x_{i_1}, \ldots, x_{i_q}\}$ does not depend on the exact assignment of these variables but only on their weight $\sum_{j=1}^{q} x_{i_j}$.

Using the first fact, the ordering heuristics can skip over the exchange of symmetric variables and so the running time decreases. However, the resulting ROBDD sizes will be the same. The second and third fact leads to a special class of variable orders. In those orders the symmetric variables are located side by side. This class will be introduced and investigated in this paper. Especially, our paper gives a negativ answer to the conjecture stated in (Panda 1994)* that for each Boolean function there is at least one order where the symmetric variables are located side by side, for which the ROBDD is of minimal size. We will present counterexamples to this conjecture.

Nevertheless, in general, it is reasonable to locate the symmetric variables side by side: To give an impressive example, consider the function $x_1x_{n+1} + \ldots + x_nx_{2n}$ (Bryant 1986). The size of the corresponding ROBDD with variable ordering $x_1, x_2, x_3, \ldots, x_{2n}$ is exponential in $n$ whereas the size of any ROBDD with an order where the symmetric variables are side by side is linear in $n$.

We present statistical facts for all partially symmetric Boolean functions with up to five input variables and experimental results of functions taken from the LGSYNTH91 benchmark set proving the new class of orders to be very efficient with respect to the ROBDD size. This leads to the idea of modifying the reordering heuristics presented in literature so that they do not reorder single variables but whole symmetric blocks. The benchmark results show that the modified algorithms outperform the original ones.

The paper is structured as follows: In Section 2 ROBDDs and symmetric functions are defined. Symmetry variable orders are introduced in Section 3 and statistical results are presented in Section 4. In Section 5 benchmark results are given. We finish with a resume of the results in Section 6.

# 2 NOTATIONS AND DEFINITIONS

We provide a short introduction to basic notions which are important for the understanding of this paper.

## 2.1 Binary Decision Diagrams

We start with a brief review of the essential definitions and properties of BDDs (see Bryant (1986)).

**Definition 1** A *Binary Decision Diagram (*BDD*)* is a rooted directed acyclic graph $G = (V, E)$ with vertex set V containing two types of vertices, *non-terminal* and *terminal* vertices. A non-terminal vertex $v$ has as label an argument index $index(v) \in \{1, \ldots, n\}$ and two children $low(v), high(v) \in V$. A terminal vertex $v$ is labeled with a value $value(v) \in \{0, 1\}$ and has no outgoing edge.

*Similar methods as those presented in this paper have independently be developed in that paper.

BDDs are also called *branching programs* (Wegener 1987). A BDD can be used to compute a Boolean function $f(x_1, \ldots, x_n)$ in the following way: Each input $x = (x_1, \ldots, x_n) \in \{0,1\}^n$ defines a computation path through the BDD that starts at the root. If the path reaches a non-terminal node $v$ that is labeled by $i$ it follows the path $low(v)$ iff $x_i = 0$ and it follows the path $high(v)$ iff $x_i = 1$. On all paths a terminal vertex is reached since a BDD is directed and acyclic. The label of the terminal vertex determines the return value of the BDD on input $x$.

More formally, we can define the Boolean function corresponding to a BDD, recursively.

**Definition 2** A BDD having root vertex $v$ *denotes* a Boolean function $f_v$ defined as:

1. If $v$ is a terminal vertex and $value(v) = 1$ ($value(v) = 0$), then $f_v = 1$ ($f_v = 0$).
2. If $v$ is a non-terminal vertex and $index(v) = i$, then $f_v$ is the function

$$f_v(x_1, \ldots, x_n) = \overline{x}_i \cdot f_{low(v)}(x_1, \ldots, x_n) + x_i \cdot f_{high(v)}(x_1, \ldots, x_n).$$

The variable $x_i$ is called the *decision variable* for $v$.

It is well-known that for each Boolean function $f$ there exists a BDD denoting $f$.

BDDs are often used as a data structure in design automation and logic synthesis. Thus there is a need of efficient manipulation of BDDs. Unfortunately, this property is not fulfilled by the general BDDs defined above (see Gergov (1992)). Therefore we need further restrictions on the structure of the BDDs.

**Definition 3** A *Reduced Ordered* BDD *(*ROBDD*)* is a BDD with the following two properties:

1. The BDD is ordered, i.e. for any non-terminal vertex $v$, if $low(v)$ ($high(v)$) is also a non-terminal vertex, then $index(v) < index(low(v))$ ($index(v) < index(high(v))$).
2. The BDD is reduced, i.e. there exists no $v \in V$ with $low(v) = high(v)$ and there are no two vertices $v$ and $v'$ with identical labels such that the sub-BDDs rooted by $v$ and $v'$ are isomorphic.

Functions denoted by ROBDDs can be manipulated efficiently (Bryant 1986). For our practical experiments we use a ROBDD package with complemented edges as described in (Brace 1990). If not other mentioned, all results are given for this kind of ROBDDs.

We associate with each ROBDD an array $\pi$ such that $\pi[i]$ denotes the variable that corresponds to label $i$. The array $\pi$ is called the *variable order* of the ROBDD. Using this notation, the function $f_v$ at the non-terminal node $v$ with label $i$ is

$$f_v(x_1, \ldots, x_n) = \overline{\pi[i]} \cdot f_{low(v)}(x_1, \ldots, x_n) + \pi[i] \cdot f_{high(v)}(x_1, \ldots, x_n).$$

## 2.2 Symmetric functions

Let $f : \{0,1\}^n \to \{0,1\}$ be a completely specified Boolean function and $\mathcal{V}_n = \{x_1, \ldots, x_n\}$ be the corresponding set of variables. The function $f$ is said to be *symmetric* with respect to a set $\lambda \subseteq \mathcal{V}_n$ if $f$ remains invariant under all permutations of the variables in $\lambda$. For completely specified functions the symmetry is an equivalence relation which partitions the set $\mathcal{V}_n$ into disjoint classes $\lambda_1, \ldots, \lambda_k$ that will be named the *symmetry sets*. A function $f$ is called *partially symmetric* if it has at least one symmetry set $\lambda_i$ with $|\lambda_i| > 1$. If a function $f$ has only one symmetry set $\lambda = \mathcal{V}_n$, then $f$ is called *totally symmetric*. For totally symmetric functions all variable orders generate ROBDDs of the same size. Thus, no reordering is necessary and we will focus on partially symmetric functions only.

Sometimes, it may be occur that $\{x_i, x_j\}$ is not a symmetric pair but $\{x_i, \overline{x}_j\}$ is one. This kind of symmetry was introduced by Hurst (1977) and named *equivalence symmetry*. In such a case, we change the phase of variable $x_j$ in the ROBDD and get symmetry in $\{x_i, x_j\}$. Note that the negation of a variable does not change the size of an ROBDD, because it exchanges the successors of the nodes with label $x_j$ only.

## 3 SYMMETRY VARIABLE ORDERS

In this section, we introduce the new class of symmetry variable orders that we will use to improve the existing reordering heuristics.

**Definition 4** Let $f$ be a partially symmetric function with the set of symmetry sets $S = \{\lambda_1, \ldots, \lambda_k\}$. A variable order $\pi$ is called a *symmetry variable order* if for each symmetry set $\lambda_i \in S$ there exists $j$ so that $\{\pi[j], \pi[j+1], \ldots, \pi[j + |\lambda_i| - 1]\} = \lambda_i$ .

By this definition, the class of symmetry variable orders consists of all variable orders where the variables of each symmetry set are located side by side. The ROBDDs that correspond to symmetry orders are called *symmetry ordered* ROBDDs. In the remainder of this section the efficiency of symmetry orders will be motivated.

It is well known that the ROBDD size of any totally symmetric function $f$ is $O(n^2)$. In a symmetry ordered ROBDD there exists a lot of sub-ROBDDs where all variables in the upper part form a symmetry set. Thus, the upper part of these sub-ROBDD forms a totally symmetric function and the size of this part is $O(k^2)$ if $k$ is the size of the symmetry set. Thus, symmetry ordered ROBDDs contain a number of small $O(k^2)$-sized sub-ROBDDs.

Furthermore, the value of a function that is symmetric in some variables $\{x_{i_1}, \ldots, x_{i_q}\}$ does not depend on the exact assignment of these variables but only on their weight $\sum_{j=1}^{q} x_{i_j}$. If one uses symmetry ordered ROBDDs, this weight is computed in neighbouring levels and no information about partial weights has to be kept over several non-symmetric levels – and keeping information may cause large ROBDD sizes. Symmetry variable orders avoid this drawback.

## 4 STATISTICAL RESULTS

Due to the remarks above, it seems that symmetry orders generate smaller ROBDDs than general orders. To check this assumption, we investigated all partially symmetric functions with three, four and five inputs. For each function we determined the number of general orders and and the number of symmetry orders that create an ROBDD with $x\%$ more

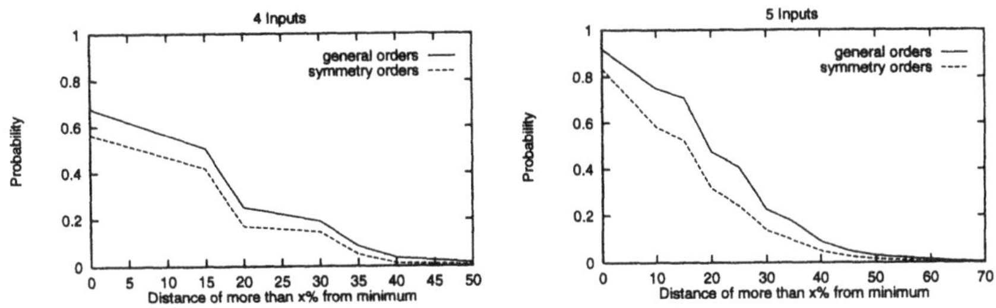


**Figure 1** Distribution of general orders and symmetry orders

nodes than the minimum ROBDD. Using these data, we computed the probability to get an ROBDD with more than $x\%$ more nodes than the minimum for an arbitrary function and an arbitrary order. Figre 1 shows the result obtained for the four and five input functions. The dashed line shows the probability that the ROBDD for an arbitrary partially symmetric function with an arbitrary *symmetry* order has more than $x\%$ additional nodes with respect to the minimum. The solid line shows the same for *general* orders. It turns out that the probability to get a $x\%$ oversized ROBDD with a symmetry order is always smaller than it is for general orders. This shows from a statistical point of view that the symmetry orders constitute an efficient subclass of variable orders. For completeness, we computed

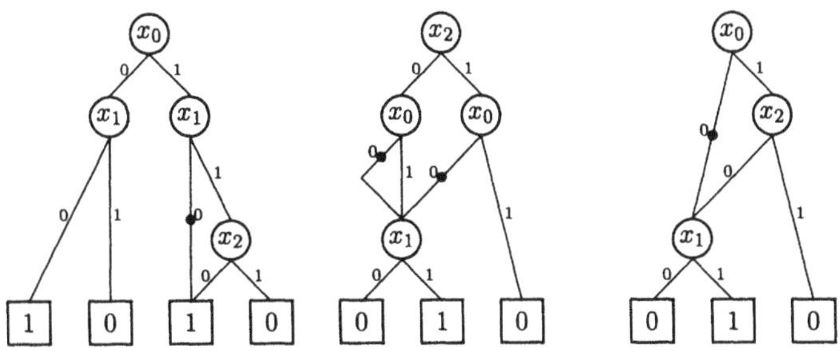

**Figure 2** ROBDDs of $f = \overline{x}_0\overline{x}_1 + x_o x_1 \overline{x}_2$

the number of partially symmetric functions for that each symmetry order results in a non-minimal ROBDD. For the 120 partially symmetric functions with three inputs there are 24 (20%) such functions. For example, the $\{x_0, x_1\}$ symmetric function shown in Figure 2 has symmetry ordered ROBDDs with 4 internal nodes while the minimum ROBDD has 3 internal nodes. For the 20.548 partially symmetric functions with four inputs there are 960 (4.7%) and for the 162.535.140 partially symmetric functions with five inputs there are only 972.280 (0.6%) such functions. The distance of the best symmetry order to the minimum was at most two nodes. We confirmed our results by performing experiments with some functions with more than five inputs. Thus, considering only symmetry variable orders seems to result in ROBDDs with sizes not larger than the minimum in almost all cases.

**Table 1** Initial ordering with symmetry orders

| Heuristic | ROBDD size | | | nodes | |
|---|---|---|---|---|---|
| | < | = | > | | |
| initial | | | | 58688 | |
| first | 9 | 5 | 3 | 58432 | 1.3% |
| median | 14 | 3 | 0 | 58020 | 1.5% |
| last | 14 | 3 | 0 | 58007 | 1.8% |
| best | 15 | 2 | 0 | 57888 | 2.5% |

| Heuristic | ROBDD size | | | nodes | |
|---|---|---|---|---|---|
| | < | = | > | | |
| initial_so | | | | 49199 | |
| first_so | 312 | 236 | 101 | 47275 | 1,2% |
| median_so | 398 | 174 | 77 | 46353 | 1.5% |
| last_so | 409 | 161 | 79 | 46362 | 1.6% |
| best_so | 534 | 112 | 3 | 45252 | 2.4% |

In (Panda 1994) it was conjectured that for ROBDDs *without* complemented edges for each functions one of its symmetry orders results in an ROBDD of minimal size. A negative answer to this conjecture was given by our experiments with these kind of ROBDDs. The four input function $f = \overline{x}_1 x_2 x_3 + x_1 \overline{x}_2 x_3 + x_1 x_2 \overline{x}_3 + \overline{x}_1 \overline{x}_2 \overline{x}_3 x_4$ which is symmetric in $\{x_1, x_2, x_3\}$ has best symmetry order of ROBDD size 9 and the minimum size is 8. For ROBDDs without complemented edges there are 80 partially symmetric functions with four inputs and 1.262.800 functions with five inputs without a minimum symmetry order.

## 5 BENCHMARK RESULTS

In this section we show the efficiency of the symmetry variable orders in practical application. For our experiments we used the CMU-BDD package contained in SIS-1.2 (Sentovich 1992) and processed 109 combinational two-level and multi-level circuits from the LGSYNTH91 benchmark set. Since the single primary outputs of a multiple output function sometimes have more symmetry we also processed each primary output of each circuit separately, to get more results about symmetric functions. Symmetry detection was executed on the ROBDDs using the algorithm proposed in (Möller 1993). We modified this algorithm to detect equivalence symmetry as well. This results in about 10 % more symmetry.

If an ROBDD is to be created from a circuit description, a heuristic, e.g. (Malik 1988), generates good initial order which is not necessarily a symmetry order. As discussed above, the size of the ROBDD may be reduced, if the initial non-symmetry order is transformed into a symmetry one. We have applied three algorithms to get a symmetry order. They differ only in the way they select the new position for a symmetry set. Heuristic first selects as position for a symmetry set the position of the first variable of the symmetry set, median selects the position of the middle variable and last selects the position of the last symmetric variable. Heuristic best calls all three methods and then selects the best order. The suffix _so denotes the methods that handle each primary output seperately. The results obtained by initial reordering are shown in Table 1. The first column gives the name of the reordering heuristic. The second, third and fourth column shows the total number of benchmark functions where the size of the symmetry ordered ROBDD is smaller, equalsized, or greater than the initial one when it was reordered with the corresponding

**Table 2** Reordering with symmetry orders

| Heuristic | ROBDD size | | | nodes | | time (sec) |
|---|---|---|---|---|---|---|
| | < | = | > | | | |
| win3 | | | | 66350 | | 14 |
| Swin3 | 25 | 29 | 2 | 64200 | 5.7% | 16 |
| sift | | | | 33878 | | 92 |
| Ssift | 26 | 26 | 4 | 33149 | 7.1% | 93 |
| win3_so | | | | 67961 | | 36 |
| Swin3_so | 693 | 1443 | 42 | 63668 | 3.4% | 41 |
| sift_so | | | | 58177 | | 116 |
| Ssift_so | 452 | 1695 | 4 | 54970 | 2.6% | 99 |

heuristic. The last column shows the total number of nodes of all ROBDDs and the average improvement over all benchmarks.

For the 109 multiple output functions we detected 56 to be partially symmetric. The initial ordering heuristic already generates a symmetry order for 39 of these functions. For more than the half of the remaining non-symmetry ordered ROBDDs the order has been improved by each of the three symmetry reordering methods and the overall number of nodes decreases. The best heuristic seems to be last and we select it for our next experiments. However, row best shows that the heuristics work well on different functions. There are only three of the single output functions for which all three heuristics generate a symmetry ordered ROBDD that is greater than the initial one. This shows that symmetry orders are also good in practice.

To reduce the size of an ROBDD several reordering heuristics have been developed. Two of them, win3 and sift (Rudell 1993) are implemented in the CMU-BDD package. To work with symmetry orders we make use of the variable blocking feature of the CMU-BDD package. Before starting reordering, we block the symmetric variables which were made adjacent by last. The modified heuristics are called Swin3 and Ssift, respectively. For all symmetric functions from the benchmark set the original heuristics win3 and sift and the modified heuristics Swin3 and Ssift were applied to the initial ROBDDs. Results are presented in Table 2. The first column denotes the reordering heuristic. The second, third and fourth column shows the total number of benchmark functions for that the modified heuristics generate a smaller, equalsized, or greater ROBDD than the original heuristic. Column nodes shows the number of nodes of all the optimized ROBDDs and the average improvement over all benchmarks. Column time shows the running time[†] of the heuristics. The additional overall running time for symmetry detection for multiple-output and single-output functions is about 88 seconds.

[†]All running times are seconds at Sparcstation 10/64 Mb

**Table 3** Benchmark results of reordering with symmetry orders

| circuit | symsets | | | init | Sinit | win3 | Swin3 | sift | Ssift |
|---|---|---|---|---|---|---|---|---|---|
| apex2 | | 1(3) | 3(2) | 2947 | 2846 | 910 | 634 | 700 | 654 |
| cps | | | 1(4) | 1455 | 1445 | 1301 | 1294 | 1035 | 991 |
| ex4 | | | 14(2) | 895 | 822 | 692 | 691 | 537 | 539 |
| seq | | | 2(2) | 5638 | 5532 | 3737 | 2586 | | |
| t481 | | | 8(2) | 63 | 33 | 33 | 21 | 33 | 31 |
| vg2 | | | 2(2) | 390 | 385 | | | 132 | 146 |
| comp | | | 16(2) | | | 146 | 128 | 146 | 107 |
| count | | | 1(2) | | | 232 | 201 | 201 | 82 |
| dalu | | | 1(2) | 4575 | 4346 | | | 1322 | 1323 |
| frg2 | | | 1(2) | | | | | 2297 | 2171 |
| i2 | 2(64) | 3(16) | 3(4) | 1586 | 1583 | 795 | 298 | | |
| i4 | | 16(3) | 50(2) | 349 | 333 | 333 | 245 | 308 | 233 |
| lal | | | 5(2) | 122 | 110 | 97 | 95 | 75 | 72 |
| my_adder | | 1(3) | 15(2) | | | 457 | 452 | 457 | 411 |
| pcler8_cl | | | 1(2) | | | 138 | 122 | 130 | 86 |
| rot | | 2(3) | 2(2) | 10224 | 10222 | 8212 | 8204 | 4574 | 4568 |
| too_large | | 1(3) | 3(2) | | | 667 | 676 | 500 | 439 |
| x1 | | | 1(2) | 1211 | 1190 | 784 | 799 | 544 | 518 |
| z4ml | | 1(3) | 2(2) | 37 | 30 | 21 | 17 | 24 | 17 |

It is shown that the heuristics that use symmetry orders generate better or same results in most cases. Swin3 saves 5.7% nodes and Ssift3 saves 7.1% nodes in average. The running time for symmetric reordering remains nearly the same. Unfortunately, there is the extra running time for symmetry detection. This increases the running time of sift in general by factor 2 and of win3 up to factor 7. One can overcome this difficulty if the symmetry detection is integrated in the reordering method following the *Idea 3* of neighbouring symmetry in (Möller 1993) as presented in (Panda 1994). Table 3 shows the effect of symmetry based reordering for some individual benchmarks. In column symsets the symmetry is given. 2(3) means that there are wo symmetry sets of three input variables. The following columns show the ROBDD size achieved by the mentioned heuristics. The leading S denotes the symmetric version. If the symmetric reordering results in the same size as the original the results are omitted.

It is shown that the symmetry modified algorithms in general outperform the original ones. Furthermore, even a small number of symmetry sets and variables can cause a great improvement. For example, for seq with only two symmetry sets of size two Swin3 saves about 30% of all nodes and for count with only one symmetry pair Ssift saves about 60%. Thus, symmetry based ordering is not only suitable for high symmetric functions but also for low symmetric functions.

# 6 CONCLUSIONS

We presented a new general method to improve the quality of ROBDD reordering heuristics for the class of partially symmetric functions by using symmetry variable orders. In this special type of orders the symmetric variables are located side by side.

Our statistical experiments as well as our benchmark experiments have shown that the efficient symmetry orders form a good subclass of all efficient variable orders. So it is useful to work only with symmetry orders when representing partially symmetric functions as ROBDDs.

It is also possible to collapse nodes of neighbouring levels corresponding to symmetric variables to *symmetry nodes*, i.e. nodes with more than two outgoing edges. This simplifies the procedure of exchanging symmetry sets. First steps in this direction can be found in (Becker 1994).

It is focus of current work to generalize the present appraoch to more powerful graph representations, like Ordered Kronecker Functional Decision Diagrams (OKFDDs) (Drechsler 1994).

# REFERENCES


S.B. Akers (1978). Binary Decision Diagrams. *IEEE Trans. on CAD*, 27(6):509–516, 1978.

B. Becker (1992). *Synthesis for Testability: Binary Decision Diagrams*, volume 577 of *LNCS*. Symp. on Theoretical Aspects of Comp. Science, pages 501–512, 1992.

B. Becker and R. Drechsler (1994). Efficient Graph Based Representation of Multi-Valued Functions with an Application to Genetic Algorithms. In *IEEE 24th International Symposium on Multi-Valued Logic*, pages 65–72, 1994.

K.S. Brace, R.L. Rudell, and R.E. Bryant (1990). Efficient Implementation of a BDD Package. In *Design Automation Conf.*, pages 40–45, 1990.

R.J. Brown, S.D.and Francis, J. Rose, and Z.G. Vranesic (1992). *Field-Programmable Gate Arrays*. Kluwer Academic Publisher, 1992.

R.E. Bryant (1986). Graph–Based Algorithms for Boolean Function Manipulation. *IEEE Trans. on CAD*, 35(8):677–691, 1986.

R.E. Bryant (1992). Symbolic Boolean Manipulation with Ordered Binary Decision diagrams. *ACM, Comp. Surveys*, 24:293–318, 1992.

D. I. Cheng and M. Marek Sadowska (1993). Verifying Equivalence of Functions with Unknown Input Correspondence. In *European Conf. on Design Automation*, pages

81–85, 1993.

R. Drechsler and B. Becker (1995). Sym*pathy*: Fast Exact Minimization of Fixed Polarity Reed-Muller Expressions for Symmetric Functions. In *European Conf. on Design Automation*, pages 91–97, 1995.

R. Drechsler, A. Sarabi, M. Theobald, B. Becker, and M.A. Perkowski (1995). Efficient Representation and Manipulation of Switching Functions Based on Ordered Kronecker Functional Decision Diagrams. In *Design Automation Conf.*, pages 415–419, 1994.

E. Felt, G. York, Brayton R., and A. Sangiovanni-Vincentelli (1993). Dynamic Variable Reordering for BDD Minimization. In *European Conf. on Design Automation*, pages 130–135, 1993.

St.J. Friedman and K.J. Supowit (1990). Finding the Optimal Variable Ordering for Binary Decision Diagrams. *IEEE Trans. on CAD*, 39(5):710–713, 1990.

H. Fujii, G. Otomo, and C. Hori (1993). Interleaving Based Variable Ordering Methods for Ordered Binary Decision Diagrams. In *IEEE Int'l Conf. on CAD*, pages 38–41, 1993.

M. Fujita, H. Fujisawa, and N.Kawato (1988). Evaluation and Improvements of Boolean Comparison Methods Based on Binary Decision Diagrams. In *IEEE Int'l Conf. on CAD*, pages 2–5, 1988.

M. Fujita, Y. Matsunaga, and T. Kakuda (1991). On Variable Ordering of Binary Decision Diagrams for the Application of Multi-level Logic Synthesis. In *European Conf. on Design Automation*, pages 50–54, 1991.

J. Gergov and C. Meinel (1992). Analysis and Manipulation of Boolean Functions in Terms of Decision Graphs. In *WG'92, LNCS*, pages 310–320, 1992.

S.L. Hurst (1977). Detection of Symmetries in Combinatorial Functions by Spectral Means. *IEE Electronic Circuits and Systems*, 1(5):173–180, 1977.

N. Ishiura, H. Sawada, and S. Yajima (1991). Minimization of Binary Decision Diagrams Based on Exchanges of Variables. In *IEEE Int'l Conf. on CAD*, pages 472–475, 1991.

B.-G. Kim and D.L. Dietmeyer (1991). Multilevel Logic Synthesis of Symmetric Switching Functions. *IEEE Trans. on CAD*, 10(4):436–446, 1991.

Y.-T. Lai, S. Sastry, and M. Pedram (1992). Boolean Matching using Binary Decision Diagrams with Applications to Logic Synthesis and Verification. In *Int'l Conf. on Comp. Design*, pages 452–458, 1992.

C.Y. Lee (1959). Representation of Switching Circuits by Binary Decision Diagrams. *Bell System Technical Jour.*, 38:985–999, 1959.

F. Mailhot and G.D. Micheli (1990). Technology Mapping using Boolean Matching and Don't Care Sets. In *European Conf. on Design Automation*, pages 212–216, 1990.

S. Malik, A.R. Wang, R.K. Brayton, and A. Sangiovanni-Vincentelli (1988). Logic Verification using Binary Decision Diagrams in a Logic Synthesis Environment. In *IEEE Int'l Conf. on CAD*, pages 6–9, 1988.

J. Mohnke and S. Malik (1993). Permutation and Phase Independent Boolean Comparison. In *European Conf. on Design Automation*, pages 86–92, 1993.

D. Möller, J. Mohnke, and M. Weber (1993). Detection of Symmetry of Boolean Functions Represented by ROBDDs. In *IEEE Int'l Conf. on CAD*, pages 680–684, 1993.

B.M.E. Moret (1982). Decision Trees and Diagrams. *ACM, Comp. Surveys*, 14(4):593–623, 1982.

S. Panda, F. Somenzi, and B.F. Plessier (1994). Symmetry Detection and Dynamic Variable Ordering of Decision Diagrams. In *IEEE Int'l Conf. on CAD*, pages 628–631, 1994.

R. Rudell (1993). Dynamic Variable Ordering for Ordered Binary Decision Diagrams. In *IEEE Int'l Conf. on CAD*, pages 42–47, November 1993.

E. Sentovich, K. Singh, L. Lavagno, Ch. Moon, R. Murgai, A. Saldanha, H. Savoj, P. Stephan, R. Brayton, and A. Sangiovanni-Vincentelli (1992). SIS: a system for sequential circuit synthesis, 1992. Department of EE and CS, UC Berkeley.

I. Wegener (1984). *Optimal Decision Trees and One-Time-Only Branching Programs for Symmetric Boolean Functions. Information and Control*, 62:129–143, 1984.

I. Wegener (1987). *The Complexity of Boolean Functions.* John Wiley & Sons Ltd., and B.G. Teubner, Stuttgart (Wiley–Teubner Series in Computer Science), 1987.

**Dirk Möller** received the diploma degree in computer science from the Humboldt-University Berlin, Germany, in 1992.
Between 1992 to 1994, he was with the Sonderforschungsbereich "VLSI Design Methods and Parallelism" at the Humboldt-University.
Since 1995 he is with the Computer Science Department at Martin-Luther-University Halle. His research interests are logic and physical synthesis.

**Rolf Drechsler** received the diploma degree in computer science from the J.W.Goethe-University Frankfurt, Germany, in 1992.
Since 1993 he is with the Computer Science Department at J.W.Goethe-University Frankfurt. His research interests are logic synthesis, testing and genetic algorithms.

**Paul Molitor** received the Diplom, Ph.D., and Habilitation degrees in 1982, 1986, and 1992, respectively, from the University of Saarland, Germany.
Between 1982 to 1992, he was with the Sonderforschungsbereich "VLSI Design Methods and Parallelism" at the University of Saarland. After being visiting professor at the University of Halle (1992/93) and the University of Freiburg i.Br. (1993), he was an associate professor in the Computer Science Department of the Humboldt-University Berlin in 1993/94. Since 1994, he is a full professor at the Martin-Luther University Halle, Germany, where he is the chairman of the Institute of Computer Science.
His research interests include logic and physical synthesis, hierarchical VLSI design, algorithms, and data structures.

PART THREE

# Partitioning and Clustering for Programmable Devices

8

# Circuit clustering and partitioning for system implementations*

*Ulrich Weinmann*[1] *and Wolfgang Rosenstiel*[2]
[1] *FZI - Computer Science Research Center*
*Haid- u. Neustr. 10-14, 76131 Karlsruhe, Germany*
[2] *FZI / University of Tübingen, Sand 13, 72076 Tübingen, Germany*
*Tel.: (++49) 721 9654 -465, Fax:-409, e-mail:weinmann@fzi.de*

**Abstract**

This paper shows several new perspectives of partitioning circuits into clusters for the implementation into programmable devices. The partitioning algorithm is based on a probabilistic node movement and is able to consider different optimization goals. Special clustering strategies based on the max flow theorem before the partitioning allow the handling of large circuits, a retiming approach following the partitioning improves the overall system performance. Several large benchmark circuits could be implemented into FPGAs with low computational costs.

**Keywords**

Maximal Flow Clustering, Partitioning, FPGA Synthesis

## 1 INTRODUCTION

The partitioning of logic structures has become a very important implementation step for the realization of circuits with programmable logic, especially FPGAs (Field Programmable Gate Arrays). These chips are used for both prototyping and the implementation of a low number of produced chips. Partitioning became necessary, because the number of equivalent gates is lower in comparison to mask-programmed gate arrays. Rapid prototyping boards have been developed to emulate large circuits with programmable logic. Here the chip utilisation is less than 40% after partitioning due to the number of external pins on a device. This paper addresses the problem by combining a new partitioning approach with technology mapping in order to achieve better implementation results.

* This research is supported by the Esprit LINK project 6855

Many approaches for partitioning circuits have been proposed to minimize the interconnections between chips. However, an exact solution to this problem can't be found in the sense that no polynomial-time algorithm for it is known. The *clustering partitioning* can be divided into seed selection, unplaced node selection and node placement. This approach is easy to implement, but the quality of the result depends strongly on the selection of the seed nodes (Murgai 1991). Ford and Fulkerson (1962) introduced the dependency between *maximum flow* and minimum cut on a graph. Besides *simulated annealing*, which uses an analogy to the annealing process in physics to avoid local minima, the *Min-Cut two-way partitioning* (Kernighan 1970) is very efficient due to group swapping. These basic partitioning algorithms have been improved in the past, mostly by reducing the calculation complexity to obtain faster partitioning results. Kuznar, Brglez and Kozminski (1993) first showed the necessity of the relationship between partitioning and technology mapping for the Xilinx architectures, using an improved Min-Cut approach.

## 1.1 Overview

In this paper we try to combine constructive and iterative partitioning approaches in order to meet the user constraints or an overall optimization goal. Figure 1 shows a general overview of the chip partitioning system.

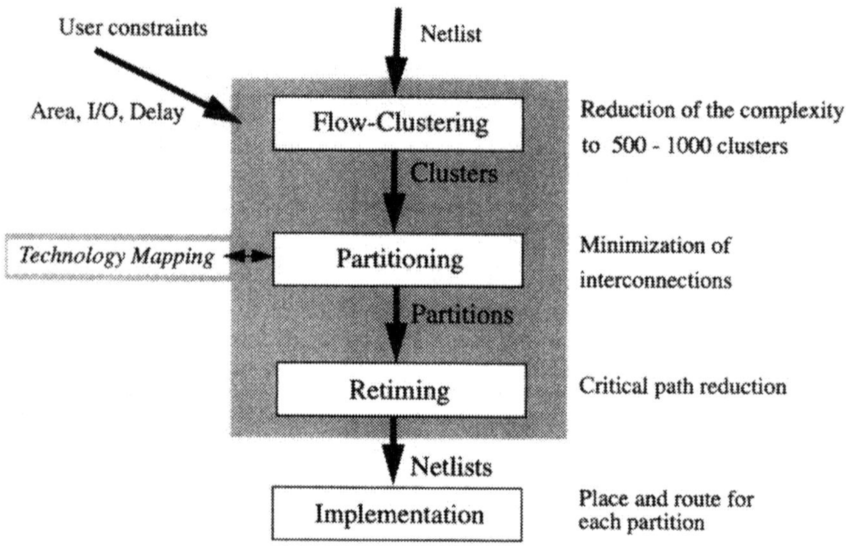


**Figure 1** Partitioning overview

Starting from a netlist description, the tool first performs a preclustering of logic structures. The goal of this partitioning step is to reduces the partitioning complexity in order to handle very large circuits. The iterative partitioning step is based on a probabilistic node exchange. Here the user requirements can be considered like partition sizes or pinning. After every optimization run, the technology mapping can be performed to obtain the actual area and delay results for each cluster. The system offers a technology independent mapping tool by Weinmann (1994), which is able to map a logic structure into any SRAM or multiplexer-based architecture.

The timing can additionally be improved by a final retiming step. This operation moves storage elements towards interconnections to reduce the length of critical paths (Leiserson 1983). Afterwards, the mapped circuits are saved and can be implemented by placement and routing tools from the FPGA vendors.

## 2 CLUSTERING

The partitioning has an exponential complexity with respect to the number of gates and connections. Therefore, larger circuits can only be efficiently partitioned by reducing the number of gates. This leads to the clustering problem, which is based on a grouping of gates (nodes) of a circuit (graph) into clusters. As a result, supernodes or hierarchical netlist descriptions can be partitioned with less effort. One drawback is the reduced search space, which now might not include the optimal result. Avoiding this problem is one goal of the clustering. Furthermore the clustering is able to connect gates, which should not be separated, such as the nodes in a cycle. The following clustering approach is the basis for an efficient partitioning.

### 2.1 Network flow based clustering

The network flow is a topic in algorithmic graph theory and can be defined as a function for every edge in the circuit. Using this definition, Fulkerson and Ford (1962) first introduced the dependency between maximal flow and minimal cut on a graph. For any network with integer edge capacities, the maximum flow value from the source to the sink is equal the capacity of a minimum cut separating the source and the sink. This theorem has not yet been considered for partitioning, because the cuts often result in very uneven partition sizes. The example in figure 2 shows this problem using a simple graph with the corresponding flows on the edges.

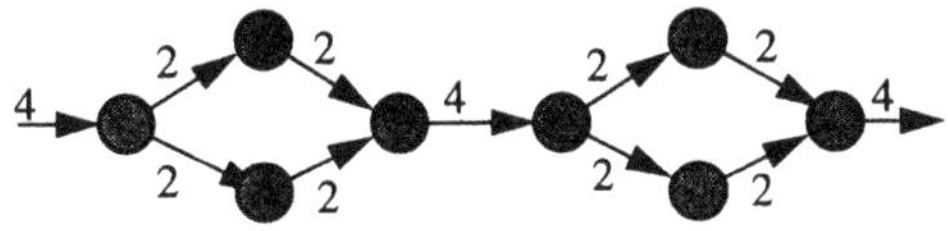


**Figure 2** Network Flow

The flow is calculated by counting all paths, which include the edge. The minimum cut can be found on the three edges with maximal flow. The first and last edge can not be used for partitioning, because the partition sizes would be very uneven. Now the goal is, using this theorem, to find edges with a high network flow, which build the basis for the iterative optimization.

The flow based clustering approach uses this information. In the first step the flow of all edges is calculated. Then these edges or their corresponding source nodes are sorted by rising flow. Starting with the node of maximal flow further nodes connected to the inputs are clustered, until their flow is larger than the corresponding output flow (this number can vary and restricts the final sizes of the clusters). The algorithm can be summarized as follows:

```
BEGIN;
FLOW_PRECLUSTERING()
     {
     Calculate Flow on all Nodes
     Sort all Nodes with rising flow in list T
     WHILE (NOT T==∅)
          {
          delete first node q from T
          Node_List(q) = {q};
          CLUSTER(q)
          }
     }

     CLUSTER(v)
          {
          FOR all input nodes vi of v DO
               IF ( flow(vi) ≤ flow(v) )
                    {
                    Node_List(q) = Node_List(q) ∪ {vi};
                    delete node vi from T
                    CLUSTER(vi)}
                    }
          }
END;
```

This can be done until all nodes are assigned to their corresponding clusters. The resulting clusters have minimal connections to their neighbouring clusters. Now the clusters have to be assigned to partitions with the goal of minimal interconnections.

Figure 3 shows a sample circuit including two sequential cycles. In order to calculate the flow on the edges, the cycles have to be cut (usually at the output of storage elements).

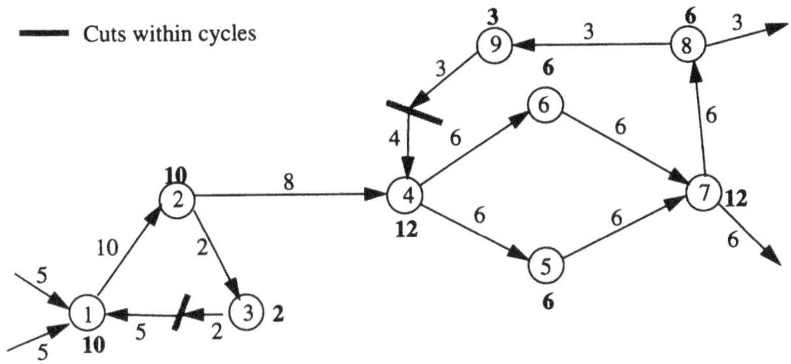


**Figure 3** Example for flow-calculation

The flow number on each edge represents the numbers of different paths including the edge. The algorithm now starts with the node of maximum flow (e.g. 7). In direction of the inputs, all nodes are clustered until the flow rises (node 4). The next iteration starts with the max flow of the remaining nodes. Figure 4 shows the resulting clusters.

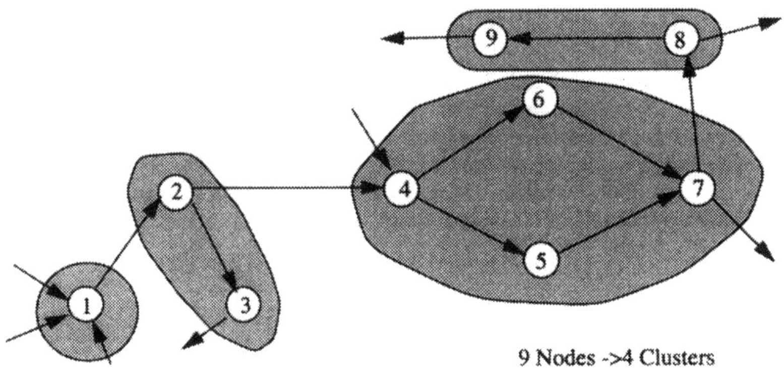


**Figure 4** Result after clustering

## 3 THE PARTITIONING ALGORITHM

This partitioning method is based on a probabilistic cluster exchange between partitions, and includes a dynamic cost function. Given is a graph G = (N,E) with a set of clusters (N) and edges (E), which will be partitioned into k partitions for the implementation into programmable logic (see figure 5).

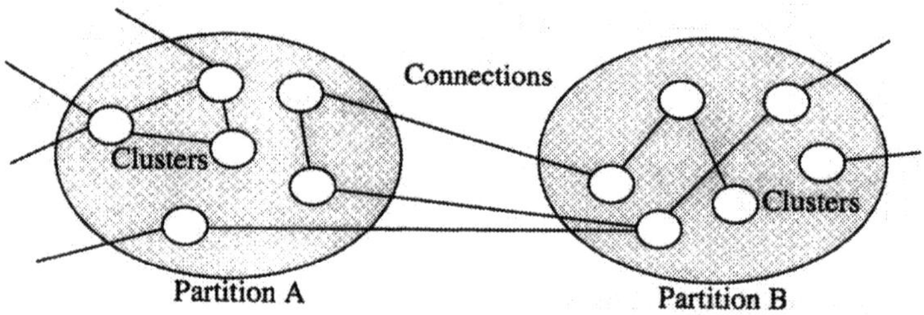


**Figure 5** Assignment of clusters to partitions

The partitioning starts with arbitrary partitions, which can be obtained by a simple next-neighbour clustering method. The iterative partitioning allows different weighted optimization factors to be added to the cost function. Using these criteria for each cluster, a *moving probability* $p_i$ of a cluster into partition i can be calculated. To illustrate the strategy, in the following a probability function including two partitioning goals will be described, which then will be extended to a generalized cost function. Here $N_i$ equals the number of clusters in partition i after each partitioning iteration, $E_i$ equals the number of inputs and outputs (I/Os) of partition i and $C_{ij}$ shows the number of connections between the two partitions i and j. These are the results after the partitioning, while $uN_i$ and $uE_i$ are user defined constraints for partition size and input/output count. The probability $p_i$ of a cluster n to move to partition i is equivalent to:

$p_i \sim e_i$ with $e_i$ = Number of interconnections with partition i

and

$$p_i \sim \frac{1}{\delta_i} \quad \text{with } \delta_i = \frac{N_i}{uN_i}$$

That means, the more connections of a cluster to another partition, the higher the probability is for the clusters to move to that partition. Furthermore this probability is inverse proportional to the number of clusters inside a partition. If both criteria have the same weight:

$$p_i \sim e_i \cdot \upsilon_i \text{ with } \upsilon_i = \frac{1}{\delta_i};$$

because of $\sum_{\forall k} p_k = 1$ follows: $p_i = \frac{e_i \cdot \upsilon_i}{\sum_{\forall k} e_k \cdot \upsilon_k}$ with k = all partitions.

The summation of the probabilities of a cluster to be moved equals 1. By calculating this equation for each cluster, a probability for movement into another partition can be defined. The general probability for cluster movement for several cost functions equals:

$$p_i = \frac{\prod_x a_{ji}^{(\gamma_x)}}{\sum_k \prod_x a_{jk}^{(\gamma_x)}}$$

$a_{xi}$ - Criteria x of cluster i
$\gamma_x$ - Weight of the criteria x

This function additionally includes weight factors, which allow the user to give different importance to the optimization criteria. If, for example, an exact pinning is important, the weight $\gamma$ of the I/O criteria would be increased. For very high weight factors the following equation applies, which gives a definite decision for movement:

$$\lim_{\gamma \to \infty} p_i = \frac{a_i^{\gamma}}{\sum_{\forall k} a_k^{\gamma}} = \begin{cases} 1 \text{ for } a_i = \max\{a\} \\ 0 \text{ for } a_i < \max\{a\} \end{cases}$$

At the beginning the weights are predefined, however, during the partitioning a change is possible. As an example, a modification would be necessary, if the allowed number of interconnections between two clusters is exceeded. Cluster movements, which cause additional chip to chip wiring, are within a small range prohibited.

The calculation of a probabilistic movement for clusters within partitions allows an easy integration of any optimization factor, which can dynamically be reconfigured. This is important for the adjustment of the partitioning to user and technology specific data. Figure 6 depicts a simple example for the criteria size and interconnections. For each cluster, k-1 calculations are needed, where k is the number of partitions. The complexity is maximal O(N(k-1)) for N clusters. This corresponds to one iteration for the partitioning. The number of iterations add a linear increase and range depending on the circuit size between 10 and 100. This estimation can be reduced, if only clusters, which are located close to the border of a partition, are considered and therefore have a high probability to be moved.

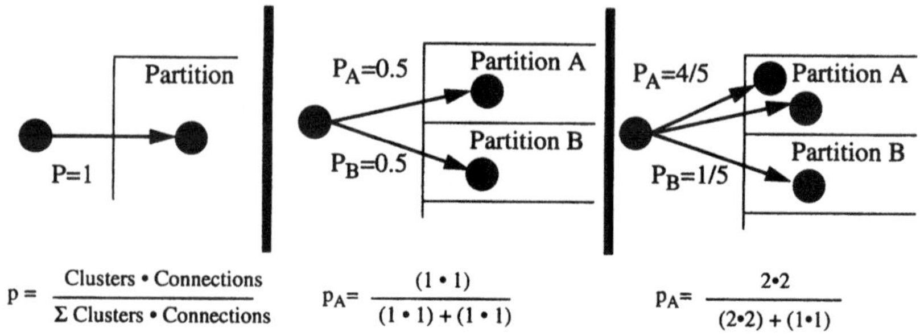


**Figure 6** Probability Calculation

## 3.1 Improved System Performance by Retiming

The retiming operation (Leiserson 1983) can be applied after the partitioning and improves the timing for FPGAs. This technique is based on the movement of storage elements within a circuit and is used to reduce the maximal delay.

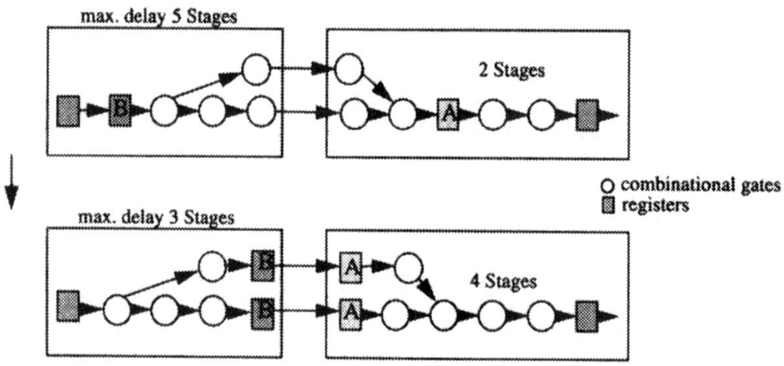


**Figure 7** Retiming after the partitioning

The drawback of this operation is a possible increase of the number of registers, if a negative register movement (Weinmann 1993) is applied. In combination with FPGA architectures with predefined storage elements (Xilinx, Actel), the timing can be improved without increasing the overall area consumption on the chip. An evaluation of benchmark circuits (Weinmann 1993) showed, that in the average less than half of the storage elements of the Xilinx architecture are used after mapping. This operation can be applied to every cluster after the partitioning, however in large designs the overall system performance is most important. Therefore the chip to chip delay has to be taken into account. The lowest delay for partitioned circuits can be achieved by putting storage elements next to the output and input pins of interconnections. This means no additional gate delay to the wiring on the board. Figure 7 depicts an illustration of the basic idea of the retiming operation in combination with the partitioning. The first solution has a maximal delay of five combinational stages in addition to one interconnection. This number can be reduced to three stages by moving register 'A' toward the input of the chip and finally to zero gate delay by moving register 'B'. The only restriction concerns divergent paths with different timing states (retiming bottlenecks), which can be avoided by duplicating gates, described in.

## 4 RESULTS

Different combinational and sequential MCNC benchmarks (Brglez 1989) were implemented and tested with the partitioning tool. The results are divided into the clustering and partitioning step. Table 1 shows the results after the flow-preclustering for the largest S-Benchmarks.

**Table 1** Results after flow-preclustering

| | | | | *Flow-Preclustering* | | | |
|---|---|---|---|---|---|---|---|
| *Bench-mark* | *#Gates* | *#Wires* | *Av. # I/O / Gate* | *#Clusters* | *#Wires* | *Av. # I/O / Cluster* | *Time* |
| s838 | 422 | 734 | 3.5 | 42 | 247 | 11.6 | <1 s |
| s953 | 424 | 801 | 3.8 | 32 | 325 | 20.2 | <1 s |
| s1196 | 547 | 1045 | 3.8 | 20 | 330 | 33.0 | <1 s |
| s1423 | 805 | 1534 | 3.8 | 126 | 584 | 9.2 | 3 s |
| s1488 | 659 | 1399 | 4.2 | 42 | 471 | 22.4 | 2 s |
| s1494 | 653 | 1405 | 4.3 | 37 | 466 | 25.1 | 1 s |
| s5378 | 2553 | 4560 | 3.6 | 136 | 710 | 10.4 | 21 s |
| s9234 | 5825 | 8427 | 2.9 | 293 | 1395 | 9.5 | 43 s |
| s13207 | 8620 | 12503 | 2.9 | 817 | 2407 | 5.9 | 77 s |
| s15850 | 10369 | 14839 | 2.8 | 798 | 2718 | 6.8 | 87 s |
| s35932 | 17793 | 31275 | 3.5 | 1616 | 7944 | 9.8 | 93 s |
| s38584 | 20705 | 35660 | 3.4 | 2235 | 9982 | 8.8 | 120 s |

Beside the number of gates and wires of the circuit, the average number of inputs/outputs per gate is given. After the preclustering these numbers can be compared with the resulting clusters and connections. For the larger circuits the number of gates is reduced to about 10% in clusters, however the number of I/Os per cluster only doubled.

Generally partitioning tools are very difficult to compare, because they depend on different parameters and cost functions. In order to achieve a realistic basis for comparison different factors can be calculated to reflect partitioning results under several optimization goals. Very important is the relationship between the number of interconnections and the deviation of the optimal cluster size after the partitioning. Of course, the number of connections can be minimized with a high deviation in the size of the clusters, however, the overall result would be rated less optimal. The average I/O count for each chip for $k$ partitions can be calculated as follows:

$$C_{av} = \frac{1}{k} \sum_{i=1}^{k} \sum_{j=1}^{k} e_{ij}$$

Each edge $e_{ij}$ between the partitions $i$ and $j$ equals a cut between chips. In order to evaluate this number, the relative factor to all edges $S$ in the circuit shows the efficiency of the partitioning.

$$C_{relative} = \frac{C_{av}}{S} \cdot 100\%$$

Together with the number of cuts the use of area in each partition has to be observed. Using an overall circuit size $B$ and a *size* $b_i$ for each partition $i$, the average deviation $a$ from the average size $\frac{B}{k}$ equals:

$$a = \frac{\sum_{i=1}^{k} \left| \frac{B}{k} - b_i \right|}{B} \cdot 100\%$$

Table 2 depicts the partitioning results for the probabilistic cluster exchange and optimization for the partitioning into two and four partitions. Even the largest partitioned circuits can be implemented in programmable logic, because the number of interconnections stays below the number of available pins of an FPGA package. For the circuit s38584 only 0.44% of the edges in the circuit graph are used for interconnections. The area deviation between the partitions was kept below 25%.

**Table 2** Partitioning results for different S-Benchmarks

| *Bench-mark* | *Two Partitions (k=2)* | | | | *Four Partitions (k=4)* | | |
|---|---|---|---|---|---|---|---|
| | *Area deviation* $a$ | *Av. # I/O* $C_{av}$ | $C_{realtive}$ | *Time* | *Area deviation* $a$ | *Av. # I/O* $C_{av}$ | *Time* |
| s838 | 0 % | 38 | 5.18 % | 23 s | 4 % | 7 | 89 s |
| s953 | 22 % | 55 | 6.87 % | 44 s | 12 % | 21 | 32 s |
| s1196 | 15 % | 39 | 3.73 % | 238 s | 23 % | 22 | 82 s |
| s1423 | 6 % | 34 | 2.22 % | 422 s | 19 % | 19 | 136 s |
| s1488 | 8 % | 55 | 3.93 % | 310 s | 12 % | 55 | 272 s |
| s1494 | 22 % | 49 | 3.49 % | 298 s | 16 % | 55 | 356 s |
| s5378 | 25 % | 147 | 3.22 % | 534 s | 13 % | 58 | 752 s |
| s9234 | 17 % | 104 | 1.23 % | 610 s | 25 % | 81 | 923 s |
| s13207 | 12 % | 121 | 0.97 % | 897 s | 24 % | 94 | 1869 s |
| s15850 | 25 % | 108 | 0.72 % | 745 s | 25 % | 83 | 2023 s |
| s35932 | 12 % | 89 | 0.28 % | 1089 s | 23 % | 103 | 1965 s |
| s38584 | 23 % | 156 | 0.44 % | 1278 s | 21 % | 112 | 2403 s |

## 5 CONCLUSION

This paper proposes a new partitioning method divided into several optimization steps. Using maximal flow calculations, a new clustering algorithm was presented. This information could be used for a probabilistic cost function to optimize area and I/O count. Additional optimization steps for delay reduction improve the performance of the partitioned system.

## 6 REFERENCES

Brglez, F., Brayan, F. and Krzysztof, D. (1989) Combinational Profiles of Sequential Benchmark Circuits. *Technical Report TR89,* Microelectronics Center of North Carolina.

Ford, L.R. and Fulkerson, D.R. (1962) Flows in Networks. *Princton University Press.*

Kernighan, K.H. and Lin, S. (1970). An Efficient Heuristic Procedure for Partitioning Graphs. *Bell System Technical Journal,* vol. 49, no. 2.

Kuznar, R., Brglez, F. and Kozminski, K. (1993) Cost Minimization of Partitions into Multiple Devices. *Proceedings of DAC 93*, 315-20.

Leiserson, C. E., Rose, F. M. and Saxe, J. B. (1993) Optimizing synchronous circuitry by retiming. *Proc. Third Caltech Conf. on VLSI.*

Murgai, R., Brayton, R. and Sangiovanni-Vincentelli, A. (1991) On Clustering for Minimum Delay/Area. *Proceedings of DAC 91*, 6-10.

Weinmann, U. (1993) FPGA Partitioning under Timing Constraints. *Int. Workshop on Field Programmable Logic and Applications*, Oxford.

Weinmann, U. and Rosenstiel, W. (1993) Technology Mapping for Sequential Circuits based on Retiming Techniques. *Proceedings of the EURO-DAC 93*, 318-23.

Weinmann, U. and Rosenstiel, W. (1994), Module Independent Mapping into TLU-FPGAs. *2nd ACM/SIGDA FPGA Workshop*, Berkeley.

9

# Circuit Partitioning For FPGAs

*G. Saucier, D. Brasen, J.P. Hiol*
*Institut National Polytechnique de Grenoble / CSI*
*46, Avenue Félix Viallet, 38031 GRENOBLE cedex FRANCE*
*Tel: (33) 76-57-46-87 Fax: (33) 76-50-34-21 E-mail: saucier@imag.fr*

**Abstract**

High-level synthesis tools have traditionally used circuit hierarchy to partition circuits into packages. However hierarchical partitioning can not be easily performed if hierarchical blocks have too large a size or too many IOs. This problem becomes more frequent with FPGAs which have small size limits and even smaller IO pin limits. An IO bottleneck often prevents the maximum FPGA package size from being reached. In this paper, two new FPGA cone partitioning algorithms are presented that have been implemented in the ASYL+ tool set. High-level synthesis is linked to cone partitioning by creating circuit hierarchy from a VHDL description.

## 1. INTRODUCTION

High-level synthesis strives to minimize the number of packages required to realize a circuit with minimum production cost at maximum frequency. The post-synthesis circuit partitioning addressed here is normally required for blocks that can not be cut based on high-level structures (architecture reorganization, data-path bit-slice partitioning, or Ram/Rom array sectioning). Such blocks are partitioned by grouping basic circuit elements like FPGA modules to meet size or IO package constraints and match the worst timing of the blocks produced during synthesis. Given that timing constraints can be met by keeping critical paths within a partition, the problem becomes finding partitions of high size to IO ratios that meet IO package constraints.

Prior work in post-synthesis circuit partitioning has primarily focused on top-down minimum-cut algorithms (Fiduccia 1982), (Hwang 1992), (Kring 1991). These algorithms recursively attempt to find the cut that minimizes the number of connections between two bipartitions (Figure 1). The min-cut of (Keringhan 1970) proposed starting with two random bipartitions followed by iterative pair swapping on all pairs of nodes. Subsequent algorithms have improved the time complexity, added a look-ahead into future moves to stabilize results, and reduced the cut size by gate replication and improved move set generation Kuznar 1993. All allow for a maximum size constraint per partition by limiting the size of a bipartition during each min-cut operation. However to meet a maximum number of IO pins constraint per partition the min-cut algorithm has to cut further, reducing the size as well as IOs of otherwise optimal partitions.Traditional min-cut approaches lack the ability to find constrained IO partitions with high Size/IO.

To meet high Size/IO requirements for specific packages, cone structure partitioning is proposed. Define a cone structure as the set of all combinational nodes that can be found between a single output and the inputs that lead to that single output (Figure 1). For sequential circuits, the inputs/outputs of the cone can also be the outputs/inputs of sequential elements. Cone partitioning selects and merges/cuts high Size/IO combinational groups within cones that exist because of low fanout. Circuits with low fanout are easily partitioned into unoverlapping cones that fit packages. Circuits with high fanout have the majority of each cone overlap several others and merging/cutting of optimal high Size/IO circuit sections is required to produce good partitions.

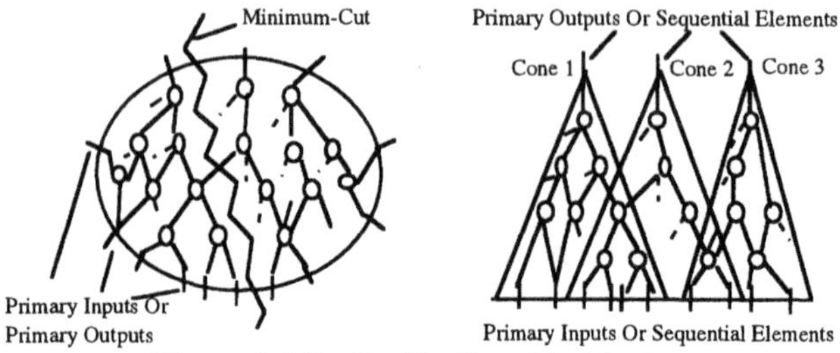


**Figure 1:** Min-Cut Vs. Cone Partitioning.

In this paper, a hierarchical partitioning algorithm is proposed that preserves high-level timing. For those nonhierarchical blocks found to be too large, flat cone partitioning is analyzed. Critical path containment within a partition is shown to be easy with cone partitioning with no loss to area. A first cone partitioning algorithm clusters by cutting each cone around highest cost cones and then cluster merges. A second cone partitioning algorithm clusters nodes found within the same cone overlapping region and then cluster merges. Experimental results using MCNC and industrial benchmarks on two different industrial FPGA packages show that the cone partitioning algorithms produce better partitions than min-cut. The hierarchical partitioning algorithm is also used to partition circuit hierarchy created from an example VHDL description.

## 2. HIERARCHICAL PARTITIONING ALGORITHM

For FPGA implementations, circuit hierarchy constrains partitioning with a prioritized list of gate clusters within clusters. Since this can guarantee the timing of high-level structures, partitioning based on circuit hierarchy can be powerful. However if the designer or high-level synthesis tool that creates the circuit hierarchy does not understand the final technology, the hierarchy overconstrains producing bad partitions with low gate utilization.

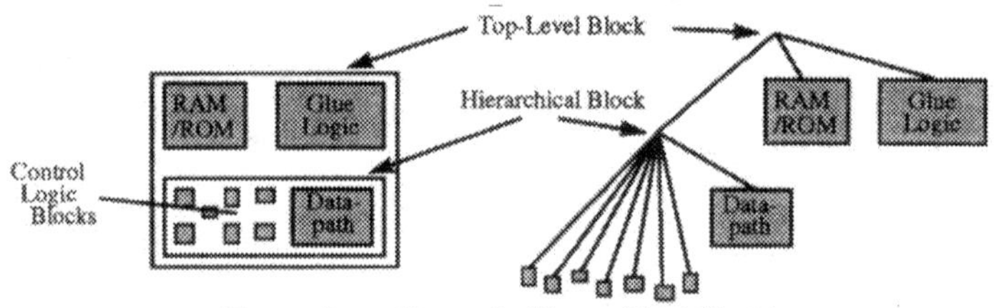


**Figure 2:** An Example Hierarchical Circuit.

For example, refer to the circuit hierarchy of Figure 2. As seen from a high-level, there is a Data-Path block, a RAM or ROM block, and Glue-Logic scattered as a controller for the data-path and a separate block of logic. At the post-synthesis level, the contents of all these blocks are mapped to FPGA library macros or modules which represent gates. The timing designed for at the high-level is preserved by implementing such blocks within one FPGA package. If the circuit hierarchy is flattened, a connected network of FPGA modules from an FPGA library results. No more Data-Path, RAM, or ROM since the typical FPGA currently has no specialized hardware sections. The only way to guarantee the desired high-level timing is to partition based on circuit hierarchy.

To exploit the hierarchy and preserve high-level timing, the following algorithm is proposed to perform hierarchical circuit partitioning.

*Hierarchical Partitioning Algorithm:*

*Step 0) Begin with a pointer to the top-level block assigned as the current BLOCK.*

*Step 1) If the current BLOCK has no hierarchical sub-blocks AND the BLOCK is larger than Size or IO constraints, perform Flat Cone Partitioning on the BLOCK. The BLOCK will be replaced by non-hierarchical sub-blocks (i.e., flat circuit partitions that meet constraints) under a new BLOCK pointer at the same level of hierarchy.*

*Step 2) If there exists hierarchical sub-blocks in the current BLOCK, recursively call the hierarchical partitioning algorithm for each sub-block (i.e., set BLOCK=sub-block and goto step 1).*

*Step 3) Perform pair-wise merging on resulting nonhierarchical sub-blocks of BLOCK to meet constraints. Merge sub-blocks (i,j) to meet MaxSize or MaxIO constraints and minimize the number of packages. For FPGAs this means to maximize the cost ratio*

$$Cost_M = \left(\frac{Size_M = Size_i + Size_j}{IO_M = IO_i + IO_j - (2 * IO_{Common})}\right)$$

*since IO pin bottlenecks cause MaxSize constraints to be rarely met. IOcommon is the number of common IO connections between the blocks. The block with largest cost=Size/IO is merged with the block that will produce a new block with largest cost. Finding larger cost merges is not a problem since size grows faster than IOs. Partitions are filled sequentially allowing for the selection of different packages. When finished merging, remove this level of hierarchy and analyze the next (i.e., move all sub-blocks of BLOCK to the parent of BLOCK, set BLOCK to its parent, and goto step 1). Exit if BLOCK is the top-level block since the circuit hierarchy has been fully partitioned.*

By recursively descending the circuit hierarchy, blocks smaller than constraints are first merged together. Such blocks connected only through the top-level are unlikely to be implemented in the same FPGA. The blocks larger than constraints are partitioned with cones.

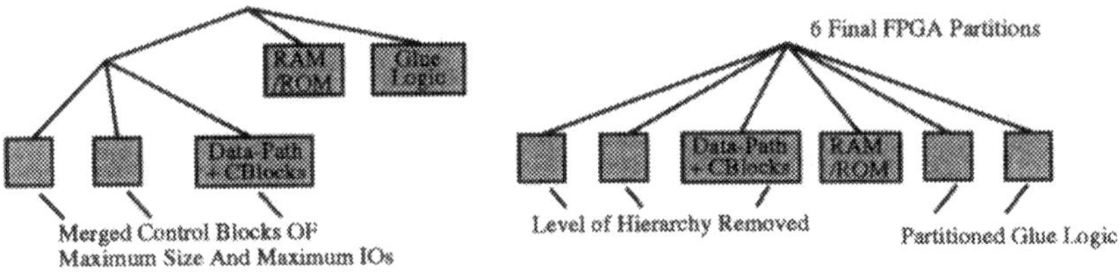


**Figure 3:** Partitioning The Example Hierarchy.

In Figure 3, the Hierarchical Partitioning Algorithm is performed on the example of Figure 2. The small control blocks of the bottom-level hierarchy are first merged, some with the Data-Path block, until MaxSize and MaxIO limits prevent further merging (Figure 3 - Left). The bottom-level hierarchy is removed and partitioning performed at the top-level (Figure 3 - Right). The Glue-Logic block was too large for constraints and was cone partitioned into two blocks. The Circuit Hierarchy is thus partitioned into 6 FPGAs with the original RAM/ROM and Data-Path blocks fully contained within an FPGA package to preserve high-level timing. Hierarchical partitioning in this way assumes some a priori rules during the creation of circuit hierarchy by designers or high-level synthesis tools to achieve good results.

• Since the target FPGAs are rarely known when designing at the high-level, nonhierarchical flat blocks should be made smaller than the smallest possible FPGA package considered. This facilitates the post-synthesis merging of the hierarchical partitioning when the target FPGA is known. Also the high-level structures of Data-Paths, RAMs, and ROMs are more properly partitioned at the high-level anyway.

• Merge together the hierarchical blocks which do not serve to preserve high-level timing. The flat cone partitioning algorithms to follow seek out highly connected sections and more accurately fill the target FPGAs since the block granularity used is the basic FPGA module. The local timing considerations of containing entire critical paths within a package are also performed by the flat cone partitioning.

These two rules are not in conflict since the first is a rule for blocks which should not be further partitioned and the second is for those that should. With hierarchical blocks designed as pre-partitioned circuit sections, the remainder of this paper will focus on the flat cone partitioning of blocks too large for target FPGA packages.

## 3. Cone Structures Within A Flat Circuit

Define a cone structure as the set of all combinational nodes that can be found connected between a single output and the inputs that lead to that single output. For sequential circuits, the inputs/outputs of the cone can also be the outputs/inputs of sequential elements. In the netlist of Figure 4 there are two cones. Cone 2 has an input pin and an output pin connected to the sequential element 2. Since sequential element 2 has only connections within cone 2, these two pins can be removed from the cone 2 pin list of IOs.

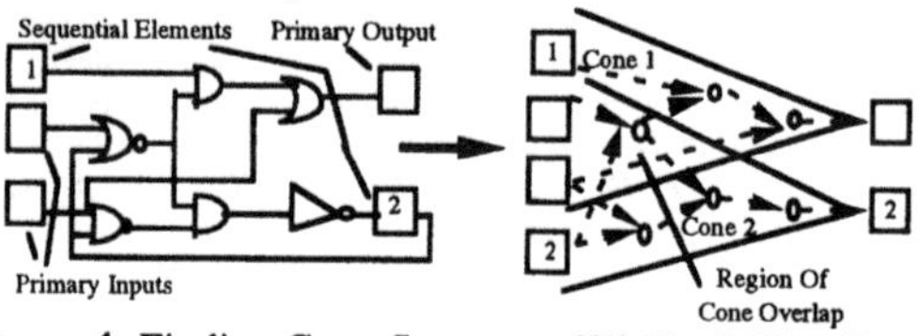


**Figure 4:** Finding Cone Structures Within A Flat Circuit.

Cone structures are minimum-cut structures because they outline regions of low fan-out. The only combinational gate node in Figure 4 with fanout greater than one results in an overlapping section between the two cones. A fanout greater than two would possibly result in an overlapping section with a third or fourth cone, making the cones harder to merge/cut. If all the nodes in cone 2 had fanout of one, the maximum Size/IO becomes 4/4 (4 nodes, 3 inputs, and 1 output). If all nodes in cone 2 had fanout of greater than one, the minimum Size/IO becomes 4/7 (4 nodes, 3 inputs, and 4 outputs). The difference between maximum and minimum Size/IOs becomes greater as the cone size increases.

## 4. Critical Path Considerations To Improve Timing

Define a critical path as a long combinational path between two sequential elements or IO pads of an original unpartitioned circuit. Classical algorithms for identifying critical paths have been published (Singh 1991). If such critical paths are cut during partitioning of the circuit into packages, resulting inter-package wire capacitances cause the circuit timing to drop. Also if only the most critical path is considered during partitioning, resulting inter-package wire capacitances can be sufficiently large along secondary critical paths to cause them to become the most critical paths. Prior work has used min-cut algorithms to create initial bipartitions that contain the n-most critical paths and then not allow gates of such paths to cross bipartition boundaries. The partitioning algorithm proposed in (Murgai 1991) clusters and merges the circuit around critical paths so that the maximum delay is minimized.

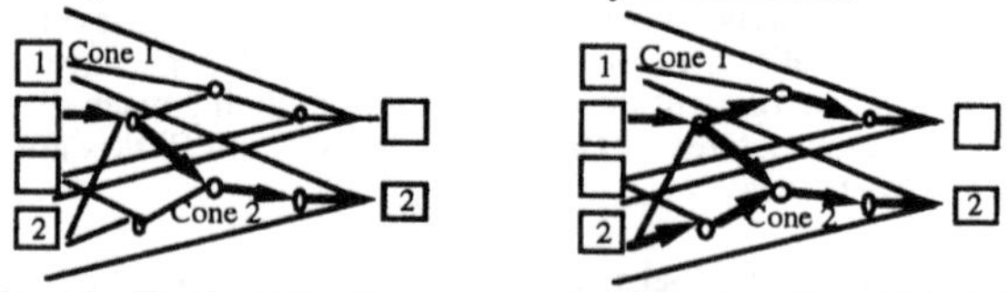


Containing The Most Critical Path Within Cone 2 And Cutting Cone 1. — Containing The N Most Critical Paths By Merging Cone 1 And Cone 2.

**Figure 5: Containing Critical Path(s) Within Cone Structure.**

Since cones are defined as starting from a primary output or sequential element and finishing at primary inputs or sequential elements, containing a critical path within a cone is relatively easy. On the left of Figure 5, containing the most critical path within cone 2 forces the cutting or immediate merging of cone 1. On the right of Figure 5, the n most critical paths are contained by forcing the merger of cone 1 and cone 2. Initial results indicate containing critical paths does not alter Size or IOs of partitions since most critical paths run through similar cones.

## 5. CUTTING FOR BEST CONES AND MERGING

In this section, the first of two cone partitioning algorithms is listed below and called *Best_Cone_Merge*. This method preserves the best cones by keeping them whole. However these whole cones are sometimes too large for package constraints with circuits of low fanout and further partitioning is required.

*Cone Partitioning Algorithm 1. Best_Cone_Merge:*

*1. Define cone structures as described in section 3.*

*2. Preserve largest cost cones and cut all others. Nonoverlapping clusters result which are the highest cost sections extracted from the original circuit. Clusters that are too large for constraints, usually the whole cones that were preserved, must be divided using min-cut.*

*3. Add sequential elements to clusters for reduced Size/IO ratio. This means adding to clusters with the largest number of common interconnections.*

*4. Merge clusters with combinational/sequential elements on the same critical path as in section 4.*

*5. Merge clusters into partitions to meet package constraints as in the hierarchical partitioning algorithm of section 2.*

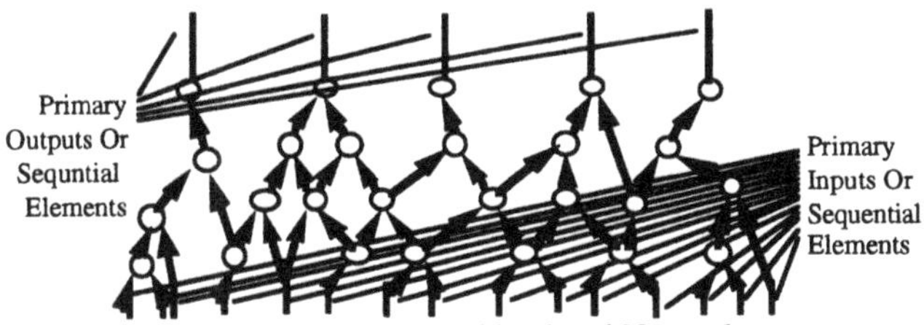


**Figure 6**: Example Combinational Network.

To illustrate this algorithm, a combinational network example is given in Figure 6. It has a total size 26 assuming each node has a size of 1 and total number of IOs 20.

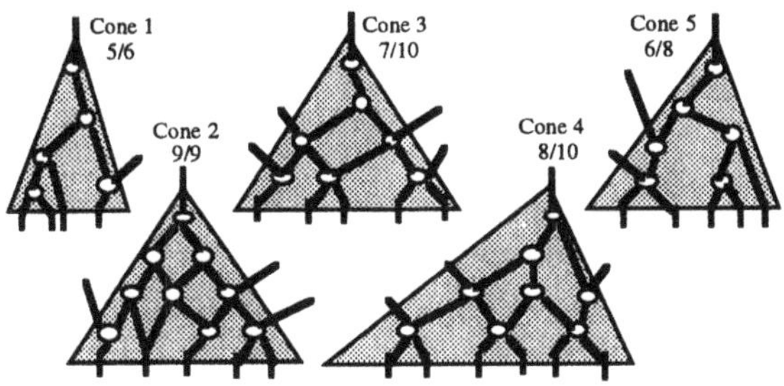


**Figure 7:** Cones Of Example Network And Associated Costs.

Figure 7 shows the network of Figure 6 after the definition of cones.

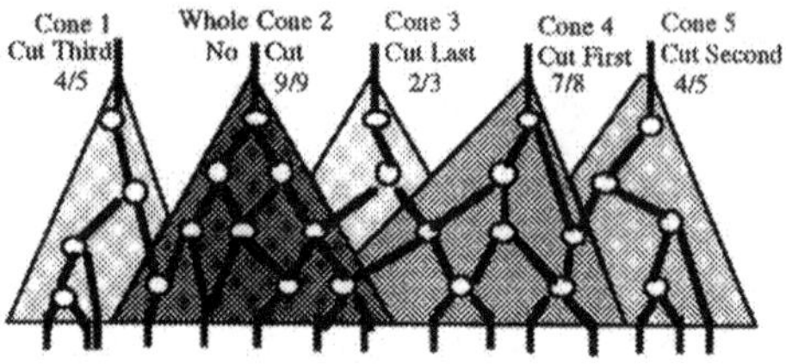


**Figure 8:** Flattened Cone Structure With Whole Best Cone 2.

In Figure 8, cone 2 with the greatest cost 1 remains whole. The other cones are cut around cone 2 producing cone 1 cost of 4/5, cone 3 cost of 4/7, cone 4 cost of 7/8, and cone 5 cost of 6/8. Of these cut cones, the highest cost cone is cone 4. Cones 3 and 5 are cut around cone 4.

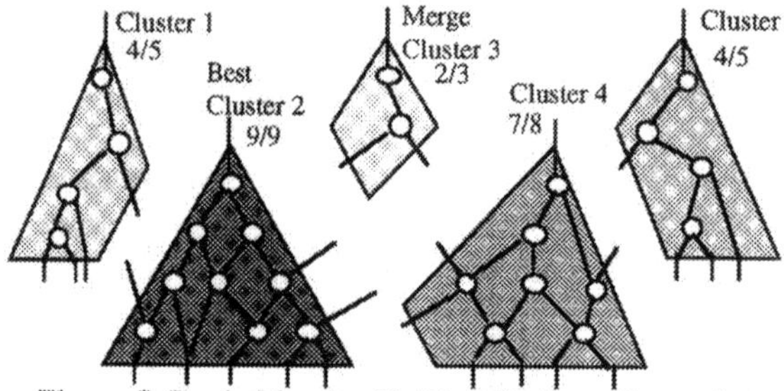


**Figure 9:** Begin Merging To Meet Package Constraints.

In Figure 9, clusters have resulted from the cut cone sections. Cluster 2 has the greatest cost 1. The costs after merging with cluster 2 are 13/12 for 2-1, 11/10 for 2-3, 16/15 for 2-4, and 13/14 for 2-5. The highest cost merge 2-3 is accepted if the merged cluster 2-3 meets package constraints.

## 6. CLUSTER ALONG CONE BOUNDARIES AND MERGE

In this section, the second cone partitioning algorithm *Cluster_And_Merge* is introduced. This algorithm clusters nodes found in similar overlapping cone sections. However these clusters often contain only one node with circuits of high fanout, forcing reliance completely on the merging operations to achieve high Size/IO partitions. Therefore this algorithm contrasts the first in that it should be used for circuits of low fanout and the first algorithm used for circuits of high fanout.

*Cone Partitioning Algorithm 2. Cluster_And_Merge:*

*1. Define cone structures as described in section 3.*

*2. Cluster Overlapping Cone Sections. Nonoverlapping clusters result which are the highest cost sections extracted from the original circuit.*

*3. Add sequential elements to clusters for reduced Size/IO ratio.*

*4. Merge clusters with combinational/sequential elements on the same critical path.*

*5. Merge clusters into partitions to meet package constraints as in cone partitioning algorithm 1.*

Figure 10 illustrates overlapping regions for the example network of Figure 6. The degree of shading in a region reflects the amount of overlap between cones. The dark region at the center is the overlap region of the three cones 2, 3, and 4. The node within this dark region has the three labels 2, 3, and 4.

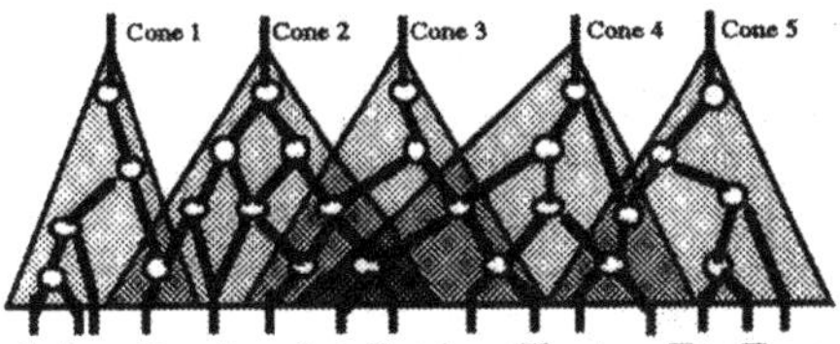


**Figure 10**: Defining Overlapping Region Clusters For Example Network.

All combinational gate nodes with the same set of labels are contained within the same overlapping region and are assigned to the same cluster in Figure 11. Note that the node at the top of clusters which were overlapping cone regions always have a fanout greater than one.

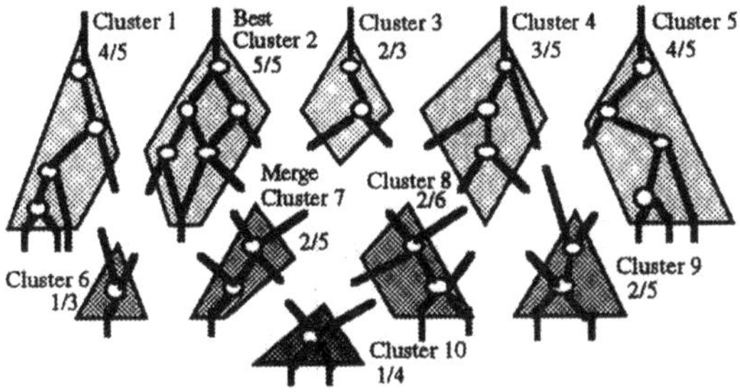


**Figure 11**: Block Clusters Formed From Cone Cuts.

In Figure 11, cluster 2 has largest cost 1. The merge costs after merging with cluster 2 are 9/10 for 2-1, 7/8 for 2-3, 8/10 for 2-4, 9/10 for 2-5, 6/6 for 2-6, 7/6 for 2-7, 7/11 for 2-8, 7/10 for 2-9, and 6/9 for 2-10. Cluster 2-7 is merged if it meets package constraints.

## 7. EXPERIMENTAL RESULTS

Table 1 lists cone partitioning information for ISCAS85 MCNC benchmarks and some real industrial circuit benchmarks. Size is the number of ACTEL2 FPGA logic modules. Primary inputs and outputs are of the initial circuit. Benchmarks with many more algorithm 2 clusters than cones indicate high cone overlap and circuits with high fanout. Cone structures are rooted at the input nodes of primary outputs or sequential element inputs. Thus for combinational circuits there can never be more cones than primary outputs, but there can be more primary outputs than cones. Since sequential elements have multiple inputs, the same does not apply.

| Circuits | Total Size | Primary Inputs | Primary Outputs | Whole Cones | Sequential Elem Size | Alg.1 Clusters | Alg.2 Clusters |
|---|---|---|---|---|---|---|---|
| c7552 | 9425 | 207 | 108 | 107 | 0 | 107 | 251 |
| c5315 | 7005 | 178 | 123 | 123 | 0 | 123 | 284 |
| c3540 | 5712 | 50 | 22 | 22 | 0 | 22 | 112 |
| c6288 | 3216 | 32 | 32 | 32 | 0 | 32 | 62 |
| c2670 | 3191 | 233 | 140 | 64 | 0 | 64 | 119 |
| c880 | 1096 | 60 | 26 | 26 | 0 | 26 | 39 |
| c499 | 912 | 41 | 32 | 32 | 0 | 32 | 43 |
| Indust1 | 7362 | 21 | 92 | 102 | 171 | 102 | 195 |
| Indust2 | 1287 | 41 | 96 | 96 | 0 | 96 | 101 |
| Indust3 | 928 | 38 | 25 | 63 | 35 | 63 | 119 |

**Table 1**: Benchmark Circuit Information.

The industrial benchmark indust1 was synthesized from a behavior level VHDL description. During the high-level synthesis process, a netlist hierarchy was created with the partitioning rules stated. Figure 12 lists a VHDL description summary of each hierarchical block and at each level of hierarchy. During the netlist partitioning, the hierarchical blocks were easily

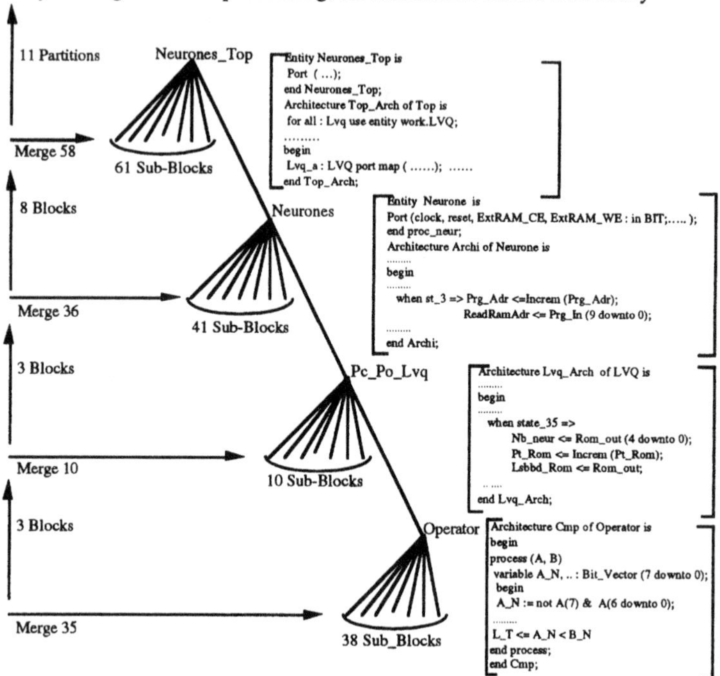


**Figure 12**: Hierarchical Partitioning Of Benchmark Indust1 With VHDL Descriptions.

merged to a minimum number of 11 packages with Size constraint 684 and IO constraint 104. This optimal result (i.e., 11 packages * 684 = 7524 is just greater than circuit size 7362) would not have been possible if the high-level synthesis had not created highly connected blocks which were smaller than the package constraints.

In Tables 2 and 3, experimental comparisons with a min-cut algorithm of (Fiduccia 1982) are presented against the cone partitioning algorithm1 of *Best_Cone_Merge* and the cone partitioning algorithm 2 of *Cluster_And_Merge*. Benchmark circuits were partitioned for two industrial packages. Table 3 lists partitioning information for a smaller package with maximum size constraint of 684 and maximum number of IO pins constraint of 104. Table 4 lists partitioning information for a larger package with maximum size constraint of 1232 and maximum number of IO pins constraint of 140. For each algorithm, Tables 2 and 3 list the number of packages into which the circuit has been partitioned (**#Pkg**), the average size of the circuit partitions (**AvSize**), the average number of IO pins of the circuit partitions (**AvIO**), and the CPU time in minutes taken during the execution of the algorithm on a Sun Sparcstation1+ (**Time**). Min-cut consistently produced bad partitions resulting in more packages because of the FPGA package IO bottleneck. Differences in results between the cone partitioning algorithms is not as clear since the typical circuits used here have average fanout (i.e, algorithm 1 clusters all met package constraints and algorithm 2 clusters all had size greater than 1).

| | Min-Cut | | | | Alg 1. Best Cone Merge | | | | Alg 2. Cluster And Merge | | | |
|---|---|---|---|---|---|---|---|---|---|---|---|---|
| Circuits | #Pkg | AvSize | AvIO | Time | #Pkg | AvSize | AvIO | Time | #Pkg | AvSize | AvIO | Time |
| c7755 | 19 | 497 | 83 | 27.8 | 15 | 628 | 67 | 4.9 | 14 | 674 | 89 | 6.4 |
| c5315 | 24 | 292 | 71 | 25.7 | 11 | 636 | 60 | 4.1 | 11 | 637 | 64 | 4.2 |
| c3540 | 13 | 440 | 70 | 8.6 | 10 | 571 | 91 | 2.6 | 9 | 635 | 75 | 1.1 |
| c6288 | 9 | 357 | 70 | 22.9 | 6 | 536 | 78 | 5.8 | 5 | 643 | 92 | 2.8 |
| c2670 | 8 | 399 | 79 | 6.9 | 6 | 532 | 50 | 1.5 | 5 | 639 | 90 | 1.0 |
| c880 | 3 | 274 | 93 | 0.1 | 2 | 548 | 78 | 0.1 | 2 | 548 | 81 | 0.1 |
| c499 | 4 | 228 | 78 | 0.1 | 2 | 456 | 52 | 5.5 | 2 | 456 | 55 | 0.1 |
| Indust1 | 17 | 430 | 78 | 32.1 | 11 | 669 | 75 | 9.7 | 11 | 670 | 88 | 6.3 |
| Indust2 | 4 | 322 | 92 | 1.8 | 2 | 644 | 89 | 1.2 | 3 | 429 | 83 | 0.3 |
| Indust3 | 4 | 232 | 67 | 1.4 | 3 | 309 | 86 | 0.6 | 2 | 464 | 98 | 1.1 |

**Table 2**: Results With Package Constraints Max Size=684 And Max IOs=104.

| | Min-Cut | | | | Alg 1. Best Cone Merge | | | | Alg 2. Cluster And Merge | | | |
|---|---|---|---|---|---|---|---|---|---|---|---|---|
| Circuits | #Pkg | AvSize | AvIO | Time | #Pkg | AvSize | AvIO | Time | #Pkg | AvSize | AvIO | Time |
| c7755 | 13 | 725 | 96 | 16.2 | 8 | 1178 | 119 | 5.2 | 9 | 1048 | 110 | 8.9 |
| c5315 | 11 | 637 | 121 | 10.6 | 6 | 1168 | 108 | 6.3 | 6 | 1168 | 90 | 6.0 |
| c3540 | 8 | 714 | 79 | 3.3 | 5 | 1143 | 96 | 2.8 | 5 | 1143 | 91 | 1.4 |
| c6288 | 5 | 644 | 87 | 8.4 | 4 | 804 | 83 | 7.4 | 3 | 1072 | 99 | 6.2 |
| c2670 | 5 | 639 | 103 | 4.5 | 3 | 1064 | 101 | 1.7 | 3 | 1064 | 119 | 2.0 |
| Indust1 | 10 | 737 | 86 | 14.2 | 6 | 1227 | 86 | 10.8 | 6 | 1227 | 99 | 7.1 |
| Indust2 | 2 | 644 | 117 | 0.8 | 2 | 644 | 74 | 1.6 | 2 | 644 | 104 | 0.3 |

**Table 3**: Results With Package Constraints Max Size=1232 And Max IOs=140.

The graphs in Figures 13 and 14 illustrate the comparisons between min-cut and the cone partitioning algorithms for the smaller and larger packages. AvSize/AvIOs (i.e., average size over average number of IO pins) are graphed against the number of partitioned packages for each benchmark circuit. Note that the AvSize/AvIOs with cone partitioning are always greater than min-cut. Also most cone partitioning AvSize/AvIOs are greater than the small package Size/IO constraint of (684/104)=6.57 and the large package Size/IO constraint of (1232/140)=8.8 (see the vertical lines shown in Figures 13 and 14). Most min-cut AvSize/AvIOs are less than both large and small package Size/IO constraints.

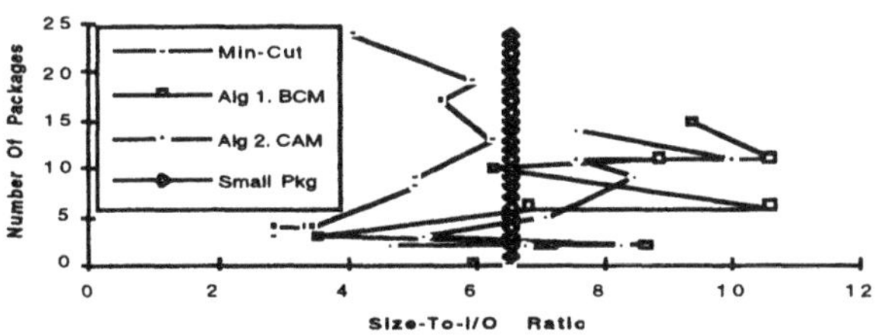


**Figure 13**: Cone Partitioning/Min-cut Results For Small Package.

Averaging the AvSize/AvIOs with the smaller package for the min-cut algorithm gives 4.4, for the *Best_Cone_Merge* algorithm gives 7.9, and for the *Cluster_And_Merge* algorithm gives 7.3. Averaging the AvSize/AvIOs with the larger package for the min-cut algorithm gives 7.0, for the *Best_Cone_Merge* algorithm gives 10.8, and for the *Cluster_And_Merge* algorithm gives 10.5. The difference between the min-cut and *Best_Cone_Merge* algorithms for the small package is 3.5 while for the large package is 3.8.

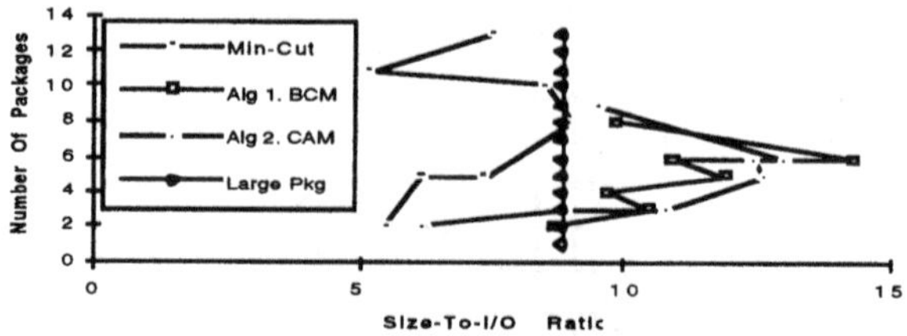


**Figure 14**: Cone Partitioning/Min-cut Results For Large Package.

The difference between the min-cut and *Cluster_And_Merge* algorithms for the small package is 2.9 while for the large package is 3.5. As the packages get larger with higher Size/IO constraints, the cone partitioning algorithms become better solutions for circuit partitioning.

## CONCLUSION

A hierarchical partitioning and two flat cone partitioning algorithms have been presented. The first cone partitioning algorithm, which should be used with circuits of high fanout, clusters by cutting each cone around the best cost cones and then merges clusters. The second cone partitioning algorithm, which should be used with circuits of low fanout, clusters nodes found within the same overlapping cone region and then merges clusters. Experimental results using two differing industrial packages show that the cone partitioning algorithms produce better partitions than min-cut. Hierarchical partitioning with a circuit hierarchy created from a VHDL description also produced an optimal number of packages.

## REFERENCES


C.M. Fiduccia et al. 1982, "A Linear-Time Heuristic For Improving Network Partitions", 19th DAC, 1982, pp. 175-181.

J. Hwang et al 1992, "Optimal Replication For Min-Cut Partitioning", ICCAD, 1992, pp.432-435.

B.W. Kernighan et al. 1970, "An Efficient Heuristic Procedure For Partitioning Graphs", Bell System Technical Journal, vol. 49, Feb. 1970, pp. 291-307.

C. Kring et al. 1991, "A Cell-Replicating Approach To Mincut-Based Circuit Partitioning", ICCAD, 1991, pp. 2-5.

B. Krishnamurthy 1993, "An Improved Min-Cut Algorithm For Partitioning VLSI Networks", IEEE Trans. On CAD, vol. CAD-33, no. 5, May 1984, pp.438-446.

R. Kuznar et al. 1993, "Partitioning Digital Circuits For Implementation In Multiple FPGA ICs", Technical Report TR93-03, MCNC, Research Triangle Park, NC, March 1993.

K.J. Singh et al. 1990, "A Heuristic Algorithm For The Fanout Problem", 27th DAC, 1990, pp. 357-360.

R. Murgai et al. 1991, "On Clustering For Minimum Delay/Area", 28th DAC, 1991, pp. 6-9.

PART FOUR

# Logic Synthesis for Programmable Devices

10

# Balanced multilevel decomposition and its applications in FPGA-based synthesis

*T. Łuba, H. Selvaraj, M. Nowicka and A. Kraśniewski*
*Warsaw University of Technology, Institute of Telecommunications*
*Nowowiejska 15/19, 00-665 Warsaw, Poland*

**Abstract**

A general decomposition concept is presented in this paper. The main strategy behind the presented Multilevel Decomposition Method is striking a balance between serial decomposition and parallel decomposition. The method is applicable to a variety of Field Programmable Gate Arrays and allows trading-off area and delay of final implementation. The results prove that the method is efficient and does not suffer from its generality.

**Keywords**

Decomposition, FPGA, technology mapping, delay minimization

## 1 INTRODUCTION

Existing FPGA-oriented technology mapping algorithms can be roughly divided into two categories according to their optimization objectives. The algorithms in the first category emphasize minimizing the number of LUTs in the solution. These algorithms include Chortle-crf (Francis et al., 1991), MIS-pga (Murgai et al., 1991), Xmap (Karplus, 1991), Trade (Wan and Perkowski, 1992) and TOS (Schlichtman, 1993). The algorithms in the second category emphasize minimizing the delay in the solution. These are DAG-Map (Chen et al., 1992) and Flow-map (Cong and Ding, 1994). Several algorithms have been implemented in both the area- and delay-minimization versions, for example MIS-pga, Chortle, and ASYL (Abouzeid et al., 1993).

Although many of the above mentioned algorithms produce encouraging results, they have a common limitation that, for a given design, a simple solution optimized with respect to a certain criterion is produced, while other solutions, preferable under different optimization objectives are ignored. Usually the logic depth (length of critical path) and the number of LUTs for these solutions vary significantly. For example, the mapping results on the Alu2 benchmark circuit, obtained by area- and delay-minimization versions of the Flow-Map and Chortle are 125 cells/12 levels or 149 cells/8 levels and 128 cells/13 levels or 227 cells/9 levels, respectively (Cong and Ding, 1994).

In general, the area-minimized designs have much larger depth, while delay-minimized designs use much more LUTs.

It is very likely that in practice the best design does not come from either of these two extreme solutions. It is important to let the designer have the flexibility to choose from a set of mapping solutions with smooth trade-off between area and depth.

In this paper we propose a balanced multilevel decomposition to solve the problem of trading-off area and depth in LUT-based FPGA technology mapping. Specifically, we are interested in obtaining a set of mapping solutions for a given design, which can meet various area and depth requirements.

Unlike the traditional approach of first performing logic minimization and then mapping the design using decomposition and other methods, the proposed method starts the mapping process straight away using different decomposition strategies described in our previous papers (Selvaraj, 1993, Luba, 1994). An important feature of multilevel decomposition is an attempt to strike a balance between the wider known serial decomposition (Luba et al. 1992, Wan and Perkowski, 1992) and the parallel decomposition (Jozwiak and Volf, 1992, Luba, 1994). Such a strategy, on one hand, is capable of detecting and utilizing the fact that a group of outputs depend on the same set of input variables and, on the other hand, eliminates redundant variables.

## 2 BASIC CONCEPTS

Taking into consideration functional dependencies (Luba, 1994), two distinct decomposition strategies can be identified for decomposing Boolean functions (Fig. 1).

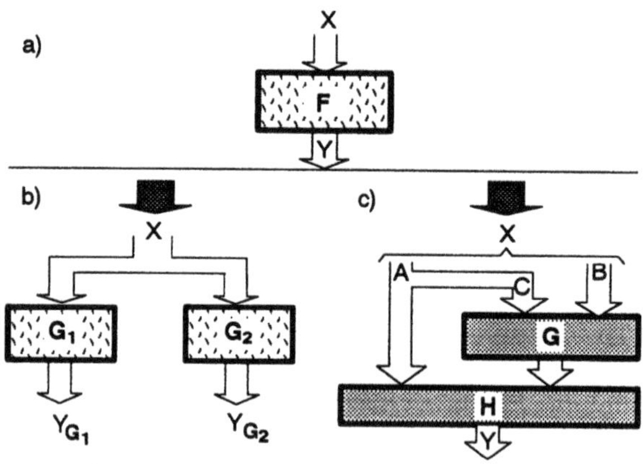


**Figure 1** Parallel and serial decomposition of function *F*

The first one, called parallel decomposition (Fig 1b), relies on partitioning the outputs into two separate groups, so that each group of outputs depend on fewer or the same number of variables as a given function $F$ i.e. $Y = Y_{G1} \cup Y_{G2}$, where for $X_{G1}$ and $X_{G2}$, the respective support sets of $Y_{G1}$ and $Y_{G2}$, we have $X_{G1} \subseteq X$ and $X_{G2} \subseteq X$.

The second decomposition strategy, called serial decomposition (Fig. 1c), relies on partitioning the input variables so that to obtain a two-level functional dependency $F = H(A,G(B \cup C))$, where $A \cup B = X$ and $C \subset A$.

The essence of the serial decomposition is to find functions $G$ and $H$, such that $G$ depends on variables in $B \cup C$, whereas $H$ depends on variables in $A$ and variables in $Z$, where $Z$ is the set of output variables of $G$.

These two strategies of decomposition play a crucial role in developing a general multi-level decomposition model.

The method is based on balancing serial and parallel decomposition, which in turn relies on certain features of minimal disjoint serial decomposition (Selvaraj, 1994).

Let $F(X)$ denote a Boolean function $F$ with a set of input variables $X$. Let $X = A \cup B$ and $A \cap B = \varnothing$, where $B$ is the set of bound variables (block $G$ inputs) and $A$ is the set of free variables (direct inputs to block $H$).

**Theorem 1.**

For every function $F$ with $|X|$ inputs, $|Y|$ outputs, $|A|$ free variables and $|B|$ bound variables, if $|X| > 2|Y|$ and $|B| > |Y|2^{|A|}$, then there exists a minimal disjoint serial decomposition such that $F = H(G(B),A)$, where the number of block $G$ outputs is $|Y|2^{|A|}$.

Theorem 1 leads to an observation that the existence of a disjoint serial decomposition $F = H(G(B),A)$, for a given number of inputs n and outputs m, is conditioned by the number of free variables (direct inputs to block $H$) and the properties of function $F$.

Using the properties of the theorem on r-admissibility (Luba, 1994) it is possible to calculate the set of free variables $A$, which determines the minimum number of block $G$ outputs (and thereby the minimum number of inputs to block $H$ or, in other words, a maximum decomposition). On the other hand, theorem 1 can be used to find whether or not a minimum decomposition exists. With the knowledge of the two extreme possibilities, it is relatively easy to find the actual maximum decomposition.

An intriguing situation arises when the process of iterative decomposition leads to a unitary function. This situation can be considered as a special case of theorem 1. By definition, a unitary function has one output and hence $|Y| = 1$. Therefore, the maximum number of block $G$ outputs is $2^{|A|}$. If the number of inputs to block $G$, $|B| > 2^{|A|}$, a minimal disjoint serial decomposition exists.

The practical importance of the above conclusion is significant as it states that a disjoint serial decomposition always exists for a unitary function if $|B| > 2^{|A|}$. Now, if minimum value is assumed for the number of free variables ($|A| = 1$), the number of bound variables $|B|$ is at least 3. This means that any unitary function with a minimum of four input variables has a disjoint serial decomposition.

When a need arises to break a function $F$ into subfunctions with two inputs, the iteration discussed so far may not lead to a complete solution. So, when a disjoint serial decomposition does not exist for a three-input unitary function, it is proposed to try non-disjoint serial (or conjunct) decomposition, $F = H(G(B),A)$, where $A \cup B = X, A \cap B \neq \varnothing$.

## 3 GENERAL AND COMPLETE DECOMPOSITION

On the basis of the presented concepts, a Multilevel Decomposition Method (MDM) has been developed. The proposed method is complete and general and can be applied to any type of FPGA Logic Block (LB) with an arbitrary number of inputs and an arbitrary number of outputs. The method looks at the Logic Block as a device capable of implementing any function with a fixed number of inputs and outputs. Unlike the traditional approach of first performing logic minimization and then mapping the design using decomposition and other methods, the proposed method starts the mapping process straight away using decomposition strategies discussed above.

Consider a function $F$: $\{0,1,-\}^n \rightarrow \{0,1,-\}^m$. Let us try to implement the function using, as a basic component, a hypothetical logic block LB with $C_{in}$ inputs and $C_{out}$ outputs, where $C_{in} > C_{out}$.

The method first tests "the sensibility" of performing disjoint serial decomposition. As LB has $C_{in}$ inputs and $C_{out}$ outputs, an ideal serial decomposition is the one where the number of block $G$ inputs, $G_{in}$, is equal to the number of LB inputs, $C_{in}$, and the number of block $G$ outputs, $G_{out}$, is equal to the number of LB outputs, $C_{out}$. If a serial decomposition is attempted with $G_{in}$ and $G_{out}$ as the number of block $G$ inputs and outputs respectively, the resulting block $H$ will have $n - (G_{in}-G_{out})$ inputs. If this number is not greater than m, then it is proposed to use parallel decomposition. In other words, if the number of inputs to the resulting block $H$ is not expected to be greater than the number of its outputs, the truth table is split into two parts using parallel decomposition. Each part is then decomposed separately using the same method.

An interesting situation arises when the number of inputs to the truth table under consideration is greater than $C_{in}$, the "sensibility" test described earlier does not recommend parallel decomposition, and the iteration process is not able to produce any result.

In such a situation, the parameters $G_{in}$ and $G_{out}$ are adjusted to $G'_{in}$ and $G'_{out}$, respectively, and serial decomposition is attempted to look for the existence of any possible function $G'$, such that: $F' = H(A, G'(x_{i1},..x_{iG'in}))$, where $F'$ is the function under consideration and $x_{i1},...,x_{iG'in} \in B$.

If a successful decomposition is found and the number of inputs to the truth table $H$ continues to be larger than $C_{in}$ (number of inputs to the given logic block), the iteration process is continued with the truth table $H$ as the new starting point after resetting the number of $G$ block inputs to $C_{in}$ and the number of $G$ block outputs to $C_{out}$.

The $G'$ function is decomposed separately, using the same iterative method, so that

$$G' = H(A, G(x_{i1},...,x_{iGin})).$$

One important aspect of the method not discussed so far is the procedure for controlling the block $G$ input/output parameters $G_{in}$ and $G_{out}$. If $C_{in}$ and $C_{out}$ are the number of inputs and outputs of the Logic Block for which the synthesis is done, initially $G_{in}$ and $G_{out}$ assume the values $C_{in}$ and $C_{out}$, respectively. Now, let us consider a situation when there is no further serial decomposition and parallel decomposition is not recommended. Under such conditions, it is proposed to increment the value of $G_{in}$ by one, i.e., $G'_{in} = G_{in} + 1$. However, before increasing the number of block $G$ inputs, it is made sure that for a given number of inputs $G_{in}$, there is no serial decomposition for $G_{out} = C_{out}, C_{out}+1,...,G_{in}-1$. For an LB with

$C_{in}-C_{out} > 1$, the existence of decomposition is verified for $G_{in} = C_{in}, C_{in}-1,...,G_{out}+1$.

For example, if the selected FPGA LB has two inputs and one output, then the consecutive block $G$ parameters, $G_{in}$ and $G_{out}$, will be 2,1; 3,1; 3,2; 4,1; 4,2; 4,3; 5,1 etc. On the other hand, if the selected FPGA LB has four inputs and one output, then the consecutive block $G$ parameters $G_{in}$ and $G_{out}$ will be 4,1; 3,1; 2,1; 4,2; 3,2; 4,3; 5,1; etc.

By relating the disjoint serial decomposition to the number of outputs, theorem 1 in practice shows the need to balance between parallel and disjoint serial decomposition strategies. When there is no possibility of finding a disjoint serial decomposition, parallel decomposition can be performed to split the truth table into two separate parts, so that the search for disjoint serial decomposition can be continued. In the view of theorem 1, parallel decomposition is a tool with which it is possible to force the condition $n > (m + 1)$ for function $F$. In the exceptional case when the process has lead to a three-variable unitary function and no disjoint serial decomposition is found for that function, conjunct decomposition is performed. If even this fails, complex decomposition is applied. Such a situation may, however, arise only if there is a need to decompose and implement a function using FPGA cells having two inputs and one output.

## 4 EXPERIMENTAL RESULTS AND CONCLUSIONS

Based on the presented general decomposition algorithm, a logic synthesis system called DEMAIN has been developed for both HP-UX and MS-DOS operating systems.

An input to DEMAIN is a truth table in Berkeley's cube-based format and the output is a network of $k$-input $l$-output cells, each realizing an $n$-variable function with $m$ outputs. The numbers $k$, $l$ can be set arbitrarily.

DEMAIN can run in either interactive or automatic mode. The automatic mode minimizes the number of Logic Blocks. In the interactive mode, the user can steer the synthesis process, according to the requirements of the chosen type of FPGA. For example, this option can be used effectively for FPGAs with Logic Blocks capable of implementing functions with different number of variables and to optimize the number of Logic Blocks or to optimize the number of levels for FPGAs with restricted interconnects. It is also possible to balance between delay and area optimization as the situation may demand.

Table 1 shows the results of decomposition of a few benchmark circuits into five-input two-output cells, which is in fact the decomposition aimed at the Xilinx Logic Blocks. The comparison of results produced by DEMAIN with the other published results shows that the proposed method does not suffer because of its universality; in fact, it provides better solutions in many cases.

Table 2 compares the minimum logic depth solutions generated by DEMAIN with those generated by other mapping algorithms designed for delay minimization.

It can be seen that, in spite of the universal character of DEMAIN, the results obtained are not worse than the results of the other programs aimed exclusively at delay optimization.

Because of arbitrary number of cell inputs accepted by DEMAIN, its applications are wider than other FPGA-oriented tools. This is because the logic cell is treated as a universal cell capable of implementing any k-input l-output Boolean function. For example, in the XILINX 3000 family, each CLB can be programmed to implement any single-output function of up to 5 input variables or any two-output function of up to 5 variables, with each output depending

**Table 1** Results on number of Xilinx LBs

| *Name* | *DEMAIN* | *TRADE* | *TOS* | *MIS-PGA* | *Chortle* |
|---|---|---|---|---|---|
| rd84 | 8 | 8 | 10 | 9 | 28 |
| rd73 | 5 | 5 | 7 | 5 | 15 |
| z4 | 4 | 4 | 4 | 4 | 4 |
| Sao2 | 23 | 27 | 23 | 28 | 26 |
| 9sym | 5 | 6 | 7 | 7 | 51 |
| root | 23 | 21 | – | – | – |
| b9 | 32 | 29 | – | 32 | 30 |
| alu2 | 17 | 22 | 47 | 96 | 94 |
| Clip | 18 | 29 | 22 | 23 | 25 |

**Table 2** Results for number of LBs/number of levels

| *Name* | *DEMAIN* | *Chortle-d* | *DAG-Map* | *Flow-Map* | *MIS-PGA* | *ASYL* |
|---|---|---|---|---|---|---|
| rd84 | 8/3 | 61/4 | 43/4 | 43/4 | 13/3 | 14/3 |
| rd73 | 5/4 | 52/4 | – | – | 8/2 | – |
| Sao2 | 23/6 | 58/4 | – | – | 45/5 | 30/8 |
| 9sym | 5/4 | 63/5 | 61/5 | 61/5 | 7/3 | 8/3 |
| alu2 | 17/3 | 227/9 | 169/8 | 162/8 | 122/6 | 60/6 |
| Clip | 18/5 | 83/4 | – | – | 54/4 | 33/6 |

on at most 4 input variables. Similarly, the ALTERA FLEX cell can implement any function of up to 4 input variables. Even the ACTEL cell which consists of three interconnected 2-to-1 multiplexers can be treated as a cell capable of implementing any Boolean function of two variables. An internal structure of the cell is irrelevant for DEMAIN. This leads us to conclude that our system is also suitable for other types of FPGAs, like those of Algotronix and Quick Logic (Selvaraj et al. 1993).

# 5 REFERENCES

Abouzeid P., Babba B., Crastes de Paulet M., Saucier G. (1993) Input-Driven Partitioning Methods and Application to Synthesis on Table-Lookup-Based FPGAs. *IEEE Trans on CAD*, **12**, No. 7, 913-925.

Chen K.C., Cong J., Ding Y., Kahng A.B., and Trajmar P. (1992) DAG-Map: Graph-based FPGA technology mapping for delay optimization. *IEEE Design and Test of Computers*, Sept. 7-20.

Cong J, and Ding Y. (1994) FlowMap: An Optimal Technology Mapping Algorithm for Delay Optimization in Lookup-Table Based FPGA Designs. *IEEE Trans. on CAD*, **13**, No. 1, 1-12.

Francis R., Rose J, Vranesic Z. (1991) Chortle-crf: Fast Technology Mapping Look-up Table-Based FPGAs, *Proc. 28-th ACM/IEEE Design Automation Conf.*, 227-233.

Jozwiak L., Volf F. (1992) An Efficient Method for Decomposition of Multiple-Output Boolean Functions and Assigned Sequential Machines, *Proc. European Conference on Design Automation*, 114-122.

Karplus K. (1991) Xmap: A Technology Mapper for Table-Lookup Field Programmable Gate Arrays, *Proc. 28th ACM/IEEE Design Automation Conf.*, 240-243.

Luba T., Markowski M., Zbierzchowski B. (1992) Logic Decomposition for Programmable Gate Arrays, *Proc. Euro-ASIC'92*, 19-24.

Luba T. (1994) Multi-level logic synthesis based on decomposition. *Microprocessors and Microsystems*, **18**, No. 8, 429-437.

Murgai R., Shenoy N., Brayton R.K., Sangiovanni-Vincentelli A. (1991) Improved logic Synthesis Algorithm for Table Look Up Architectures, *Proc. IEEE International Conf. on Computer Aided Design*, 564-567.

Schlichtmann U. (1993) Boolean Matching and Disjoint Decomposition for FPGA Technology Mapping. *IFIP Workshop on Logic and Architecture Synthesis*, 83-102.

Selvaraj H., Czerczak A., Kraśniewski A., Łuba T. (1993) A Generalized Decomposition of Boolean Function and its in FPGA-Based Synthesis. *IFIP Workshop on Logic and Architecture Synthesis*, 147-166.

Selvaraj H. (1994) FPGA-based Logic Synthesis. *Ph.D. Dissertation*. Warsaw University of Technology.

Wan W., Perkowski M.A. (1992) A New Approach to the Decomposition of Incompletely Specified Multi-Output Function Based on Graph Coloring and Local Transformations and Its Application to FPGA Mapping, *Proc. European Design Automation Conf.*, 230-235.

# 6 BIOGRAPHIES

All the authors are from the Warsaw University of Technology, Institute of Telecommunications. Currently Tadeusz Łuba and Andrzej Kraśniewski are Professors. Henry Selvaraj has just received his Ph.D. degree from this University and Mirosława Nowicka is a member of research staff in the Institute.

11

# Disjoint Decomposition for LUT FPGA Synthesis

*Ulf Schlichtmann*
*Siemens AG, Semiconductor Group*
*81549 Munich, Germany*
*Tel.: +49-89-4144-4823. Fax: +49-89-4144-8159.*
*email:* ulf@hl.siemens.de

**Abstract**

Field-Programmable Gate Arrays (FPGAs) enjoy increasing popularity in the design of digital systems. It is imperative to support the design process with CAD tools if quality designs are to be obtained. However, the structure of FPGAs forbids the use of many algorithms developed for other types of chips. New algorithms are required. Decomposition techniques are especially well-suited for Look-Up-Table (LUT) FPGAs. This paper describes how to perform decompositions efficiently using ROBDDs. The quality of a function's decomposition depends on the selected partitioning of the function's variables. An effective heuristic to partition variables for decomposition is introduced in this paper.

**Keywords**

Functional decomposition, disjoint decomposition, ROBDD, variable partitioning

## 1 INTRODUCTION

Field Programmable Gate Arrays (FPGAs) are a class of integrated circuits that has seen rapid gains in usage and popularity in the recent past. Originally introduced about a decade ago, nowadays they are used extensively for prototyping and for production runs of small to medium volumes. The general architecture of all currently available FPGAs consists of a regular array of identical basic logic modules connected by routing resources. Both the logic module and the routing resources are user-programmable.

In Look-Up Table (LUT) FPGAs the logic module is dominated by a $2^q$ bit memory, thus allowing for the implementation of any Boolean function of at most $q$ inputs. The logic module is referred to as a q-LUT. Typical examples are the Xilinx FPGAs.

Logic synthesis and technology mapping are two key steps in the design of a digital circuit. The task of logic synthesis is to optimize a circuit, described by a set of Boolean equations. Generally, this optimization depends heavily on the identification of common subexpressions and on decomposition of Boolean functions into less complex functions (Brayton 1990). Traditionally, the quality of logic synthesis has been measured in the number of literals required to represent the circuit after optimization. The task of technology mapping is to transform the optimized set of Boolean equations, usually represented as a Boolean network, into a netlist such that each element of the netlist can be implemented by a cell that is available in the given target technology.

Techniques for these tasks have been developed with the fine-grained cell libraries for mask-programmable ASICs in mind. While some of the techniques remain applicable for FPGAs, significant optimization potential would be wasted if these techniques were simply adapted for FPGAs. Instead, new techniques need to be developed.

Functional decomposition of Boolean functions is an optimization technique that is very effective both in logic synthesis and in technology mapping for LUT FPGAs. While the theory of functional decomposition of Boolean functions was originally developed about fourty years ago (Ashenhurst 1952), due to the high complexity of computing decompositions, it has not been put to use widely. However, recently is has been shown how to perform functional decompositions more efficient than previously possible, thus reviving interest in decomposition theory (Lai 1993b, Schlichtmann 1993). The quality of a functional decomposition depends on the selected partitioning of the function's variables. So far, no efficient partitioning strategies have been reported.

The organization of this paper is as follows. Following some definitions, Section 3 describes how to compute decompositions using ROBDDs. An effective heuristic to determine variable partitionings is introduced in Section 4. Results are given in Section 5.

## 2 DEFINITIONS

A Boolean variable $z$ can have two values, here denoted by 0 and 1. By $\overline{z}$ the complement of the variable $z$ is denoted. Both $z$ and $\overline{z}$ are called literals.

Let $\underline{z}$ denote a vector of Boolean variables, $\underline{z} = (z_1, z_2, \ldots, z_n)$. With $|\underline{z}|$ the number of components of the vector is obtained. A fully specified, single output Boolean function, $F(\underline{z}) : B^{|\underline{z}|} \rightarrow B$ is a function of Boolean variables. A multi-output Boolean function is written as a vector $\underline{g}(\underline{z})$ of single-output Boolean functions. The number of outputs is obtained by $c = |\underline{g}(\underline{z})|$. Alternatively, $(g_1(\underline{z}), \ldots, g_c(\underline{z}))$ will also be used instead of $\underline{g}(\underline{z})$. The number of elements in a set $S$ are obtained by $|S|$.

Reduced, Ordered Binary Decision Diagrams (ROBDDs) as proposed by Bryant (Bryant 1986) are employed. The expression $ROBDD(f)$ refers to the node $v$ that roots the ROBDD of $f$. By $index(v)$ the position of the variable associated with ROBDD-node $v$ in the variable ordering is obtained. To obtain the position of a variable $z$ in the variable ordering, $index(z)$ will be used as well. The first variable in the ordering is assigned $index = 0$.

## 3 DISJOINT DECOMPOSITION USING ROBDDS

Given a Boolean function $F(\underline{z})$ and a partitioning of the variables into *bound* variables $\underline{x}$ and *free* variables $\underline{y}$, $F(\underline{z}) = F(\underline{x}, \underline{y})$, an expression of $F$ having the structure $F(\underline{x}, \underline{y}) = f(\underline{g}(\underline{x}), \underline{y})$ is termed a *decomposition*. The function $\underline{g}$ is called *decomposition function*, $f$ is called *composition function*. The value $c = |\underline{g}(\underline{x})|$ is referred to as the *communication complexity* of the decomposition. If no variables are shared between $\underline{x}$ and $\underline{y}$, the decomposition is said to be *disjoint*, otherwise *non-disjoint*. This paper focusses

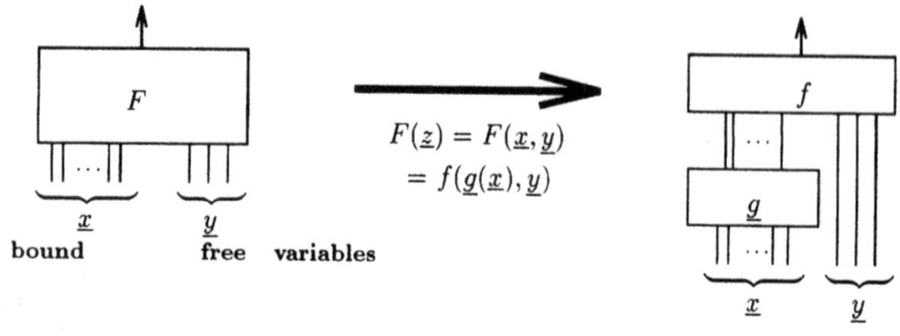


**Figure 1** Structure of a disjoint decomposition

on disjoint decompositions. The structure of such decomposition is shown graphically in Figure 1. The *decomposition problem* is: Given $F(\underline{z})$ and a partitioning of $\underline{z}$ into $\underline{x}$ and $\underline{y}$, determine the minimum communication complexity.

Previously, methods have been suggested to determine $\underline{g}(\underline{x})$ based on decomposition charts (Ashenhurst 1952) and on the on- and off-covers of Boolean functions (Roth 1962). While these methods solve the decomposition problem, they are not as efficient as the ROBDD-based approach discussed next.

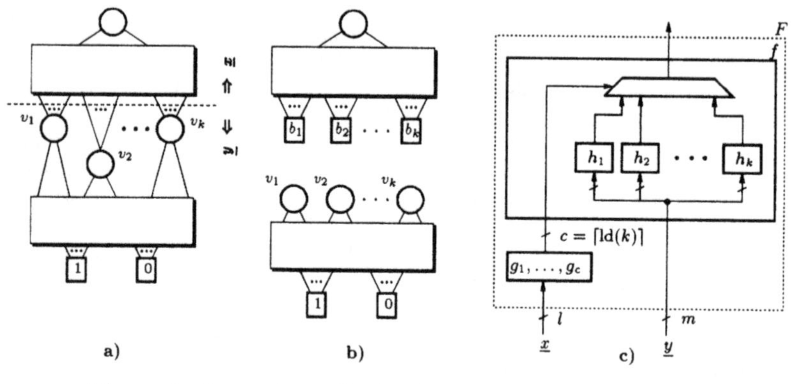


**Figure 2** Computing a disjoint decomposition using ROBDDs

The basic idea of the ROBDD-based approach is shown in Figure 2. An ROBDD is built for function $F(\underline{x}, \underline{y})$. The bound variables are arranged before the free variables in the variable ordering. Assume $l = |\underline{x}|$. If a *cut-line* is drawn at level $l$ in the ROBDD, i.e. separating the BDD-nodes $v_b$ with $index(v_b) < l$ from the nodes $v_f$ with $index(v_f) \geq l$, then $cut\_nodes(ROBDD(F(\underline{x}, \underline{y}), l))$ denotes the set of nodes that are located below the cut-line and have at least one incoming edge that originates above the cut-line. The ROBDD is then cut at the position of the cut-line. Each of the nodes $v_1, \ldots, v_k$ in this set now becomes a root node of an ROBDD. The functions $h^1(\underline{y}), \ldots, h^k(\underline{y})$ represented by ROBDDs rooted by $v_1, \ldots, v_k$ only depend on the free variables $\underline{y}$. They can be thought of as being connected to the data inputs of a $k$-to-1 MUX. Thus, the communication complexity $c$ is easily determined from a function's ROBDD. Then $c = \lceil ld\ k \rceil = \lceil |cut\_nodes(ROBDD(F(\underline{x}, \underline{y}), l))| \rceil$ control inputs are needed to select one of the data inputs. The functions $\underline{g}(\underline{x})$ computing these control inputs only depend on the bound variables $\underline{x}$. This way, in Figure 2c, the decomposed structure shown on the right-hand side in Figure 1 is obtained.

Computing a decomposition using an ROBDD results in speedups ranging from 17 to 35 compared to previously known methods on a set of benchmark examples (Lai 1993b).

# 4 PARTITIONING VARIABLES TO MINIMIZE COMMUNICATION COMPLEXITY

The problem of selecting a variable partitioning has received only little attention so far. In (Lai 1993) all partitions with $|\underline{x}| = l$ ($l = 4$ or $5$) are enumerated and evaluated. This approach can be quite costly for Boolean functions that depend on many variables. Additionally, since $|\underline{x}|$ is constrained, only a limited solution space is evaluated. It may instead be desirable in the first steps to decompose a large function into intermediate functions that themselves are still too large to fit into a given $q$-LUT. In (Sasao 1993) only decompositions with $|\underline{y}| = 2$ are considered, thus significantly limiting the solution space. In (Schlichtmann 1993) it is shown that it is beneficial to select symmetric variables as bound variables. Many Boolean functions, however do not have symmetric variables.

Therefore, a heuristic is presented here to adequately address the problem of determining an effective partitioning. It proceeds by first determining an initial solution followed by an iterative improvement phase. These steps result in a variable *ordering.* Only then a cut in the ordering is selected to obtain the final variable *partitioning.* A cost function is presented that guides this selection of the cut position. No restriction is placed on the number of variables in either bound set or free set.

## 4.1 Initial Variable Ordering

This heuristic is based on the assumption that generally an ROBDD that is small in the number of nodes will also allow decompositions with a small communication complexity. It proceeds as follows:

1. Search for a variable $z_d$ such that $F(z_1, \ldots, z_d = 1, \ldots, z_n) = 1$, i.e. $z_d$ can be disjunctively split from $F(\underline{z})$. If found, delete this variable from $F(\underline{z})$, remember it and continue with step 1.
2. Search for a variable $z_c$ such that $F(z_1, \ldots, z_c = 0, \ldots, z_n) = 0$, i.e. $z_c$ can be conjunctively split from $F(\underline{z})$. If found, delete this variable from $F(\underline{z})$, remember it and continue with step 1.
3. Arrange the variables found so far in reverse order at the end of the variable ordering. The variable found first will be the last variable and so on.
4. On the remaining function, apply the procedure reported by Wu and Marek-Sadowska (Wu 1993).

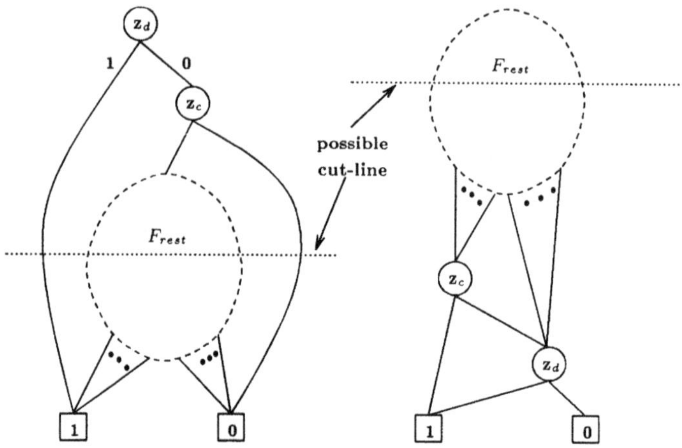


**Figure 3** Ordering of variables that can be split off disjunctively or conjunctively

The rationale for step 3 in the above procedure is illustrated in Figure 3 for the ROBDD of a function $F(\underline{z}) = z_d + z_c \cdot (F_{rest}(\underline{z}'))$ where $\underline{z}'$ does not contain $z_d$ and $z_c$ anymore. If the variables $z_d, z_c$ were put at the top of the variable ordering as shown in the left part of the Figure, the 1-successor of $z_d$ would lead directly to the terminal node 1, the 0-successor of $z_c$ would lead directly to the terminal node 0. Then, for every cut-line below $z_c$ both terminal nodes would be contained in the set *cut_nodes*. If the variables are put into the last position of the variable ordering, as shown in the right part of the Figure, this problem is remedied. The ROBDD of $F_{rest}$ (dashed in the Figure) remains unchanged.

## 4.2 Iterative Improvement

The variable ordering so obtained will usually not represent an optimum. Therefore, an iterative improvement heuristic is employed next.

The basic idea of the method is to swap variables across a given cut-line. Whereas up to now variable orderings have been considered without regard to a specific location of a cut-line, now variables need to be swapped with regard to a specific cut-line. However, the best location for the cut-line can only be determined after iterative improvements of the variable ordering have been performed. Therefore, variable swapping needs to be performed for every possible cut-line. Thus now $|\underline{z}| - 1$ ROBDDs will be considered, one for each possible cut-line.

The algorithm *variable_exchange* is shown in Figure 4. The cut-lines are considered sequentially. The $level(cut)$ of $cut$ is the minimum $index(v)$ of any node $v$ located below $cut$. $ROBBD^l(F(\underline{z}))$ denotes the ROBDD of $F(\underline{z})$ that is being treated with respect to $cut$ with $l = level(cut)$. $ROBDD_{z_u \leftrightarrow z_l}(F(\underline{z}))$ denotes the ROBDD with variables $z_u$ and $z_l$ exchanged in the variable ordering. Variables are always swapped in pairs, i.e. for $cut$ at level $l$ a variable $z_u$ with $index(z_u) < l$ is swapped with another variable $z_l$

```
variable_exchange(ROBDD(F(z))) {
  for cut = 1 to |z| - 1 {
    do {
      beneficial = FALSE
      z_u = determine_upper_swap_var(ROBDD^cut(F(z)), cut)
      z_l = determine_lower_swap_var(ROBDD^cut(F(z)), cut)
      if |cut_nodes(ROBDD^cut_{z_u<->z_l}(F(z)), cut)| <= |cut_nodes(ROBDD^cut(F(z)), cut)| {
        ROBDD^cut(F(z)) = ROBDD^cut_{z_u<->z_l}(F(z))
        beneficial = TRUE
        for oc = index(z_u) + 1 to index(z_l) {
          if |cut_nodes(ROBDD^oc_{z_u<->z_l}(F(z)), oc)| <= |cut_nodes(ROBDD^oc(F(z)), oc)|
            ROBDD^oc(F(z)) = ROBDD^oc_{z_u<->z_l}(F(z))
        }
      }
    } while (beneficial)
  } }
```

**Figure 4** Iterative improvement of variable ordering based on variable exchanges

with $index(z_l) \geq l$ in the variable ordering. A swap is considered to be *beneficial*, if it decreases $|cut\_nodes(ROBDD(\underline{z}), l)|$. The communication complexity is obtained easily as $\lceil ld(|cut\_nodes(ROBDD(\underline{z}), l)|)\rceil$. For a given *cut* at level $l$, variable swaps are performed as long as beneficial swaps exist. Pairs of swap variables are determined by tentatively swapping each variable $z_u$, $index(z_u) < l$ below *cut* and selecting the variable that yields an optimum of $|cut\_nodes(ROBDD(\underline{z}), l)|$. This is done by *determine_upper_swap_var*. The respective procedure is performed for all variables $z_l, index(z_l) \geq l$, by swapping them above *cut*. The best variables are then permanently swapped, if the swap is beneficial.

To consider more than just one cut-line at a time, after each swap of $z_u$ and $z_l$ it is checked if this swap would improve the solution for other cut-lines also. Only other cut lines *oc* in the range $index(z_u) < level(\text{oc}) \leq index(z_l)$ need to be considered.

## 4.3 Determining Partitioning

Now the variable ordering has been improved for every possible cut-line. In order to actually perform the decomposition, one of the possible cut-lines needs to be selected to determine a partitioning of $\underline{z}$ into $\underline{x}$ and $\underline{y}$.

To this end, a cost function is proposed that estimates the number of logic modules required to implement a Boolean function $f$ in $q$-LUTs. It is based on the assumption that a function will be decomposed according to the Shannon expansion $f(\underline{z}) = z_i \cdot f(\underline{z}) + \overline{z}_i \cdot f(\underline{z})$. While this leads to a pessimistic estimation of implementation cost, empirically it has yielded good results in guiding the choice of a partioning of the variables.

The cost of implementing a single-output function $g_i(\underline{x})$ is given as:

$$\text{cost}(\underline{g}(\underline{x})) = \begin{cases} 1 & \text{for} |\underline{x}| \leq q \\ 2^{(|\underline{x}|-q)+1} - 1 & \text{for} |\underline{x}| > q \end{cases} \tag{1}$$

The total cost $C$ of a decomposition of a function $F(\underline{x}, \underline{y}) = f(\underline{g}(\underline{x}), \underline{y})$ then is

$$C = \text{cost}(f(\underline{g}(\underline{x}), \underline{y})) + \sum_{i=1}^{|\underline{g}(\underline{x})|} \text{cost}(g_i(\underline{x})) \tag{2}$$

Of all possible variable orderings associated with different cut-lines *cut*, the cut-line that results in a decomposition minimizing $C$ is selected.

## 5 RESULTS FOR TECHNOLOGY MAPPING ON XILINX

The presented decomposition method has been implemented in the synthesis tool TOS. It is embedded into a design flow that consists of preprocessing the initial Boolean network by assuring that no two nodes in network compute the same function, either globally or locally. A partial collapsing is performed where nodes are collapsed into their successors, if these will then still have no more than $q$ fanins. After decomposition, the same operations are performed as a postprocessing step.

Finally, nodes in the network are merged into logic modules (Configurable Logic Blocks, CLBs) as permitted by the technology, using cardinality matching.

| | Random | | BDD-Min | |
|---|---|---|---|---|
| | CLBs | CPU [s] | CLBs | CPU [s] |
| NO iterative improvement | 2927 | 57.2 | 1774 | 27.9 |
| WITH iterative improvement | 1773 | 113.8 | 1715 | 72.9 |

**Table 1** Effects of various options of the presented algorithm

Table 1 permits an evaluation of the presented methods for variable partitioning. TOS optimized for number of CLBs on the XC3000 architecture. The total number of required CLBs for a set of 20 MCNC benchmark circuits is given. Four data points are presented. Variables were ordered both randomly (Random) and using the heuristic presented in Section 4.1 (BDD-Min). Additionally, the iterative heuristic of Section 4.2 was employed to improve on the initial results.

Two facts are obvious from Table 1: the iterative improvement heuristic is very effective. The better it's starting situation, the better it's final results. Also, the heuristic proposed to select an initial variable ordering is very effective. Just by itself it results in almost exactly the same number of required CLBs as using the iterative improvement heuristic on a random variable ordering.

The required CPU time increases by a factor of 2 to 3 when the iterative improvement heuristic is used. Interestingly, using the BDD-Min heuristic results in less CPU time than using a random ordering. This is because with a random ordering, the number of decompositions that need to be performed rises significantly.

# 6 CONCLUSION AND FUTURE WORK

A method to compute disjoint decomposition using ROBDDs has been presented. Also, an effective method to partition variables for decomposition has been introduced. It has been shown that these methods yield good results using only little CPU time.

Future work will focus on non-disjoint decomposition, decomposition of multi-output Boolean function and improved methods for variable partitioning.

# 7 REFERENCES

Ashenhurst, R. L. (1952) The Decomposition of Switching Functions. *Bell Laboratories' Report No. BL-1.*

R. K. Brayton, R. K.; Hachtel, G. D. and Sangiovanni-Vincentelli, A. L. (1990) Multilevel Logic Synthesis. *Proceedings of the IEEE*, **78**, 264 - 300.

Bryant, R. E. (1986) Graph-Based Algorithms for Boolean Function Manipulation. *IEEE Transactions on Computers*, **35**, 677 - 91.

Lai, Y.-T.; Pan, K.-R. R.; Pedram, M. and Sastry, S. (1993) FGMap: A Technology Mapping Algorithm for Look-Up Table Type FPGAs Based on Function Graphs. *Workshop Notes International Workshop on Logic Synthesis IWLS*, 9b1 - 4.

Lai, Y.-T.; Pedram, M. and Vrudhula, S. B. K. (1993b) BDD Based Decomposition of Logic Functions with Applications to FPGA Synthesis. *30th Design Automation Conference DAC*, 642 - 7.

Roth, J. P. and Karp, R. M. (1962) Minimization over Boolean Graphs. *IBM Journal of Research and Development* **6**, 227 - 38.

Sasao, T. (1993) FPGA Design by Generalized Functional Decomposition, in *Logic Synthesis and Optimization* (ed. T. Sasao), Kluwer Academic Publishers, Boston / London / Dordrecht.

Schlichtmann, U. (1993) Boolean Matching and Disjoint Decomposition for FPGA Technology Mapping. *Proceedings IFIP International Workshop on Logic and Architecture Synthesis*, 83 - 102.

Wu, Y.-L. and Marek-Sadowska, M. (1993) Efficient Ordered Binary Decision Diagram Minimization Based on Heuristics of Cover Pattern Processing. *European Conference on Design Automation EDAC*, 273 - 7.

# 8 BIOGRAPHY

Ulf Schlichtmann joined the Institute of Electronic Design Automation (Prof. Dr.-Ing. K. Antreich) in 1990 to start the synthesis research group. Together with Michael Hermann he laid the foundation for the FPGA synthesis tool TOS. In 1991, he spent 2 months as a visiting scholar with the Microelectronics Center of North Carolina, Research Triangle Park, USA. His dissertation focusses on core algorithms for FPGA oriented logic synthesis and technology mapping. Since 1994, he is a member of the technical staff in the Siemens semiconductor group, mainly involved with macrocell based IC design and design reuse.

# 12

# Performance Comparison of Logic Block Units Implemented in FPGA's Families

*M.ROBERT, L.TORRES, F.MORAES and D.AUVERGNE*
*UNIVERSITE MONTPELLIER II*
*LIRMM, 161 rue Ada 34392 Montpellier cedex 5, FRANCE*
*Fax : ( 33 ) 67–41–85–00, e–mail : robert@lirmm.fr*

**Abstract :**

Due to the different solutions used to implement combinatorial logic blocks, comparison of performances of circuit mapping, on different FPGA's target, is quite hard to be analyzed. In order to accurately compare the performances of the logic blocks architectures used in FPGA's families we use facilities in automatic layout synthesis to compare the performances of a set of logic functions implemented with different logic block alternatives. Comparisons are obtained from the evaluation of the performances of the layout extraction of these implementations. As an application of the results obtained a methodology is proposed to optimize the size of transistor of the transmission gates used to implement Look–up Tables.

**keywords :**

Programmable logic blocks, Layout synthesis, transistor sizing.

## 1– INTRODUCTION

FPGA architectures often make use of complex cells to efficiently implement circuitry with the help of logic synthesis tools. Among the several FPGA technologies available today *(Xilinx, Actel, Altera, etc...)*, the comparison of timing performances is always device *(XC3000, XC4000, ACT1,...)* and technology (CMOS process and programming technique) dependent. For a given FPGA device, the delay of a path depends on the logic block architecture, the number of the logic depth of the mapping, the routing delays associated and the I/O cells. The resulting delay information is directly given by the manufacturers without any technological information on the process parameters used. Consequently, the effect of the logic block structures on the speed of FPGA is quite hard to be analyzed.

In order to focus on the effect of logic block architecture on FPGA timing performances, it is necessary to select a common CMOS process, and to implement a set of logic functions using the different logic blocks alternatives. A first complete study of logic blocks has been proposed in Singh (1992,1991): an experimental approach is taken, in which a set of benchmark logic circuits is synthesized into different FPGA's using a same CMOS process. It is shown that the fine grain blocks (Nand2,...) are slow, because they require many levels of logic and consequently a large routing delay. Five and six input look up tables and Actel muxes give the best performances because of the balance obtained between routing and logic delays. This study covers a large selection of combinational circuits, and show the effect of granularity on performance taking into account an approximation of the routing delays. However the limitations come from the evaluation performed at the transistor level without any consideration on the layout and the transistor sizes (always minimum). Moreover the routing delay is taken as a constant similar for each connection.

The effect of logic block layout on the speed of FPGA's is studied in this paper. Using a layout generator, we evaluate the post–layout performances of the basic cells taking into account the load capacitances and the transistor widths. To generate the different CMOS logic functions, we use an efficient layout methodology, where the cells are implemented as an array of rows with different heights (Moraes 1993). Optimized cells are generated with a two pass procedure : first the automatic generator transforms the electrical netlist in a symbolic description, then this description is translated into a layout after technology interpretation and compaction procedures. The main differences with the traditional cell based approach are : no explicit routing channel between rows, variable row heights, all cells are vertically transparent to the second metal layer used to vertically cross the cells, technology independence, and optimization of the transistor widths depending on the user constraints.

This paper is divided into 4 sections. Section 2 presents the logic block investigated, and the layout synthesis approach used. In section 3 we describe the experimental procedure used to evaluate the post layout performances of the different logic blocks, and the results obtained. In section 4, we discuss on the improvement of the look up table logic blocks using a methodology to size the transmission gates.

## 2 – DESCRIPTION OF LOGIC BLOCKS

The FPGA architectures consists in an array of identical cells separated by wiring channels. The Actel architectures implement 4–to–1 muxes in a single level (mux A, Figure 1) Gamal (1989). The logic block of Xilinx XC series uses a programmed look up table ( LUT ) to implement boolean functions : 4 input look up table for the XC2000 series LCA (*figure 2*), 5 input look up table for the XC3000 and XC 4000 series.

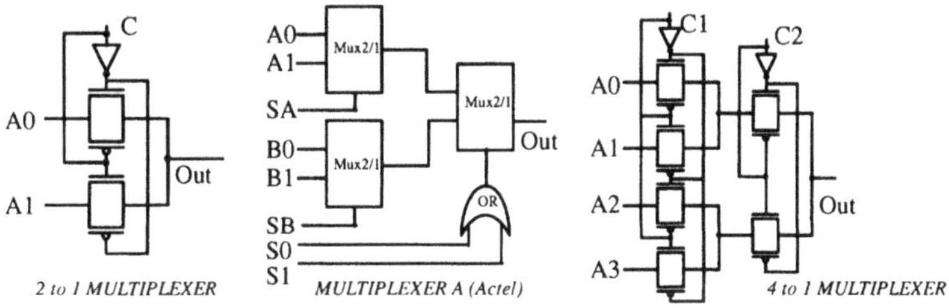


**Figure 1 :** Multiplexer basic modules

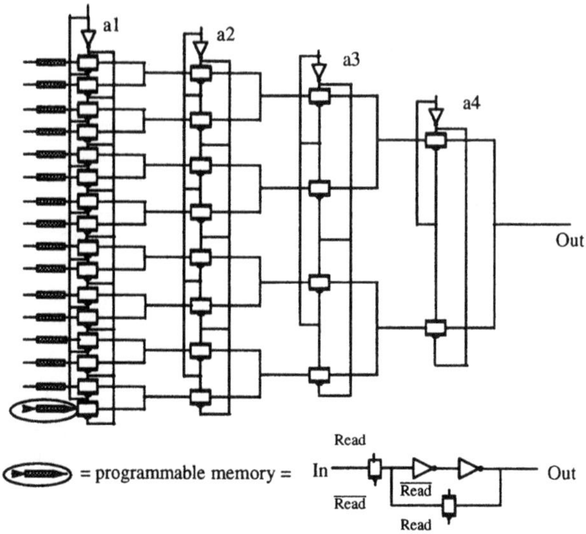


**Figure 2 :** Four–input Lookup Table (LTK4)

## 2.1 Layout methodology

To generate the different CMOS logic functions, we use an efficient layout methodology, where the cells are implemented as an array of rows with different heights *(Tropic* generator) (Moraes 1993, 1993bis). Optimized cells are generated in a two pass procedure : first the automatic generator transform the electrical netlist into a symbolic description, then this description is translated into a layout after technology interpretation and compaction procedures. Interconnection is realized with two metal layers for intra–cell and inter–row routing.

The general procedure of the layout synthesis (TROPIC) covers 5 mains steps : basic cell extraction, partition of the set of cells into rows, basic cell generation, placement – routing and layout compaction.

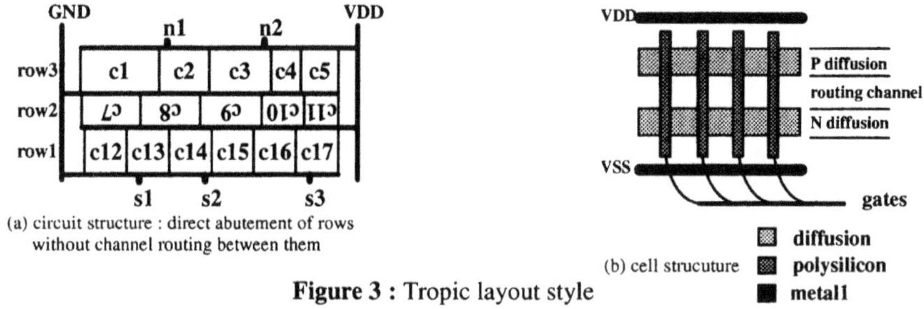


**Figure 3 :** Tropic layout style

Figure 3a illustrates the organization of the layout, as a direct abutment of cells. As shown in Figure 3b, each cell is divided into 5 parts : two parts dedicated to the diffusion rows for the transistor implementation, and three parts devoted to the routing regions.

The proposed layout style (Moraes 1993) is characterized by the placement of cells in horizontal rows, where the even rows are horizontally mirrored (Figure 3a). The main differences with the traditional cell based approach are : no explicit routing channel between rows, variable row heights, all cells are vertically transparent to the second metal layer used to vertically cross the cells, technology independence, and optimization of the transistor widths depending on the user constraints.

The main characteristics of this layout style are as follows:

•gate terminals of complementary transistors are imposed to share the same polysilicon column, this configuration avoids vertical cycles in the routing problem (the only exception is for the transmission–gates) ;

•polysilicon layer connects complementary gates (vertically) ;

•metal 1 layer is used for drain/source power connections (vertically) and for equipotential nets in the routing channel (horizontally) ;

•metal 2 layer is used for vertical connections between rows, and to connect the N and P parts of each cell ;

•diffusion layer is only used horizontally for transistor implementation and neighbour drain/source terminals;

•multiple contacts are used in drain/source terminals, this allows to reduce parasitic capacitances and resistances ;

•orientation of the input/output pins of the blocks (north south, east, west), this option is very useful for the floor planning phase of the chip.

Performance driven layouts are obtained through a complete automatic optimization sequence between independent modules as illustrated in Figure 4. The automatic generator, ***TROPIC***, transforms the electrical netlist into a symbolic description, translated after technology interpretation and compaction procedures, ***PRINT*** (Cathebras 1990) or CADENCE compactor, into a layout with minimum width transistors. After layout extraction, a timing or electrical simulation of the macro–cell in its environment (load, input pulses, ...) is performed (***PATH RUNNER*** (Deschacht 1990), HSPICE or ELDO). A post layout sizing applied to the extracted electrical description generates a new netlist with optimized transistor widths, ***PSIZE*** (Azemard 1992). Using this description, an optimized layout is generated.

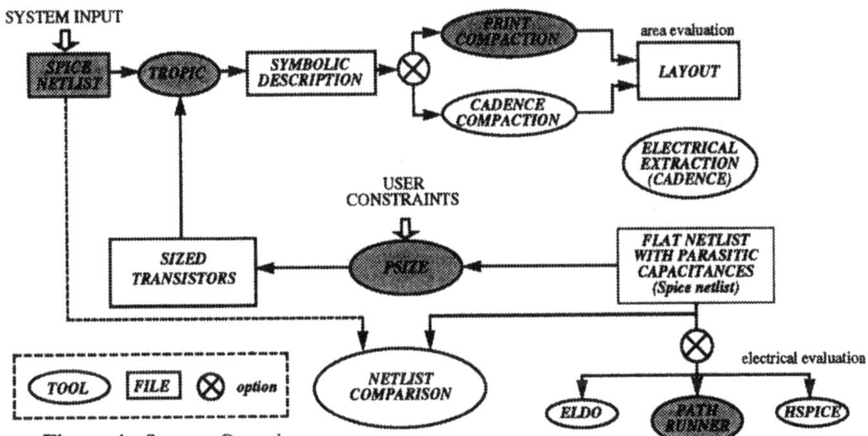


**Figure 4 :** System Overview

Figure 5 illustrates the automatic implementation of the logic block structures obtained with the *Tropic* generator from an electrical netlist (SPICE FORMAT).

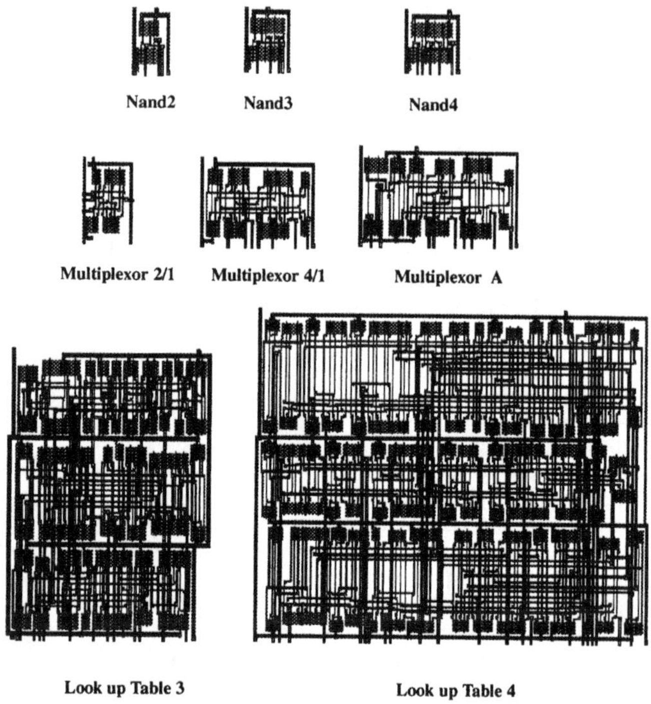


**Figure 5:** Layouts of logic blocks obtained with the tropic generator

## 3 – EXPERIMENTAL PROCEDURE

The experimental procedure is described in Figure 6. Using the transistor level schematics (Spice netlist) of the logic blocks presented previously (Figures 1,2,4) we have implemented two logic functions :

$f1 = abc+ab\bar{d}+ac\bar{d}$, proposed in Singh (1992)

$f2 = ab + fc\bar{d} + \bar{f}be+\bar{b}d\bar{c}$

The choice of simple functions avoid a shift in the results due to the quality of the logic synthesis tools. Our main objective in this first study is to analyze the influence of the layout level. For more complex functions the results obtained at the electrical level in Singh (1992) can be used as a reference.

As described in Figure 6, different types of transistor widths have been analyzed (W minimum, and W = 16 µm ) and the output load has been fixed to Cin and 10 Cin, where Cin represent the input capacitance of an inverter). After layout generation, the extracted netlist contains all the parasitic capacitances (diffusion, routing,...).

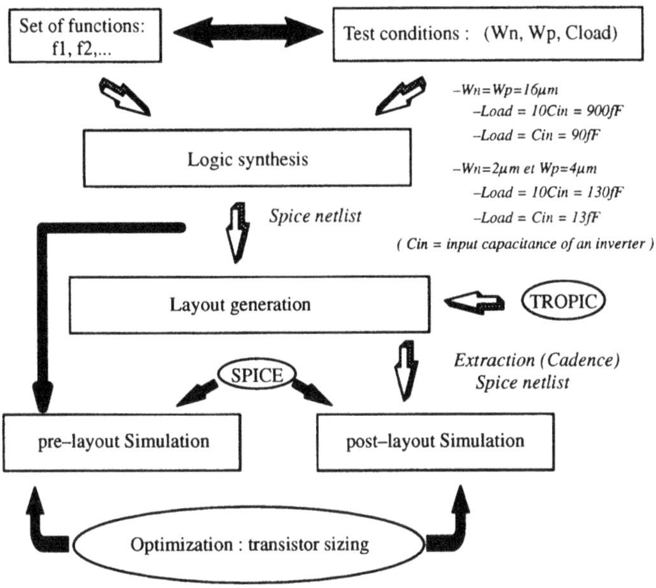


**Figure 6 :** Summary of the experimental procedure

## 3.1 Results

In order to don't overload this section, we present here some of the main results obtained with these experiments. As shown in the results given in Figures 7 and 8, the Mux 2/1 architecture exhibit the lowest delay before and after layout. The effect of the layout capacitances reverse the ranking between Nand gates and Mux A, Mux 4/1 gates.

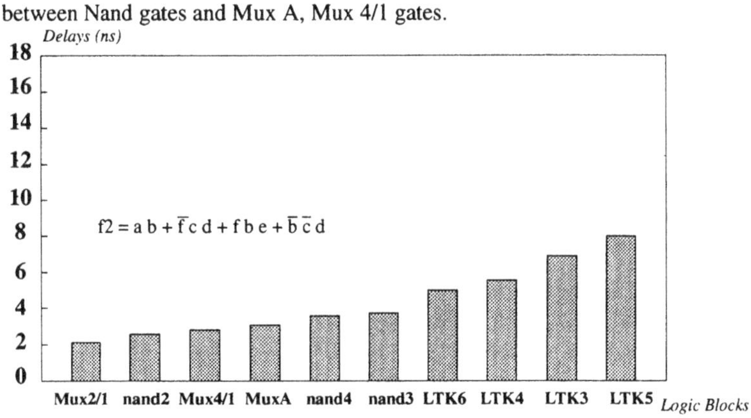


**Figure 7 :** pre–layout evaluation of the delays for the logic function f2.
(Wn=2µm and Wp=4µm and CL=130fF= 10 Cin)

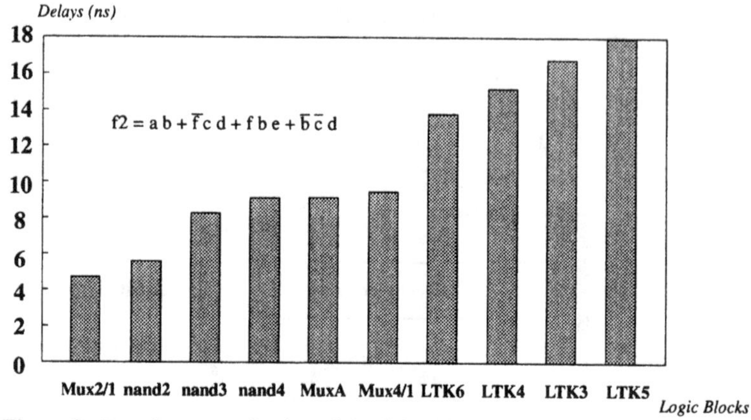


**Figure 8 :** Post–layout evaluation of the delays for the logic function f2. (Wn=2μm and Wp=4μm and CL=130fF= 10 Cin)

The timing performances of the look up tables implementation is not as good as reported in Singh (1992). One of the reason is that these modules are not optimally used in this test configuration. For more complex functions, the full capacity of the look up table will be used. However, at this level it is interesting to analyze the effect of the transistor sizing on the timing performances of the look up table modules, this is the object of the next part.

## 4 – LOOK UP TABLES OPTIMIZATION

Transistor sizing at layout level is necessary to improve the overall performance of integrated circuits. Based on a local optimization defined through an explicit formulation of delays (Auvergne 1991), a sizing methodology has been applied to the look up table cells. We defined local sizing laws under the constraint of symmetrical rise and fall times for the devices considered. For each structure, the solutions of the sizing problem involve the determination of the size of the transistors of the controlling devices and the size of the N and P transistors of the gate under consideration. A Typical critical path is represented in Figure 8 where we separate a control block (Structure I) and a data block ( Structure II ). The optimization strategy begins by the sizing of the structure II from the output (where the load capacitance $C_L$ is fixed) to the data inputs. The local sizing rules used for this "ANDORI" structure are given as follow (Auvergne 1991) :

$$\left\{ \begin{aligned} XN &= \frac{\sqrt{Y}\,(1 + 12K'(n_2-1))}{\left[1 + \frac{\mu n}{\mu p} + 12K'\left[(n_2-1) + \frac{\mu n}{\mu p}(n_1-1)\right]\right]^{\frac{1}{2}}} \\ XP &= \frac{\mu n}{\mu p}\,\frac{\sqrt{Y}\,(1 + 12K'(n_1-1))}{\left[1 + \frac{\mu n}{\mu p} + 12K'\left[(n_2-1) + \frac{\mu n}{\mu p}(n_1-1)\right]\right]^{\frac{1}{2}}} \end{aligned} \right.$$

Where :

- $Y = \frac{C_L}{C_{ref}}$ is the output load, with respect to a reference capacitance $C_{ref} = C_{ox}L_{min}W_{min}$ .

- $X_N = \frac{W_N}{W_{ref}}$, $X_P = \frac{W_P}{W_{ref}}$, represents the transistor size, with respect to a reference width.
- $n_1$ and $n_2$ respectively represents the numbers of parallel transistors in P and N array
- $K' = \frac{1}{24K} \frac{\frac{Vcc}{2} - V_T}{\frac{Vcc}{2} - V_T^*}$ with $K = \frac{8Vcc(Vcc - V_T)}{7Vcc^2 + 4V_T^2 - 12VccV_T}$ where K is a slowly varying technological coefficient.

For the 1.5μm CMOS process under consideration the value of these coefficients are : $C_{ref}$=4.4fF, K= 1.274 and K' = 0.049 .

Depending on the logic level to be transmitted the structure of the look up table has been replaced by an equivalent "ANDORI" configuration allowing full application of the sizing procedure, as follows : low or high levels to be passed transform the structure in equivalent N or P serial arrays (Figure 10), directly sized as for general "ANDORI" configuration (Auvergne 1991). Then the P or N companion transistor is sized in the ratio of mobility of its partner ($\mu_n/\mu_p$=2.4).

For example, as shown in Figure 9, with a given load of 130 fF the transistor width obtained for the transmission gates TG5, TG6, TG7, TG8, of the structure II are : $W_N$ = 5 μm and $W_P$=13μm.

After optimization of the structure II the input load of structure I is known (this is the input of *inv3* – node "6" – in Figure 9 ). The same backward process is applied to the array of structure I from the output ( here node "6" ) to the input ( node "IN" ). Figures of the resulting transistor sizes are given in the corresponding buble of Figure 8.

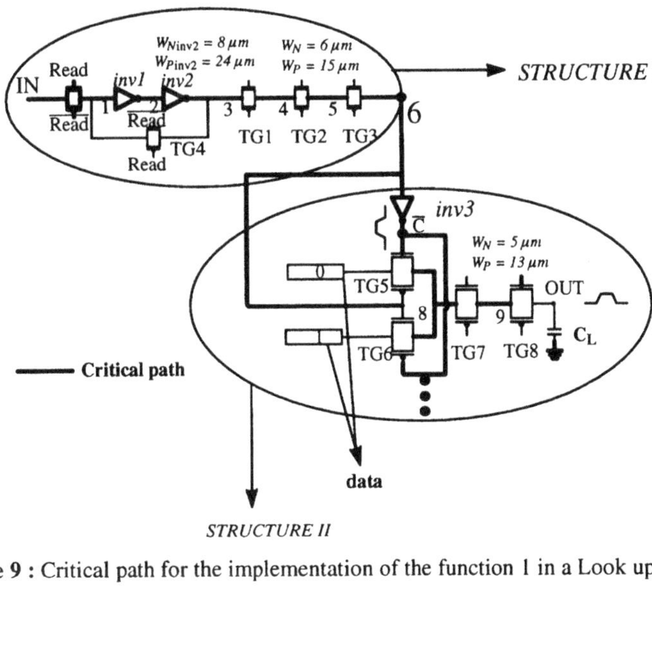


**Figure 9 :** Critical path for the implementation of the function 1 in a Look up Table 3

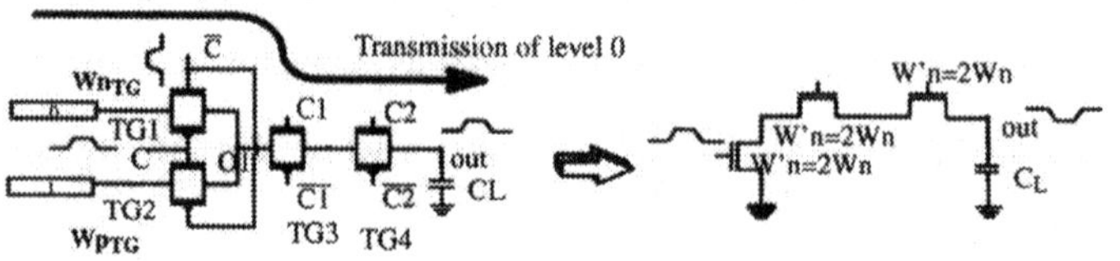


(a) "NAND3" configuration

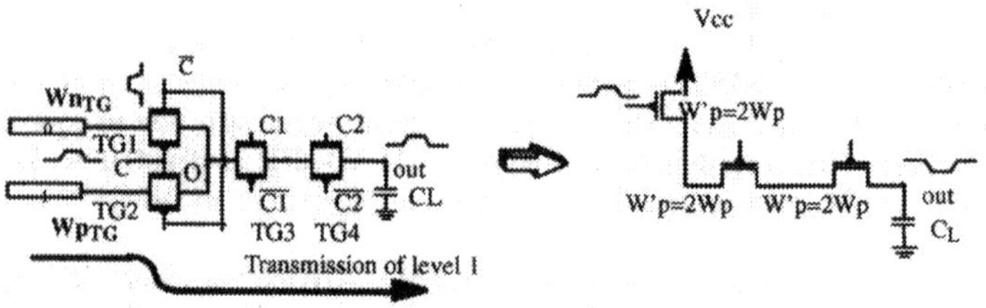


*(b) "NOR3" configuration*

**Figure 10 :** Optimization procedure

As shown in *Table 1*, where we compared Look up Table implementations of the functions f1 and f2 with different transistor size alternatives, it appears clearly that sized solutions result in the fastest implementations of the look up table cells (up to 50% reduction of the delay). The comparison with the other logic block structures (Figures 11 and 12) shows that after optimization of the transistor widths the look up tables exhibit very good speed performances.

| | | function1 | | | | | | | | function2 | |
|---|---|---|---|---|---|---|---|---|---|---|---|
| | | LTK3 | | | | LTK4 | | | | LTK6 | |
| | | $C_L$=130fF | | $C_L$=900fF | | $C_L$=130fF | | $C_L$=900fF | | $C_L$=130fF | |
| | | $t_{HL}$ | $t_{LH}$ | $t_{HL}$ | $t_{LH}$ | $t_{HL}$ | $t_{LH}$ | $t_{HL}$ | $t_{LH}$ | $t_{HL}$ | $t_{LH}$ |
| Fixed width | Wn=16µm Wp=16µm | | | 5.40 | 2.20 | | | 3.96 | 2.25 | | |
| | Wn=2 µm Wp=4 µm | 5.33 | 3.6 | | | 3.53 | 2.87 | | | 5.06 | 4.34 |
| Optimization of transistor width | | 3.27 | 3.07 | 3.51 | 3.44 | 1.77 | 1.64 | 2.78 | 2.61 | 2.24 | 2.21 |
| Benefits | | 39% | | 35% | | 50% | | 30% | | 50% | |

**Table 1 :** Optimization results for the functions f1 and f2 implemented with different look up tables (delays in ns) and transistor sizing alternative (minimum and constant sizes, optimized sizes for a given load).

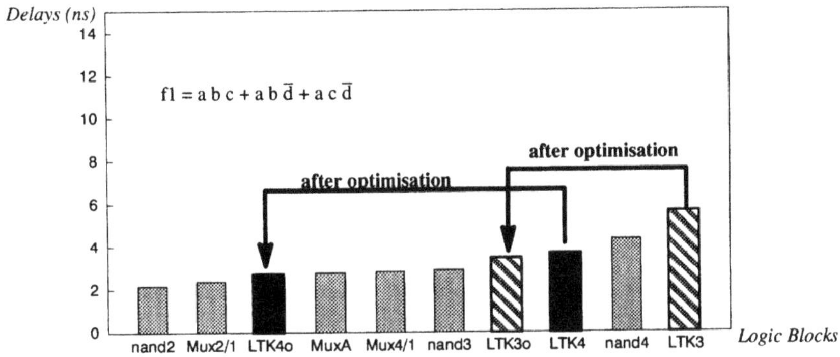


**Figure 11 :** delays of the function f1 (with Wn=2μm and Wp=4μm and CL=130fF) and performance improvements of the look up table 3/4 implementation after transistor sizing

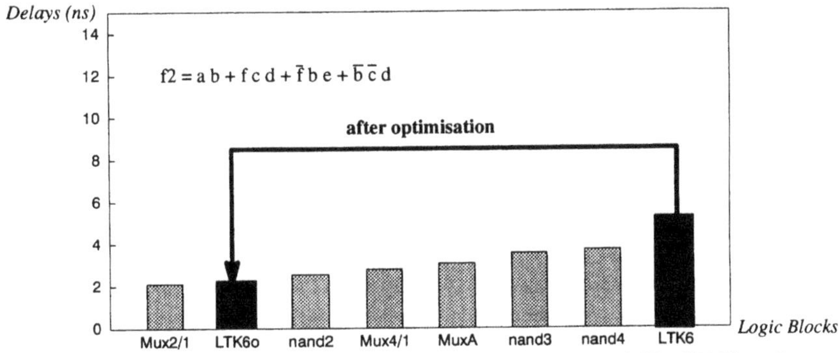


**Figure 12 :** delays of the function f2 (with Wn=2μm and Wp=4μm and CL=130fF) and performance improvements of the look up table 6 implementation after transistor sizing

# 5- CONCLUSION

We explored in this work the comparative performances of the different logic blocks used in FPGA mapping taking into account the layout evaluation. The main conclusion is the confirmation of the good performances of the look up table and multiplexer modules. A significant reduction in delay is obtained by optimizing the transistor widths of the look up tables. The use of a layout generator is of great help to evaluate architectural alternatives as well as the fast migration of circuits in different processes. The high regularity of the layout style used, gives the possibility to parameterize the layout capacitances, allowing an accurate prediction of performances for logic synthesis tools.

Selection of FPGA structures as a technology management alternative for performance driven design is a widely opened problem. Using automatic layout generator combined to transistor sizing and performance evaluation we were able to compare different logic blocks unit used to implement FPGAs. Examples are obtained in different logic paths and compared in terms of speed and area. Evidence is given of LUT based high performance implementation, through the definition of optimal sizing for logic units.

## 6–REFERENCES

D.Auvergne, N.Azemard, V. Bonzom, D. Deschacht, M.Robert, (1991), Formal sizing rules of CMOS circuits, The European Design Automation Conference, Amsterdam, February 1991.

N.Azemard, V.Bonzom, D.Auvergne, (1992) P.SIZE : a sizing aid for optimized designs, in Proc of EURO–DAC 92, Hamburg (Germany), pp. 160–165.

G.Cathebras, (1990), Contribution à la compilation structurelle des circuits intégrés CMOS, Phd Thesis, LIRMM, Montpellier, France, 90.

D.Deschacht, P.Pinede, M.Robert, D.Auvergne, (1990), PATH RUNNER: an accurate and fast timing analyse, in Proc. of EDAC 90, Glascow (Scotland), pp. 529–533.

A. E. Gamal (1989), An architecture for electrically configurable gate arrays, in IEEE Journal of Solid State Circuits, Vol. 24 , No 2, April 1989.

F. Moraes, N.Azemard, M.Robert, D. Auvergne, (1993), Flexible macrocell layout generator, in Proc. of 4th ACM/SIGDA physical Design Workshop, Los Angeles, 1993, p. 105–116

F. Moraes (bis 1993), N.Azemard, M.Robert, D.Auvergne (1993), Tool box for performance driven macrocell layout generator, in Fourth Eurochip Workshop on VLSI design training, Toledo, September 1993.

S. Singh, J. Rose, P. Chow, D. Lewis (1992), The effect of logic Block architecture on FPGA performance, in IEEE Journal of Solid State Circuits, Vol. 27 , No 3, March 1992.

S. Singh, (1991), The effect of logic Block architecture on the speed of field programmable gate array, M.A. Sc. thesis, Dept. Elect. Eng., Univ. of Toronto, Ont. Canada, Aug. 1991

## 7–BIOGRAPHY

**Daniel AUVERGNE** since 1981, is Professor at the Laboratory of Computer Sciences, Robotics and Microelectronics of Montpellier, France, developing Microelectronics in the engineering program of the university. Current field of research includes the various aspects of VLSI design such as timing models, automatic layout generation, optimization and transistor sizing. Author or coauthor of more than 110 publications in reviews or communications in international conferences in the field of solid state physics and VLSI design. He has been a member of the program committee of EUROASIC and EUROCHIP workshop , ESSIRC, EDAC, PATMOS. He has been a reviewer of contributions submitted to VLSI, EUROASIC, EDAC, ISCAS, IEE Electronic circuits and systems, IEEE trans. on CAD, IEEE journal of Solid State Circuits

**Fernando MORAES** was born in Passo Fundo, Brazil. He received the B.S. degree in electrical engineering in 1987 and the M.S. degree in computer science in 1990, both from Federal University of Rio Grande do Sul (UFRGS), at Porto Alegre, Brazil. He obtained the Ph.D. degree in microelectronics in 1994 at the LIRMM laboratory, at Montpellier, French. Currently he his working as researcher at UFRGS. His research interests include layout synthesis, timing analysis and timing optimization at layout level.

**Michel ROBERT** is Professor at the University of Science of Montpellier (France), where he develop microelectronics in the engineering program. He was born in France in 1957. He received the M.S. and ph.D. degrees in electronics from the University of Montpellier, in 1980 and 1987. From 1980 to 1984 he was employed by Texas Instruments in the Semiconductor division. He then joined the University of Montpellier as a researcher in the Laboratory of Computer Sciences, Robotics and Microelectronics (LIRMM). Author or coauthor of more than 80 publications in the field of VLSI design, his research concerns the design methodologies of MOS integrated circuits, layout synthesis, and the design of real time image processing integrated circuits.

**Lionel TORRES** was born in France in 1970. He received the M.S degree in electronics from the University of Montpellier (France) in 1992. He prepare, since 1993 a ph.D degrees in microelectronic at the Laboratory of Computer Sciences, Robotics and Microelectronics (LIRMM). His research interest include image processing, embedded system and FPGA .

PART FIVE

# Structural Optimization

13

# Area Optimization of Bit-Parallel Custom Data Paths

*Z. Sahraoui, L. Rijnders, J. Vanhoof, I. Bolsens, H. De Man*
*IMEC*
*Kapeldreef 75, B-3001 Heverlee, Belgium*

**Abstract**

This paper presents an area optimization method for large and highly customized bit-parallel data paths. Area optimization of the data paths is achieved by identifying and removing redundancies. The method uses reduction tables to achieve an efficient transformation of multi-output gates during redundancy removal. It can handle both stuck-at and duplicate nets. Stuck-at redundancies are identified using a novel approach in test pattern generation. Duplicate nets are identified by constructing the ROBDDs of the corresponding net functions. The method takes also into account interaction through exchange of information between high level synthesis and data path synthesis.

Experiments done on industrial examples show that area optimization performed with the method achieves results comparable to those obtained with logic synthesis, but the run times are reduced by at least one order of magnitude.

## 1 INTRODUCTION

To satisfy the tight throughput constraints of real-time digital signal and image processing applications, the silicon compiler CATHEDRAL-2/3 [Lanneer,1991][Note,1991] maps algorithms onto a set of customized, lowly multiplexed and cooperating data paths. These data paths go through a number of optimization steps in order to fulfill the area and timing requirements. Among these steps *redundancy identification and removal* is performed for area optimization.

The task of redundancy identification and removal in CATHEDRAL-2/3 is viewed both as a quality and a complexity problem. Due to the area and timing requirements of the application domain, the quality of the data paths is important. The best possible implementation of a custom data path is achieved by composing hierarchically a set of RT-operators [Katsadas,1992], such as adders, subtractors, comparators, multipliers, multiplexers and shifters, retrieved from a generic library of module generators. These operators are carefully designed, hence they do not contain any redundant hardware, and also guaranty an optimal mapping of their function onto a given standard cell library. However, the hierarchical composition of these operators, combined to data path customization, may introduce redundancies in originally redundancy-free regions. The presence of redundancies impacts negatively the area and makes the implementation of the data paths sub-optimal. The presence of redundancies also deteriorates the testability of data paths, i.e. makes them difficult to test. The bad testability raises the need to introduce extra hardware to make the data path testable, and consequently impacts negatively the area even more. Therefore it is important to identify and remove such redundancies.

As for the complexity problem, because of the application domain characteristics, the hierarchical composition of the RT-operators produces large data paths which have important logical depths. The previous cited characteristics concur to make area optimization techniques like logic synthesis not well adapted to perform its task on bit-parallel custom data paths. Experimental results obtained with industrial examples confirm these ideas.

In this paper we present an alternative area optimization method for bit-parallel custom data paths. The method allows to handle large bit-parallel custom data paths, which can contain more than one hundred operators, within practical run times, and produces good results by taking into account the characteristics of the application domain and of the data paths. Area optimization is achieved through the identification of redundancies and their removal. Redundancy identification is performed using test pattern generation algorithms. The removal of a redundancy is directly performed on the data structures of the data path netlist, in a hierarchical manner. Areas not concerned by the redundancy are not modified, hence no remapping is required. Once identified, a redundancy is removed and the effect of its removal is propagated using rules stored in a *reduction table* [Vandeweerd,1989]. Each gate in the hierarchical structure has a reduction table which specifies how the gate must be transformed. The combination of reduction tables and hierarchy allows us to perform redundancy removal on combinational as well as sequential circuits described at the gate level or at the macro block level, or a mixture of both levels. Experiments realized with a tool based on this method demonstrate the viability of the idea.

This paper is divided into three sections. First, we introduce a set of selected examples, taken from real-life applications of audio and video digital signal processing, to give an idea of the kind of data paths we are dealing with. Next, we introduce our method based on the idea of identifying and removing redundancies using reduction tables. Finally, we present experimental results obtained by our method and how they compare to those obtained with logic synthesis.

# 2 EXAMPLES IMPLEMENTATION

## 2.1 Data path generation

A CATHEDRAL-2/3 architecture consists of data paths whose composition is tailored toward the algorithm of the application at hand. The data paths are synthesized through a number of steps [Note,1991], starting from the signal flow graph (SFG) of the algorithm:

1. During operation clustering, atomic operations of the signal flow graph are grouped into clusters.
2. In the data path assignment step, the clusters are assigned to dedicated data paths in such a way that the total area is minimized under the constraint of a given global cycle budget and I/O constraints.
3. In the data path definition step, the internal composition of each data path is constructed so that it can execute all clusters assigned to it.

These steps have the goal of optimizing data path area by sharing as much as possible the available hardware and without violating the timing constraint. This resource sharing is done at the RT-level, since the atomic operations in the SFG are RT-operations. During the data path definition step, atomic SFG

**Table 1** Example parameters

| Example | Data path | Gates | Area | Delay | Clock rate |
|---|---|---|---|---|---|
| RGB-YUV codec | dp-0 | 6622 | 17.235 | 117 | 27 |
| | Clipper | 495 | 1.343 | 44 | 27 |
| Cordic processor | dp-0 | 432 | 1.119 | 63 | 16 |
| Mobile terminal | TX | 3086 | 8.272 | 54 | 6.9 |
| | RX | 2318 | 7.041 | 152 | 8.6 |
| Reed-Solomon decoder | BM | 5943 | 9.902 | 79 | 10 |

operations get assigned to operators. These operators are retrieved from a generic library of high quality module generators described manually by designers well versed in the design of arithmetic blocks. Manual design of the operators allows, among other advantages, an efficient use of multi-output gates such as full adders and half adders. This type of gates represents in general a large share of the gates used in a data path. Therefore it is important to use them if they are provided in a standard cell library. In summary, the operators implement their operations efficiently, guaranty an optimal mapping of their function onto a given standard cell library and contain no redundancies. The resource sharing performed at the RT-level, in addition to the high quality of the operators motivated the idea to preserve the connectivity structure of the data paths during area optimization, and to operate only on those regions concerned by the redundancies. Another reason to preserve the original connectivity structure is that some high level tools rely on information derived during the high level synthesis step to perform their task. This information is often presented in the form of relationships between signals in the SFG of the algorithm and nets in the structure of the data path. So, if the structure of the data path is in some way altered then the information provided by the high level synthesis step becomes irrelevant. An example of such a tool is the SFG Tracing method for formal verification [Claesen,1991], which traces SFG reference signals back to the netlist of the implementation to formally prove that the latter corresponds to the initial specifications.

## 2.2 Examples from the application domain

In order to assess of the potential of our method, we selected a set of four industrial examples representing a wide range of complexities that CATHEDRAL-2/3 can synthesize. The applications were selected from the audio and video DSP application domains. The examples are listed in Table 1. The area is given in $mm^2$, the delay in ns and the clock rate in Mhz.

From the four examples we selected six data paths:

- The RGB to YUV codec digitally converts a 24 bit RGB video signal at a rate of 27 MHz to black-and-white images at 13.5 MHz, and samples down the color information 2-fold to a rate of 6.75 MHz. A separate data path, the clipper (Figure 7), takes care of the clipping. The chip can also execute the opposite interpolation operation to reconstruct the original RGB signal. The system clock is 27 MHz.

- The Cordic algorithm is a wide-spread iterative approximation algorithm for the trigonometric functions. The Cordic data path allows to execute one Cordic iteration in one clock cycle. For application in a digital FM-demodulator, the clock rate is 16 MHz.
- TX and RX are the matched FIR filters in the transmitter, resp. receiver, of a 6.8 kbit/s CDMA satellite mobile terminal [Philips,1993], using a 255-bit spreading code length. Both 8-fold oversampling filters have 32 taps and operate at a sample rate of 6.936 MHz. The TX has a system clock of 6.936 MHz, and RX has a system clock of 8.67 MHz.
- The Berlekamp-Massey algorithm is the most complex part of a 80 Mb/s Reed-Solomon decoder. It iteratively computes from a syndrome the error locator polynomial, of which the roots indicate the error positions. This data path also supports premultiplication for code shortening, and a polynomial multiplication to compute the error values. It operates in the 8-bit Galois field GF(256) at a symbol rate of 10 MHz using a system clock of 10 MHz.

## 2.3 Implementation of the RGB-YUV codec

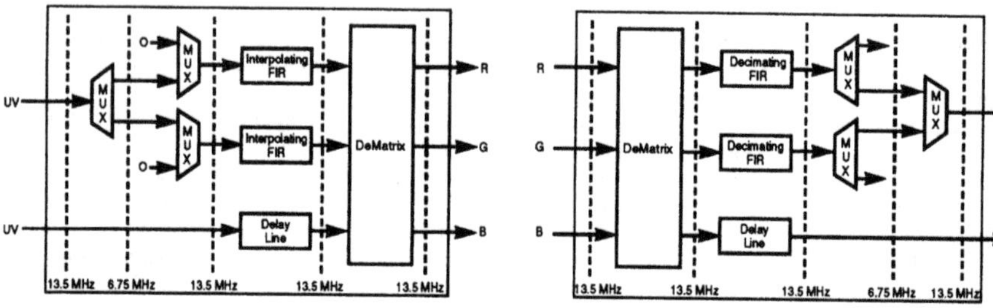


**Figure 1** RGB-YUV coder and decoder.

In a first phase, the multi-rate YUV-RGB algorithmic description has been expanded to a single-rate description. The resulting description went through a number of transformation and synthesis steps which produced a processor with a data path that executes the conversion at a clock rate of 27 MHz. The single-rate coder and decoder architectures are shown in Figure 1. All multiplications were expanded to add-and-shift operations before data path definition was performed.

An implementation of the architecture of Figure 1. produced a data path of 136 operators. It contains 6468 gates, occupies 17.23 $mm^2$ and has a clock period of 117 ns (the data path needs to be pipelined). The core of the data path, i.e. the data path without the register files implementing the delay lines, has an area of 7.2 $mm^2$ and contains 2672 gates, of whom 759 are full adders and 1569 are EXOR and multiplexer gates. Full adders, EXOR and multiplexer gates alone occupy 92.4% of the area of the data path core.

To perform area optimization on the data path, logic synthesis techniques were tried first. We used SIS1.1 [Brayton,1987] to achieve that. After 33 hours of CPU time, SIS produced a structure netlist which has 18.13 $mm^2$ of total area and has a delay of 229 ns. We note an increase in area and delay which goes contrary to our expectations. We also note the large run time.

The explanation of these negative results resides in the composition of the structure netlist. The synthesized structure contains 12164 gates among which there is no full adder. This means that the 759 full adders present in the original structure were transformed to a set of smaller gates, with a corresponding increase in area and delay. Moreover, all EXOR and multiplexer gates that were in the original structure were replaced in the synthesized one, inducing also an extra increase in area and delay. Finally, another increase in area and delay should be observed after placement and routing due to the larger routing area caused by the important increase in the number of gates (more than two times). The total increase in area and delay makes obvious the need of an alternative method of performing area optimization. This other method must be able to accommodate all the previous observations concerning area, delay and run time.

# 3 TABLE DRIVEN REDUNDANCY REMOVAL

## 3.1 Redundancies in a data path

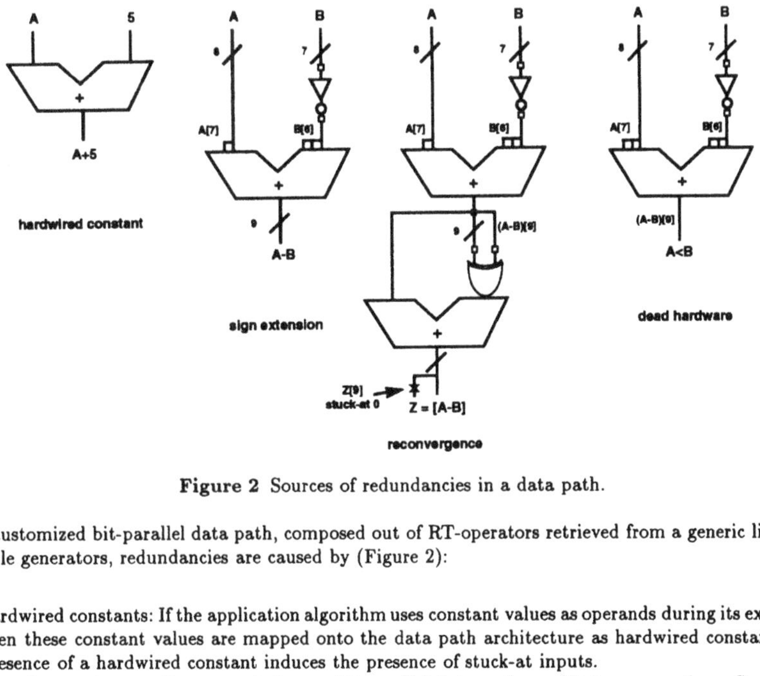


**Figure 2** Sources of redundancies in a data path.

In a customized bit-parallel data path, composed out of RT-operators retrieved from a generic library of module generators, redundancies are caused by (Figure 2):

- hardwired constants: If the application algorithm uses constant values as operands during its execution, then these constant values are mapped onto the data path architecture as hardwired constants. The presence of a hardwired constant induces the presence of stuck-at inputs.
- bus alignment: operations executed on a bit-parallel data path are bit true operations. Sometimes, a bit-true operation needs a casting operation to produce the correct result. The casting is done by a bus alignment which may create duplicate bit-values. An example is a sign extension needed by a two's complement operation.

- reconvergence: the hierarchical composition of the operators may create a reconvergence. Since path reconvergence is a necessary condition for the presence of stuck-at nets, it is possible that redundancies are introduced in regions previously non-redundant.
- unexploited functionality: operators are generated by module generators. A module generator can provide operator instance with various functionalities. Sometimes, not all functionalities are needed, or the full precision of an operator is not used, hence hardware may be generated but is not exploited. In such cases "dead" hardware is created (vs. dead code in software compilers). For instance, a comparator can be built out of a subtractor which produces the necessary comparison flag but has its result disregarded.

The above cited causes create three types of redundancies in a data path:

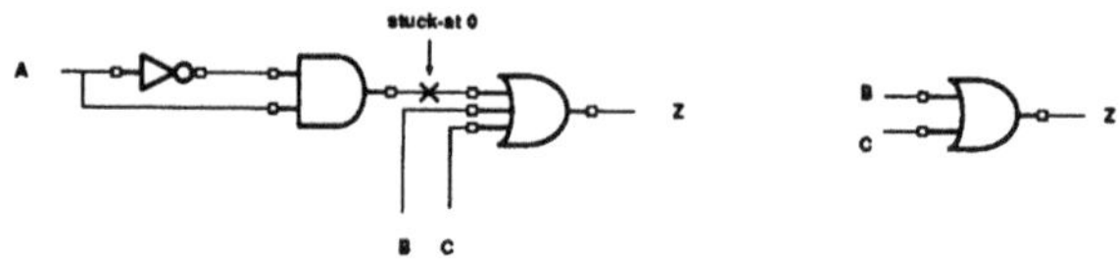


**Figure 3** Removing stuck-at nets.

- stuck-at nets: a stuck-at net can be replaced by its stuck-at value without changing the function performed by the data path. If such a redundancy is identified, it is removed by performing a "constant propagation" (vs. constant propagation in software compilers) on its fanout and dead hardware removal on its fanin (Figure 3).

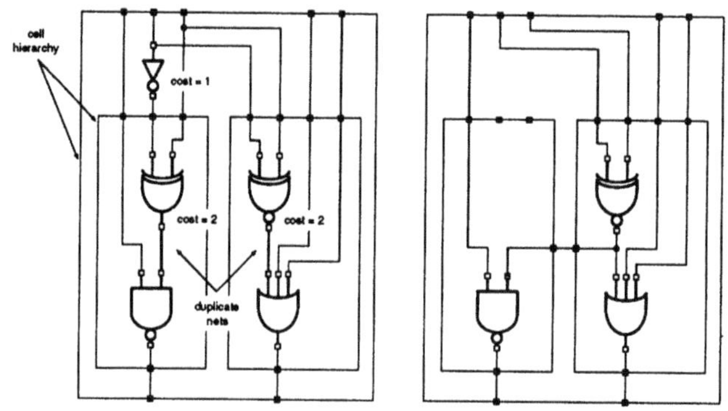


**Figure 4** Removing duplicate nets.

- duplicate nets: two or more nets which have the same Boolean representation for all input combinations are called duplicate nets. In general, the presence of duplicate nets means that only one net is necessary to produce, while the others can be simply removed. For that, it is necessary to select the net to keep, and the ones to remove. A greedy selection is adopted. The net that needs the least hardware to produce is kept, while the other net or nets are removed. An example of such a case is shown in Figure 4. The net which has cost 2 is distributed to the required fanins, while the one of total cost 3 is removed.

- dead hardware: it is the hardware that is not necessary to produce the required functionality. This kind of redundancy does not need an identification phase, and is easy to remove.

## 3.2 Redundancy identification

We have two types of redundancies to identify: stuck-at nets and duplicate nets. There exists a number of methods of identifying stuck-at redundancies. Most of these methods are based on automatic test pattern generation algorithms. An ATPG algorithm finds a stuck-at redundant fault by failing to generate a test. An important factor of performance in test pattern generation is the number of decisions in the search space. The more decisions are made in order to find a test the larger is the search space. ATPG is performed employing an algorithm which reduces the number of decisions [Sahraoui,1993]. Unlike classical test generation methods which perform their search on a netlist of gates, this algorithm works at a higher level. The gates of the circuit to test are partitioned into clusters. The search performed at the cluster level becomes a progressive translation of a set of value assignments to another set of value assignments that steps over cluster boundaries. To be able to step over cluster boundaries ***sensitization*** and ***propagation*** conditions are employed. Sensitization and propagation conditions play the same role as respectively the Boolean function associated to a gate and the non-controlling values of the gate inputs. The identification of duplicate nets necessitate the computation of the associated Boolean functions. An effective way of representing Boolean functions is reduced and ordered binary decision diagrams (OBDDs). Two nets can be proven to be duplicate if they have the same OBDD representation.

## 3.3 Reduction tables

After its identification, a redundancy is removed by performing a constant propagation on its fanout and dead hardware removal on its fanin. These two procedures obey rules that can be specified for each type of gate. We associate with each type of gate a set of propagation and removal rules grouped in a table called *reduction table* [Vandeweerd,1989]. The reduction table of a gate (Figure 5) specifies how the behavior of that gate changes if its inputs are stuck-at and/or if one of its outputs is not used. In this case, the gate is replaced by another gate which performs the same function but with a cheaper area and/or delay cost. In that circumstance the gate is said to have been reduced. A reduction table contains a set of entries corresponding to all possible combinations of gate inputs and gate outputs. A reduction table entry is a rule which has two fields (Figure 5). The first field, the IF part of the rule, specifies the input combination. The second field, the THEN part of the rule, indicates the type of transformation.

A gate input can have one of three values: "0","1" or "x". The "0" or "1" value indicates that the corresponding input is stuck at that value. The "x" value indicates that the corresponding gate input has no fixed value. If the gate has $n$ inputs then its corresponding reduction table can contain at most $3^n$ entries. In case the gate has more than one output, a reduction table entry can also specify what kind of transformation must be applied to the gate if one of its outputs is not used. Consequently, a gate output can also have two states, used or not used. the possible input and output combinations can produce $3^n \times 2^m - 1$ entries, where $n$ and $m$ are respectively the number of gate inputs and outputs. For

```
(setq and2 '( (
;;; inputs and outputs
( (a b) (z) )
;;; inputs that are commutative
( (a b) ) )
(
( (0 0) (x) nil () ( (z 0) ) )
;;; IF a = 0 AND b = 0
;;; THEN the cell is removed and its output z is stuck-at 0

( (0 1) (x) nil () ( (z 0) ) )
( (0 x) (x) nil () ( (z 0) ) )
( (1 1) (x) nil () ( (z 1) ) )
( (1 x) (x) nil () ( (z b) ) )
) ) )
```

**Figure 5** Reduction table of 2-input AND gate.

each gate input/output combination there is a transformation to accomplish. Such a transformation is not *aware* of what happens to the gate. The transformation specifies how a gate is reduced, regardless of the function of this gate. The transformation establishes therefore a general relation between a gate and its replacement, valid under certain conditions. Therefore, reduction tables allow the transformation of all sorts of gates like e.g. ANDs, ORs, full adders and memory elements, and hence redundancy removal is possible on combinational as well as on sequential circuits.

The generation of the reduction tables is automated by means of logic synthesis. After their generation, the reduction tables may be manually updated to accommodate gates that require special attention. These gates have in general complex functions, whom logic synthesis has difficulties in reducing. For example, a full adder with a stuck-at 0 input is best reduced to a half adder, a transformation logic synthesis is unable to perform. Manual updates can be kept to the minimum since there are only a few types of these special gates. The number of special gates is small, so the cost of a good efficiency in redundancy removal is not as high as one can fear. The automatic generation of the reduction tables, combined to adequate manual updates for a limited number of specific gates permits to reach a high level of efficiency during redundancy removal without paying off a high cost.

The automatic generation of reduction tables is performed as follows (Figure 6):

1. First, we generate all possible combinations of gate inputs and outputs. Each input can have a "0", "1" or "x" value. Each gate output, if there are more than one output, can have a "x" or "()", i.e. unused value. We exploit gate input commutativity to reduce the number of combinations and hence the number of entries (see figure). For that, the reduction table also contains a list of gate inputs that can commute.
2. With each combination we associate a transformation to apply, i.e a replacement gate to determine, if there is one. The replacement gate is determined by its Boolean equation, which is derived from the Boolean equation of the original gate to which we apply the conditions set by the input/output combination.

```
function red_table_gen (gate, commut_list))
/* gate = (name, Boolean equation) */
/* commut_list = list of commutative inputs */

begin
      /*
      calculate all input/output combinations
      taking into account the commutativity list
      */
      combi_list = calculate_combi_list(gate, commut_list);

      while(combi_list != EMPTY)
      begin
            current_combination = first item of combination list;
            determine Boolean equation of replacement gate;
            generate structure netlist;
            reduction_table = add entry to reduction table;
      end
      return reduction_table;
end
```

**Figure 6** Automatic generation of reduction tables.

3. In case the replacement gate is not a leaf cell, its structure netlist is synthesized from its Boolean equation.
4. Finally, the corresponding table entry is constructed from the information obtained in the previous steps.

## 3.4 Mixed style design and hierarchy

Because of efficiency reasons, it is common practice to mix different layout generation styles to generate the layout of a chip. For instance, regular structures like register files, multipliers, ROMs and RAMs may be generated using procedural module generators, while less regular structures may be implemented using standard cells. This layout strategy imposes new constraints on redundancy removal. For instance, a situation can happen where the process of constant propagation pushes a stuck-at value to the input of a macro cell. In this case the propagation must stop since the internal structure of the macro cell cannot be altered, and therefore the input in question must be tied to power or ground depending on the stuck-at value. Due the preceding, redundancy removal, as it is implemented in our system, is able to deal with mixed style designs.

In addition, since we consider user interactiveness important for subsequent use of the tool, and since an important means of dialogue between the user and the tool is hierarchy, the tool performs its redundancy removal task on hierarchical netlists. Hierarchy is a powerful means to understanding the complexity of a design, which permits the user to give in an easy way hints to the tool and to steer the optimization

**Table 2** Experimental results–area optimization.

| Example | Data path | redundancy removal | | | | logic synthesis | | | |
|---|---|---|---|---|---|---|---|---|---|
| | | Gates | Area | Delay | CPU | Gates | Area | Delay | CPU |
| RGB-YUV codec | dp-0 | 6468 | 16.481 | 117 | 332.3 | 12164 | 18.136 | 229 | 120628.7 |
| | Clipper | 288 | 0.273 | 31 | 61.6 | 193 | 0.162 | 16 | 58 |
| Cordic processor | dp-0 | 390 | 0.782 | 53 | 46.1 | 840 | 0.802 | 40 | 866.4 |
| Mobile terminal | TX | 2800 | 7.402 | 46 | 159.6 | 3400 | 7.567 | 47 | 198.4 |
| | RX | 2301 | 6.806 | 152 | 57.4 | 5549 | 7.491 | 179 | 14436.9 |
| Reed-Solomon decoder | BM | 5825 | 8.575 | 79 | 801.1 | 9030 | 8.675 | 82 | 34262.8 |

process towards directions that the tool would be otherwise unable to find. That is also why the user is provided with the possibility to feed directly the redundancy removal procedure with information about redundant nets in the data paths at any hierarchical level, in addition to the information provided by the redundancy identification procedure. The user is left with full control on the whole process of identifying and removing redundancies if he/she desires that. In the contrary case, redundancy removal can be operated automatically.

# 4 EXPERIMENTAL RESULTS

The method presented in this paper has been implemented in C running under UNIX. The goal of the implementation was to build a fully operational tool, based on the previous ideas and able to handle large data paths in practical run times.

We performed area optimization of the data paths presented in Section 2.2 with both SIS and our method. The results of these experiments are reported in Table 2. The area is given in $mm^2$, the delay is in ns and the CPU time is given in seconds on a DEC-5000.

Some interesting remarks can be made. First, two data paths had an increase of their area and delay after area optimization performed with SIS. One is the data path "dp-0" of the RGB-YUV codec, which has already been presented in Section 2.3. The second data path is the receiver of the mobile terminal example. We can also note for these two data paths very large SIS run times; several orders of magnitude larger than those achieved with our method. Area optimization performed with our method on the same two examples produced an improvement in area, gate count and no increase in the delay. The improvement in gate count over SIS is noteworthy. The area figures are those of the cell areas, hence we expect an even better improvement with our method after placement and routing because of the smaller routing area. We also note for the rest of the data paths, except the clipper of the RGB-YUV codec, comparable area and delay results after area optimization performed with SIS and our method. However, our method achieves much smaller run times, and for most data paths one order of magnitude smaller. However, it can be noted that our method obtained a worse delay for the data path of the Cordic processor. This delay improvement achieved with SIS is nevertheless unnecessary since the data path fulfills the timing constraint of 10 Mhz. It is interesting to note that our method of performing area optimization cannot cause the delay to increase. It can only improve it or yield the same delay. This characteristic is important because area optimization is performed independently of delay optimization, and hence no negative impact on delays is feared if area optimization is performed with our method.

```
function clipper ()

begin
       input yx0, ux0, vx0;
       output y0, u0, v0;

       if (yx0 < 16)
         then y0 = 16;
         else  if (yx0 > 235)
                  then y0 = 235;
                  else  y0 = yx0;
       if (ux0 < 16)
         then u0 = 16;
         else  if (ux0 > 240)
                  then u0 = 240;
                  else  u0 = ux0;
       if (vx0 < 16)
         then v0 = 16;
         else  if (vx0 > 240)
                  then v0 = 240;
                  else  v0 = vx0;
end
```

**Figure 7** Behavioral description of the clipper.

It can be noted that our method yields better area results for all examples except the clipper of the RGB-YUV codec. As shown in Figure 7, the clipper guarantees that the input is bounded within a given range. Its function hence is pure control. In addition, the presence of many constants in the description implies that the original structure was completely altered during the removal of hardwired constants (the big difference between the area of the original netlist and the optimized one supports this idea). However, we explained that one of the characteristics of our method is that it tries to preserve the original structure of the data path, something that is not possible with the clipper. Logic synthesis techniques are also particularly suited for problems where the structure to be generated is not known beforehand, which is the case for the clipper. The conjugation of the previous reasons led to the better results obtained with SIS.

## 5 CONCLUSION

Experimental results obtained with industrial examples demonstrate that the objectives we set for our method were achieved. Area optimization of CATHEDRAL-2/3 data paths is performed without violating the timing constraints, and within practical run times. The amount of optimization is comparable to that obtained with logic synthesis, if not better, but with run times at least one order of magnitude smaller. The set of selected examples put forward the strength and the weakness of our method. Our method deals best with large data paths, but produces less good results with smaller, and control and constant dominated ones.

## ACKNOWLEDGEMENTS

The authors wish to thank Paul Six, from IMEC, and Jef Van Meerbergen and Albert Van der Werf, from Philips, for many useful discussions.

## REFERENCES

[Brayton,1987] R. Brayton, R. Rudell, A. Sangiovannin-Vincentelli, A. Wang, "MIS: A Multi-Level Logic Optimization System," IEEE Trans. on CAD, November 1987.

[Claesen,1991] L. Claesen, M. Genoe, F. Proesmans, E. Verlind, H. De Man, "SFG-Tracing: a Multi-Level Formal Verification Methodology," Invited presentation, Workshop on Formal Verification for Hardware and Software, IMAG, Grenoble, October 1991.

[Katsadas,1992] E. Katsadas, Z. Sahraoui, M. Wouters, V. Derudder, I. Bolsens, P. Six, H. De Man, "Two Alternative Implementations for a Library of Functional Building Blocks," IFIP Workshop on Synthesis, Generation and Portability of Library Blocks for ASIC Design, Grenoble, March 1992.

[Lanneer,1991] D. Lanneer et al, "Architectural Synthesis for Medium and High Throughput Signal Processing with the New CATHEDRAL Environment," in *Trends in High-Level Synthesis*, editors R. Camposano and W. Wolf, Kluwer, 1991.

[Note,1991] S. Note, W. Geurts, F. Catthoor, H. De Man, "CATHEDRAL-III: Architecture-Driven High-Level Synthesis for High Throughput DSP Applications," Proc. DAC-91, San Francisco, 1991.

[Philips,1993] L. Philips, I. Bolsens, A. Rabaeijs, B. Vanhoof, J. Vanhoof, "Silicon Integration of Digital User-end Mobile Communication Systems," Proc. ICC 93 International Conference, Geneva, Switzerland.

[Sahraoui,1993] Z. Sahraoui, "A Method for Extracting Propagation and Sensitization Conditions in Redundancy Identification," Esprit 2260 Report, June 1993.

[Vandeweerd,1989] I. Vandeweerd, et al., "REDUSA: Module Generation by Automatic Elimination of Superfluous Blocks in Regular Structures," IEEE Trans. on CAD, September 1989.

14

# DataPath regularity extraction

*R.X.T. Nijssen and J.A.G. Jess*
*Eindhoven University of Technology, Dept. of Electrical Engineering, Design Automation Section/ES, P.O. Box 513, 5600 MB Eindhoven, The Netherlands,*
R.X.T.Nijssen@ele.tue.nl

**Abstract**

An efficient method is introduced to extract datapath regularity from circuit netlists so as to speed up and improve placement of datapaths. A search-wave traverses the circuit constructing the datapath by choosing the most regular extensions as indicated by a local regularity index measuring the uniformness of the neighborhood of known datapath stages.

**Keywords**

Regularity Extraction, Datapath Placement, Layout Generation

## 1 INTRODUCTION

The inherent regularity of data-path circuitry can be exploited to obtain high density layouts. This can be achieved by aligning, ie. placing linearly, cells associated with the same bit-slices in one direction, and by aligning the identical (or highly similar) bit slices perpendicularly. This regularity comes from the use of bit-wise parallel operators at the architectural design level. These operators are implemented by repeating single-bit operator cells to form a multi-bit module. By interconnecting the corresponding bit-cells of different modules, a datapath is formed consisting of a number of highly similar bit*slices*. The cells within the same multi-bit module comprising a *stage* are identical and can hence be aligned. By imposing geometrical regularity upon the size of one side of the bit-cells in a slice, eg. they span the same number of rows in all stages, alignment in the other direction is also provided for. Figure 1a shows an abstracted representation of an ideal datapath aligned in 2-directions. Furthermore, datapath regularity is made up by the fact that the majority of nets in datapath circuitry is local to either one single slice or one single stage, so there are no or just a few nets that are connected to cells which are connected to both multiple slices and stages.

Combined exploitation of these two characteristics, alignability of cells in two perpendicular directions and coinciding perpendicular locality of nets, yields dense placements. The information on alignability consists of an assignment to one stage and, at the same time, one slice for every component. With this information, the task of generating a good relative placement for datapath circuitry in terms of density and wire length is decomposed from a general 2-dimensional placement problem of all cells of the datapath to one linear arrangement of the stages of only one single slice, and one linear arrangement of the slices of only one stage. These two orderings can be solved using eg. [GCT77][NKH+93] or [Asa82]. This complexity reduction follows directly from the properties of the connectivity and geometry of stages and slices. In particular, the arrangement of the stages within a slice does not depend

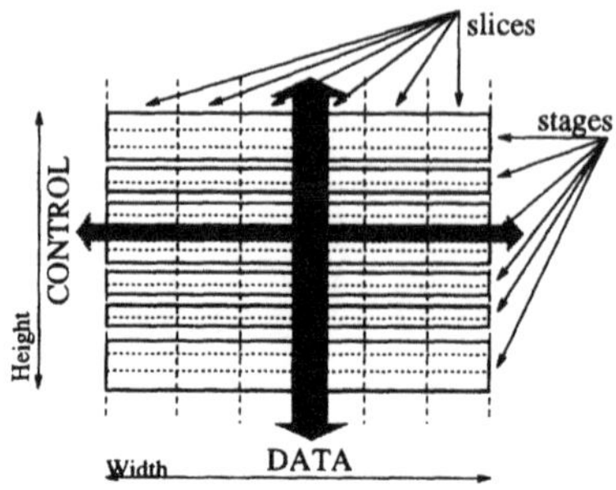


**Figure 1** ideally aligned datapath

on the arrangement of the slices within a stage, hence these two tasks can be carried out separately. In addition, the latter ordering task is trivial in most cases. The total solution space of the problem is thereby reduced by many orders of a magnitude, allowing for the size of a datapath circuit to be placed in one run to be much larger than with conventional methods which have no explicit notion of regularity. The paradigms they are based on will fail to implicitly work towards perpendicular alignment of datapath cells, despite long run-times.

Special-purpose CAD environments, often called *data-path compilers* (DPCs), have been developed that do take advantage of these characteristics. Like generic tools, DPCs are incapable of extracting information from a circuit on its regularity. Thus there is a need for a general datapath regularity extraction algorithm.

The problem formulation is then as follows: Given a network, the problem is defined as tagging as many network entities as possible with both a slice tag and a stage tag. The goal is to determine this labeling such that these slices and stages describe a suitable bit-cell alignment in both directions. In other words, the individual slices should be as similar as possible.

Previous work on datapath regularity extraction is based on a regularity guided partitioning method presented in [OHN87], like [HS88][CH93]. None of these methods however are designed to extract complete datapaths, they rather form a number of datapath chunks called location macros, ie. clusters to be used as preplacement seed units for conventional placement tools. They do not determine slice and stage assignments to the contained cells. Only a few datapath regularity characteristics are used, and the extraction procedures wrongly identify certain common regularity aberrations as regular.

## 2 REGULARITY CHARACTERISTICS

Three basic categories of data path regularity characteristics are distinguished, namely *structural* ie. the minimal representation of the circuit entities and their connectivity, *functional* ie. annotations to circuit entities regarding their function or usage, eg. signal flow direction, the types or type equivalences of terminals and their purpose, eg. clock, etc. *explicit* ie. data (partially) specifying the data-path, eg. a group of nets tagged as a bus. In most cases, the basic structural data already provides sufficient information to accurately quantify the regularity of most data

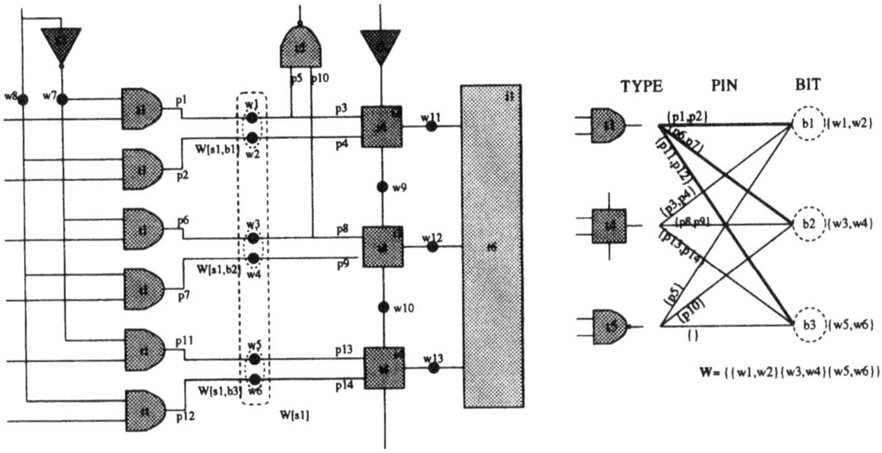


**Figure 2** a)Part of a datapath b) Regularity Graph of a datapath

paths. This type of information is always available for any circuit, therefore we henceforth focus on it. The other categories are optional, but may be used to enhance the amount of detail so as to be able to distinguish between cases that are structurally identical. The structure of a circuit is completely defined by a set of module instances, the type of a module instance, a set of nets, and a set of pins, ie. the connectivity between modules and nets through specified terminals. The type of a wire is taken to be its degree.

The key idea of our modeling is that we observe the amount of regularity in the neighborhood of one already known datapath stage, the *vantage* stage. We call this *local relative datapath regularity*. A stage that consist of wires (instances) will henceforth be referred to as a *Wstage* (*Istage*). Consider figure 2b showing a part of an example datapath circuit. In this example, it is assumed that the wires $\{w_1, w_2\}$, $\{w_3, w_4\}$, and $\{w_5, w_6\}$ are each in a distinct slice, and that together, they form one stage $s_1$, denoted as $\{\{w_1, w_2\}, \{w_3, w_4\}, \{w_5, w_6\}\}$. Wstage $s_1$ is outlined in the figure. A candidate Istage, say $s_2$, is formed by the cells of type $t_4$.

The structural regularity characteristics used as criteria to measure the extent of regularity are the presence of:

**W-T** *multi-slice module types* like $T(i_2) = T(i_3) = T(i_4) = t_4$ relative to vantage Wstage $s_1$. Such module types occur across the entire width of the datapath, as seen from a certain vantage Wstage like $s_1$.

**W-I** *multi-slice modules* like instance $i_1$ if the vantage Wstage is $\{\{w_{11}\}, \{w_{12}\}, \{w_{13}\}\} = s_3$ These are modules that are adjacent to all slices of the vantage Wstage. Instances like $i_1$ may be a non-expandable block or a block possibly even containing another datapath.

**I-T** *multi-slice wire types* like $T(w_{11}) = T(w_{12}) = T(w_{13}) = 2$ relative to vantage Istage $s_2$. This is a wire type attached to all slices of the vantage stage.

**I-W** *multi-slice wires* like $w_7$. Such wires are common for all slices of the vantage Istage. This category mainly consists of control signals for all bits, like enable signals, function selectors etc, largely constitute this category.

For conciseness, we will henceforth focus on the first criterion W-T, viz. the presence of multi-slice module types. This explanation can straightforwardly be generalized to the other criteria.

## 3 DEFINITIONS

Let $I$ denote the set of *instances*, $W$ the set of *wires*, and $P \subset I \times W$ denote the set of *pins*. The bipartite graph $C = (I, W, P)$ is the usual network graph describing the structure of any circuit. Let $E = I \cup W$ be the set of circuit entities. The function $I : P \rightarrow I$ maps each pin to its connected instance, and $P : E \rightarrow 2^P$ maps each wire or instance to the set of pins through which it is connected. Note that the *function* is denoted as $I()$ unlike the *set* $I$, etc., and that for brevity of the notation, a function name may be overloaded unambiguously depending on its argument. Further, let $T$ denote the set of circuit entity types. The function $T : E \rightarrow T$ maps each entity to its type. We define the type of a wire $w \in W$ as the number of its terminals, $T(w) = |P(w)|$.

A *stage* is a set of entities of the same type, and that are to be aligned in one direction. A *slice* is a set of entities that are to be aligned in a direction perpendicular to the stages. Ideally, a slice contains all nets corresponding with one bit of the datapath, and nets crossing slices are contained in one stage.

The assignment of circuit entities to stages and slices is denoted by labels. All entities associated to the same stage have the same *stage-label* being one of $S = \{s_1, s_2, ..., s_{|S|-1}, s_\infty\}$. Entities for which stage label is not known (yet) are labeled $s_\infty$. Entities of the same slice have the same *slice-label* being one of $B = \{b_1, b_2, ..., b_{|B|-1}, b_\infty\}$. Entities for which a slice label is not known (yet) have slice label $b_\infty$.

We use following functions mapping entities to stages or slices: The function $\varsigma : E \rightarrow S$ maps entities to stage labels. Note that by this definition of $\varsigma$, an entity can be member of at most one single slice. Also recall that by definition of a stage, the types of all entities in the same stage are the same. The function $\beta : E \rightarrow B$ maps entities to slice labels. In the example of figure 2, $\varsigma(w_1) = s_1$ and $\beta(w_3) = b_2$.

With these definitions, the alignment of the components of a datapath can be fully described by $\beta$ and $\varsigma$. The goal of our regularity extraction method boils down to constructing a $\beta$ and $\varsigma$ mapping defined for as many circuit entities as possible.

The functions $\varsigma$ and $\beta$ are not injective, so they have no inverse. Instead, we define the function $\varsigma^{\leftarrow} : S \rightarrow 2^B$ of $\varsigma$ returning the set of slices of stage $s \in S$ as

$$\varsigma^{\leftarrow}(s) = \{b \in B | \exists e : e \in E : \varsigma(e) = s \wedge \beta(e) = b\} \tag{1}$$

Function $\beta^{\leftarrow} : S \times B \rightarrow 2^E$ returns the set of entities of which a slice $b \in B$ of stage $s \in S$ consists:

$$\beta^{\leftarrow}(s, b) = \{e \in E | \varsigma(e) = s \wedge \beta(e) = b\} \tag{2}$$

For example, in figure 2, wireslice $\beta^{\leftarrow}(s_1, b_1) = \{w_1, w_2\}$ is encircled by a dotted oval, and wirestage $\varsigma^{\leftarrow}(s_1) = \bigcup_i \beta^{\leftarrow}(s_1, b_i) = \{\{w_1, w_2\}, \{w_3, w_4\}, \{w_5, w_6\}\}$ is enclosed by the dashed frame.

We define the notion of local relative datapath regularity as the extent to which the appropriate criteria like W-T are present between a known stage $s$ and one of its adjacent candidate stages. An *adjacent candidate stage* $s' \in S$ of $s$ consists of a number of entities of the same type of which at least one is adjacent to an entity of $s$. We represent this adjacency with a *regularity graph*.

# 4 REGULARITY GRAPH

For every criterion, there is a relation between it and its stage $s$ reflected by a complete bipartite graph:

$$\Gamma_{W-T}(s) = (\varsigma^{\leftarrow}(s), T, B \times T) \tag{3}$$

$\Gamma_{W-T}(s)$ has a vertex for every slice of stage $s$, and a vertex for every instance type of any adjacent instance. The type-vertices of graph $\Gamma_{W-T}(s)$ are obtained from $C$ by removing all instances from it that are not adjacent to $s$, then collapsing all remaining instances of the same type onto one type-vertex. The edges $\epsilon \in B \times T$ of $\Gamma_{W-T}(s)$ are tagged with a set of pins realizing the respective adjacencies between the elements in a slice of $s$ and a criterion. The attribute function $\gamma_{W-T}(s) : B \times T \rightarrow 2^P$ returns the set of pins between stage $b \in B$ and type $t \in T$:

$$\gamma_{WT}(s, \epsilon) = \{p \in P(w) | \epsilon = (b, t) \wedge w \in \beta^{\leftarrow}(s, b) \wedge T(I(p)) = t\} \tag{4}$$

Figure 2 shows the wirestage-to-celltype relation graph $\Gamma_{W-T}(s_1)$ for the stage $s_1$ of figure 2. All cell types of cells that are not adjacent to any wire in $s_1$ are omitted from the graph. Note that the fact that $b_3$ has no connection to a cell of type $t_5$ is represented by edge $(b_3, t_5)$ having an empty associated set of pins.

# 5 RELATIVE REGULARITY METRIC

The proposed *relative local regularity metric* $\rho_{W-T}(s)$ is based on a statistical analysis of the cardinalities of the pin sets on the edges of every celltype node of $\Gamma_{W-T}(s)$. This metric is inversely proportional to the extent of regularity from wirestage $s$ with respect to the types of its adjacent cells.

We denote the score vector made up of the cardinalities of the pin sets on the edges of node $t \in T$ of $\Gamma_{W-T}(s)$ as $X_{W-T}(s, t)$. Thus, the elements $x_{W-T}(s, t)[i]$ of $X_{W-T}(s, t)$ are defined as

$$x_{WT}(s, t)[i] = |\gamma(s, (b_i, t))| \tag{5}$$

In figure 2, the sizes of the pin sets on the edges of node $t_1$ are $X_{W-T}(s, t_1) = [|\{p_1, p_2\}|, |\{p_6, p_7\}|, |\{p_{11}, p_{12}\}|] = [2, 2, 2]$. The extent of regularity between vantage stage $s$ and adjacent instance type $t$ is obtained by interpretation specific statistical properties of $X_{W-T}(s, t)$. Ideally, $X_{W-T}(s, t)$ has a uniform distribution, corresponding with a maximally regular candidate stage of type $t$.

To quantify the extent to which the distribution of $X_{W-T}(s, t)$ matches a uniform one, we compute the *range* and the *symmetry* of $X_{W-T}$. If the range, defined as the difference between the largest and the smallest values of $X_{W-T}$, is 0, then $X_{W-T}$ is uniformly distributed. If it is larger, but not too large, we look at the the skewness of the distribution of $X_{W-T}$, expressing the symmetry of $X_{W-T}(s, t)$. The skewness is given by the *third absolute central moment* $\mu_3$ about the average of the vector. The $n$-th absolute central moment of a distribution about its average is defined as $\mu_n = \mathrm{E}[|X - \overline{X}|^n]$This measure increases as the symmetry decreases. A completely regular environment of $s$ is reflected by $\mu_3 = 0$. The relative local regularity $\rho_{W-T} : S \times T \rightarrow \mathbf{R}$ from stage $s \in S$ relative to cell type $t \in T$ is defined as

$$\rho_{W-T}(s, t) = R(X(s, t))(\mu_3(X(s, t)) \tag{6}$$

This function is a suitable metric because $\rho_{W-T}(s) = 0$ implies maximal regularity, $\rho_{W-T}(s) = \infty$ maximally irregular while the value of $\rho_{W-T}(s)$ increases monotonically as the regularity decreases. These properties allow comparison of the extent of regularity at different stages of the datapath and with respect to different types.

# 6 REGULARITY EXTRACTION ALGORITHM

The objective of regularity extraction is to maximize the number of circuit entities that have a stage label. Our regularity extraction algorithm works by expanding a search-wave through the network. The search-wave consists of stages that have already been discovered. The search-wave is initialized with one or more *seed* stages. In practice, these can be found easily eg. from attributes at the circuit's primary terminals, or they may be specified explicitly.

The search-wave is extended from one of its stages $s$ with one suitable not yet discovered adjacent stage $s'$, over its full width at a time. Extensions are chosen from a set of candidate extensions. This set is ordered by the value of the regularity metric associated with the respective each candidate extensions. This process is iterated until there are no more suitable wave extensions left. With every extension, the newly added stage is tagged with a new stage label from $S$. The slices of the new stage $s'$ inherit the slice labels of the slices of $s$ via the pins that realize the adjacency of $s'$ and $s$. In case of maximally regular datapath circuitry, all entities have a unique stage label $\varsigma^{\leftarrow}(s_\infty) = \emptyset$ and the slices of every stage $s \in S$ have the same number of elements $\forall b : b, b' \in \varsigma^{\leftarrow}(s) : |\beta^{\leftarrow}(s,b)| = |\beta^{\leftarrow}(s,b')|$. The optimization goal $f_c$ can thus be defined as "minimize $f_c = |\varsigma^{\leftarrow}(s_\infty)|$". The algorithm described here tries to achieve this by greedily selecting the most regular candidate extension known at any moment. Particularly when the circuit is highly regular many choices will be equivalent because they have the same regularity value.

In the algorithm outlined below, the wirestage seed of the search-wave is given as SEED. Priority queues are used for efficiency. Two different threshold values further control the expansion process. An upper limit $\rho_{tr}$ filters candidate extensions leaving those that are sufficiently regular. A lower bound $\mathrm{w}_{tr}$ prevents the algorithm from selecting a candidate extension that is regular, but too narrow in terms of the number of slices. If the thresholds are set very tight, say $\rho_{tr} = 0$ and $\mathrm{w}_{tr} = |\varsigma^{\leftarrow}(\mathrm{SEED})|$, the algorithm will only find slices that are completely identical.

Stage $s := s_0$; Queue $Q_{W-T} := \{0, \mathrm{SEED}\}, Q_{I-T} = \emptyset$;
$\rho_{W,\min}, \mathbf{P}_W := Q_{W-T}$.front(); $\rho_{I,\min}, \mathbf{P}_I := Q_{I-T}$.front();

analyze_wire_stage(s) { /* fill $\gamma_{W-T}$ */
    $\gamma_{W-T}[s,] := \emptyset$;
    **forall** $b \in \varsigma^{\leftarrow}(s)$ **do**
        **forall** $e \in \beta^{\leftarrow}(s,b)$ **do**
            **forall** $p \in P(e)$ **do**
                $\gamma_{W-T}[T(I(p)), b] \cup = \{p\}$;
    /* compute $\rho_{W-T}(s,t)$ for all types */
    **forall** $t \in T$ **do**
        **forall** $b$ **do**
            $\mathbf{P}(s,b) := M[t, \mathbf{W}(s,b)]$;
        **if** $|\mathbf{P}(s)| \geq \mathrm{w}_{tr} \rightarrow$
            $\rho_{W-T}(s,t) := \mu_3(\mathbf{P}(s,b))$;
            **if** $\rho_{W-T}(s,t) < \rho_{tr} \rightarrow Q_{W-T} \cup = \{\rho_{W-T}(s,t), \mathbf{P}(s)\}$ }

**do**{   s := new S ;
    **if** $\rho_{W,\min} < \rho_{I,\min} \rightarrow$ mark$_W(\mathbf{P}_W, s)$; **analyze_wire_stage**(s) ;
    **else** mark$_W(\mathbf{P}_I, s)$; **analyze_inst_stage**(s) ;
    $\rho_{W,\min}, \mathbf{P}_W := Q_{W-T}$.front(); $\rho_{I,\min}, \mathbf{P}_I := Q_{I-T}$.front(); }
**while** $(|Q_{W-T}| > 0) \vee (|Q_{I-T}| > 0)$;
post_process();

The function **analyze_inst_stage** is analogous to **analyze_wire_stage**. The tagging procedure mark assigns the new stage label $s$, and the slice labels to the entities of $s$ inherited via the pins in $\mathbf{P}$. The main loop in the algorithm ends if there are no wave continuations with sufficient regularity left from sets of wire-strands or sets of instance-strands. The post-processing phase resolves undefined stage and slice labels for entities for which, looking back, a clear choice can now be made. This means that if for a module a slice and stage label is induced by its environment with limited ambiguity, and if the conditions for it being part of a datapath are met, it will be added to the datapath.

The algorithm has a run-time complexity of $\mathcal{O}(|P|)$ since every pin is analyzed at most twice. In practice, only pins in regular areas are considered, so actual run times are smaller.

# 7 RESULTS AND CONCLUSIONS

We implemented the algorithm in C++. The current implementation uses a more or less greedy strategy in that it favors wirestage-to-celltype wave expansions over others. The expansion process ends if all candidate expansions have been exhausted. Our results, which could not be included in this paper due to space limitations, show that the algorithm correctly extracts slices and stages from a number of circuits of different sizes and moderate irregularity. The results were evaluated by comparing the output of the extraction with explicit regularity information provided by module generators. The extraction performance is then given by the percentage of entities of which the extractor found the same stage/slice membership. The method has been applied to a number of large examples from various circuit synthesis environments such as a 8048 microprocessor core. In all cases, the datapath in the netlist was regained within a few seconds of CPU time. The threshold values proved to be an effective means to control the susceptibility for irregularity.

The datapath extraction algorithm presented in this paper enables datapath placement of circuits that were generated by general synthesis frontends. The algorithm is suitable for large circuits which are at least partially regular only to a some extent.

# REFERENCES


Asa82] T. Asano. An optimum gate placement algorithm for mos one-dimensional arrays. *Journal of Design Systems*, 6(1):1–27, 1982.

CH93] C.E. Cheng and C.-Y. Ho. Sefop: A novel approach to data path module placement. In *Proceedings of the International Conference on Computer Aided Design*, pages 178–181. IEEE, Nov 1993.

GCT77] S. Goto, I. Cederbaum, and B.S. Ting. Suboptimum solution of the back-board ordering with channel capacity constraint. *IEEE Transactions on Circuits And Systems*, 24(11):645–652, Nov 1977.

HS88] M. Hirsch and D. Siewiorek. Automatically extracting structure from a logical design. In *Proceedings of the International Conference on Computer Aided Design*, pages 456–459. IEEE, 1988.

NKH+93] H. Nakao, O. Kitada, M. Hayashikoshi, K. Okazaki, and Y. Tsujihashi. A high density datapath layout generation method under path delay constraints. In *Proceedings of the Custom Integrated Circuits Conference*, pages 9.5.1–9.5.5. IEEE, 1993.

OHN87] G. Odawara, T. Hiraide, and O. Nishina: Partitioning and placement technique for cmos gate arrays. *IEEE Transactions on Computer Aided Design*, CAD-6(3):355–363, May 1987.

15

# EPR - a Synthesis Tool for Speed Optimization

*E. Fehlauer, St. Rülke and G. Franke*
*FhG IIS Erlangen - Department EAS Dresden*
*Zeunerstr. 38*
*D-01069 Dresden*
*Germany*
*phone: (+49) 351 4640 728* *fax: (+49) 351 4640 703*
*email: ruelke@eas.iis.fhg.de*

**Abstract**

We propose a novel and practicable approach to speed up sequential circuits with pipeline structure at gate level which is called Extended Peripheral Retiming (EPR). Key to EPR is the idea to compute optimal register positions in an iterative manner and only for selected circuit partitions at a time. Different sequential and combinational optimization techniques are combined together. Using EPR, complexity of the optimization task is broken down and bottlenecks of other approaches are removed. Experimental results show the benefit of EPR.

**Keywords**

Sequential digital circuit, pipelining, optimization, peripheral retiming

## 1 INTRODUCTION AND OVERVIEW

Methods to optimize a structural description of a synchronous sequential circuit at gate level may be subdivided into combinational, sequential, and combined approaches.

Combinational techniques are the traditional approaches to optimize a circuit but they are limited to circuit partitions between register boundaries only (DeMicheli, 1994). Sequential techniques are more sophisticated but they are usually limited by their restricted scope. Most of them treat only sequential elements or apply a restricted set of transformations to special partitions of the circuit. Besides, techniques like Retiming (Leiserson, Rose and Saxe, 1983) require run time intensive computations; the optimal register position is determined by a polynomial time procedure. Most interesting are approaches which combine different sequential and/or combinational techniques together. So they augment the optimization potential to consider an enlarged range of optimization goals or to enhance the quality of the solution like (Malik et.al., 1991) and (Martin, 1993). However, up to now, no well defined common method is available taking into account the simultaneous optimization of both, the sequential as well as the combinational part of the circuit in a global manner.

We want to contribute to combined techniques. Our work was inspired most of all by both Peripheral Retiming (Malik et.al., 1991) and Extended CombToPipe (ECTP) (Fehlauer, Rülke and Franke, 1993). EPR (Extended Peripheral Retiming) is a novel and practicable approach to speed up sequential circuits with pipeline structure at gate level taking into consideration area penalties. The original approach Peripheral Retiming performs the following consecutive tasks: Firstly, registers are moved temporarily to the periphery of the circuit (primary inputs and outputs). Then, the resulting enlarged logic between the registers is optimized globally using combinational techniques. Finally, the registers are replaced using Retiming. Unfortunately, this complex approach requires costly computations. Moreover, global combinational optimization may create also additional paths (pseudo dependencies) such that the final register replacement sometimes may not be possible.

The idea of EPR is to perform register replacement in an iterative manner and only for selected partitions of the circuit. The iterative nature of the approach enables us to combine Peripheral Retiming with other techniques gaining timing and/or area optimization in each of the iterations. If pseudo dependencies occur, a backtracking technique ensures that they will be eliminated. The iterative register replacement in combination with the local optimization of logic between the currently created register boundaries utilizes our method ECTP. Based on heuristics, ECTP generates only a minimal number of registers, and ensures that the existing clock cycle constraints will be met - if necessary by generating additional pipeline stages.

In summary, the complexity of the optimization task is broken down, bottlenecks of the original approach are removed, and different sequential and combinational optimization techniques are combined together.

EPR is designed to be a value-added tool to synthesis systems (Synopsys, SIS). So, the user always operates in familiar surroundings. The synthesis tools provide us not only with required algorithms for timing and area optimizations but also with facilities to obtain design analysis.

## 2 GOALS AND DESIGN OBJECTS

We consider the problem of performance-driven optimization of sequential circuits at gate level, especially the post-synthesis timing improvement if performance (clock cycle) constraints are violated. The goal is to speed-up an initially synthesized circuit without a complete redesign - only by transformations of the structure obtained in the initial design.

Design objects we are focusing our work to are sequential circuits with extended pipeline structure and single-phase edge-triggered clocking. A circuit with pipeline structure consists of a sequence of k processing stages $P_1, ..., P_k$ separated by registers (e.g. clocked DFF) and has an acyclic topology. Each processing stage is formed by a combinational block which may have different granularity (single gate or complex module of connected gates). The combinational block of stage i communicates with the adjacent block of stage i+1 only via registers.

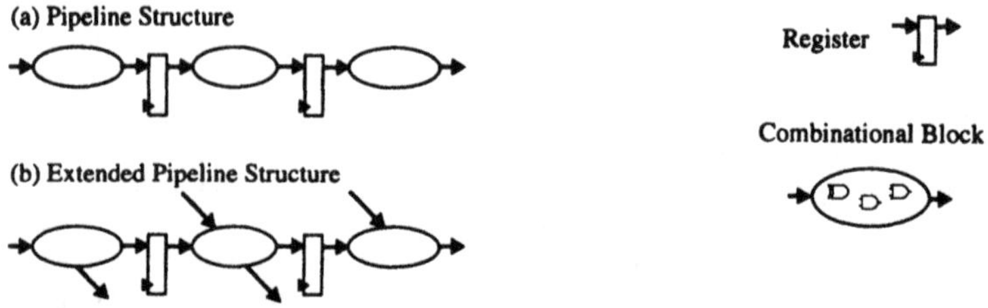


**Figure 1** Classification of pipelined circuits.

In case of pure pipeline structures, all primary inputs of the circuit are connected to the combinational block of the first stage and all primary outputs of the circuit are connected to the combinational block of last stage (see Figure 1a). Therefore, on each path between any primary input and any primary output the same number of registers occur.

More flexible structures with additional primary inputs and primary outputs connected to combinational blocks of any stage, but not to two or more blocks of different stages, are called extended pipeline structures (see Figure 1b). Consequently, paths between a primary input $PI_x$ and different primary outputs $PO_{y1}$, ..., $PO_{ym}$ may have different numbers of registers. Note that if two or more paths between a certain primary input $PI_x$ and a certain primary output $PO_y$ exist then all of this paths do have the same number of registers.

## 3 METHODOLOGY OF EPR

Analogous to original Peripheral Retiming, the activities done by EPR can be assigned to three consecutive main steps (see Figure 2):

1. Step S1 moves registers temporarily to the periphery of the circuit.
2. Step S2 optimizes the resulting enlarged combinational logic between the registers globally.
3. Step S3 replaces the registers iteratively (S3a through S3d).

### *Register shifting to the periphery (step S1)*

In a more detailed view, only the idea to move registers after the PIs and/or before the POs in order to enlarge the potential for combinational optimization is common in both approaches Peripheral Retiming and EPR. But unlike Peripheral Retiming, EPR does not require the numbers of registers at the periphery. EPR only detects efficiently, if such a register shifting is possible. If this detection succeeds, the circuit is proven to have an extended pipeline structure and consequently, our approach can be performed successfully. Note that the information about the sequential properties of the circuit is stored in the path weight matrix PWM which can be created tracing the circuit paths between PIs and POs. For the following computation steps in EPR it is enough to consider only the combinational circuit (where all registers are removed) and PWM as the fundamental.

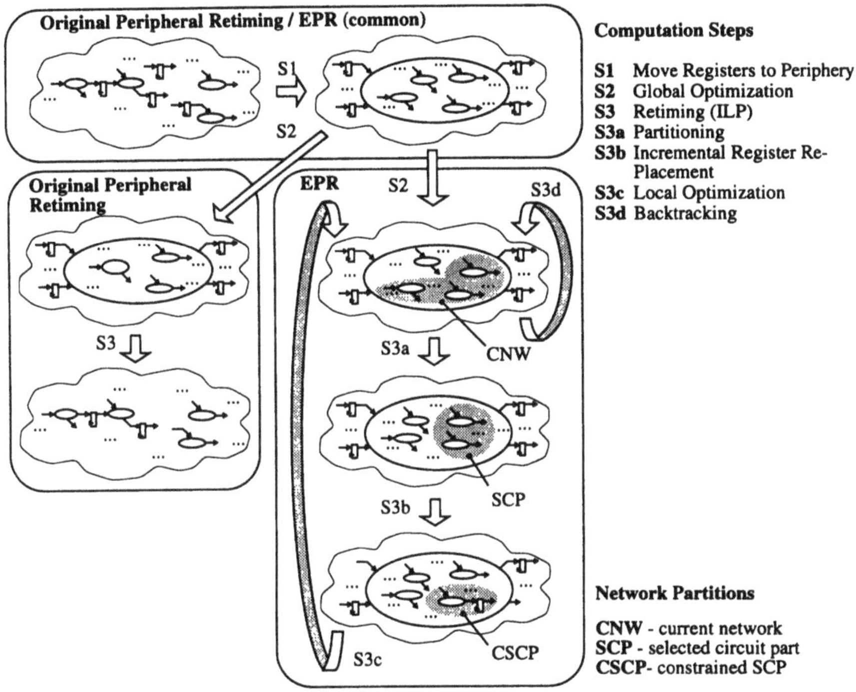


**Figure 2** Comparison of Peripheral Retiming and EPR

### *Global optimization (step S2)*

Different to Peripheral Retiming, EPR controls by parameters whether the global optimization should be performed or not. Additionally to the classical approach, EPR is able to handle pseudo dependencies with negative path weight which may result from the optimization. The related backtracking technique is quite simple. After having restored the circuit to the state before optimization there are two types of further treatment (controlled by parameters):

- No global combinational optimization is performed but local optimization will be done in the further EPR steps.
- The optimization is repeated using another optimization method (e.g. within SIS a lot of different optimization scripts are available or you can draw up your individual scripts).

### *Register replacement (steps S3a through S3d)*

In EPR, register replacement is determined by the idea to cope with the inherent complexity of this process by an iterative approach, by partitioning techniques, and by heuristics. Combining different optimization techniques gains different optimization goals simultaneously. Goal is not only to fit the given clock cycle constraints, but also to consider the related area penalties.

At the start of each iteration of register replacement, a special partition of the circuit called SCP (selected circuit part) has to be selected (step S3a of Figure 2). To compute SCP, in a first step all paths of the circuit which currently do not suffice PWM have to be chosen.

Next, SCP will be reduced further by considering limitation of pipeline stages and timing constraints as shown in step S3b of Figure 2. This will be done in two steps:

- Based on the pipeline stage of each PI and each PO as well as considering the relative position of each gate to the pipeline stages of neighboring PI and PO (see Figure 1), we compute the range of possible pipeline stages for each gate. This range is given by an early pipeline level and a late pipeline level.
- Based on the delay of each gate and the relating arrival and required times of input and output signals of these gates, this range is to be restricted further.

All gates having the same range are comprised to a circuit part called CSCP (constraint selected circuit part). In summary, only such circuit parts remain in a CSCP which have to carry at least one register on each of their paths to meet the timing constraints.

During CSCP computation it can be decided whether pipelining is necessary to fit the timing constraints or not. Pipelining is necessary if the number of required registers on a particular path exceeds its capacitance in the current PWM. In this case, the restrictions resulting from the delay of a gate are not compatible to the restrictions resulting from the pipeline stages of the neighboring PI and PO. Treatment of pipelining in the further running of our approach is very simple. It is enough to increment PWM but all other computation steps are not influenced.

In the partitioning process, usually more than one circuit part fit the stated properties of a CSCP. Therefore, the most promising one is selected using heuristics. Two different heuristics provided good results in experiments. Controlled by parameters it can be selected whether to choose the circuit part with the greatest number of gates or the circuit part with the fewest gates on its critical path.

Only for this selected circuit part an optimal register position will be computed. Based on ideas of (Münzner and Hemme, 1991) and related extensions in (Fehlauer, Rülke and Franke, 1993), we exploit a modified MaxFlow algorithm for this purpose. As a precondition, the selected circuit part has to be transformed into a weighted MaxFlow graph MFG. The modified MaxFlow algorithm cuts MFG completely on a minimal number of edges. This cut represents the optimal register position. The problem to compute the weights of particular edges in MFG will be the objective of an other paper and will not be handled here in detail. It depends on both, the type of the connection which is represented by the edge as well the goal to cut MFG completely by the algorithm used.

According to the determined optimal register position, the related registers have to be inserted in the circuit. Afterwards, combinational timing and/or area optimization can by applied locally for all combinational blocks as shown in step S3c of Figure 2. If pseudo dependencies occur for a particular combinational block, the handling (step S3d of Figure 2) is similar to that in the global optimization step explained above.

The current iteration of register replacement is closed by correcting PWM considering the included registers and possibly the changed (removed or added) paths.

As mentioned above, a couple of options ensure the flexibility of EPR. Briefly summarized, there are options to maintain the global optimization, to handle pseudo dependencies, to control partitioning, and to select heuristics.

# 4 EXPERIMENTAL RESULTS

Since there is a lack of sequential benchmarks having extended pipeline structure we use the following three different circuits to show results of our approach:

- *exam1* is a simple example with 8 pipeline stages. The 9 PIs as well as the 6 POs are connected to different stages of the pipeline. All combinational blocks in the particular pipeline stages are arbitrary combinations of ANDs, ORs, and INVs.
- *exam2* is the VHDL description of a real designed controller. For experiments *D* and *F*, first we cut all loops, then we optimized the circuit, and finally we connected the loops again.
- *exam3* is a cascade of unrolled FSM circuits. It was obtained using the MCNC benchmark ex3 after having applied the following operations: a) cutting all loops behind the latches which results in additional inputs and outputs, b) tripling the resulting circuitry, c) connecting the additional outputs of one subcircuit with additional inputs of the other subcircuit.

To set up unique test conditions, we worked on Sun Sparc10 workstation using the Berkeley system SIS for logic optimization and the LSI Logic library LCA10k for timing analysis. Handling these examples in the unique environment, we performed the following experiments:

- *A*: no explicit optimization,
- *B*: combinational timing optimization,
- *C*: Retiming for minimal clock period,
- *D*: Peripheral Retiming (we have added a tool to SIS which performs Peripheral Retiming using routines and data structures of the SIS system),
- *E*: varying sequences of *A* through *D*,
- *F*: EPR with different options.

We analyzed the final timing of the particular circuits and we compared clock cycle (Clock), number of registers (#REGs), and combinational area cost (CAC). The results of our experiments are comprised in Table 1. To illustrate experiment *E*, we show the results of applying *C* behind *B*.

**Table 1** Experimental Results

| | | *A* | *B* | *C* | *D* | *E* | *F* |
|---|---|---|---|---|---|---|---|
| *exam1* | Clock | 5.20 | 3.78 | 4.95 | 4.31 | 3.78 | 3.51 |
| | #REGs | 13 | 13 | 10 | 10 | 12 | 10 |
| | CAC | 43 | 40 | 40 | 32 | 21 | 19 |
| *exam2* | Clock | 11.55 | 11.59 | 10.93 | 8.01 | 7.63 | 8.67 |
| | #REGs | 30 | 30 | 30 | 51 | 71 | 52 |
| | CAC | 528 | 503 | 526 | 259 | 259 | 184 |
| *exam3* | Clock | 8.28 | 8.02 | 7.25 | 9.59 | 8.02 | 8.14 |
| | #REGs | 7 | 7 | 15 | 9 | 7 | 8 |
| | CAC | 189 | 189 | 190 | 213 | 189 | 134 |

Table 1 shows the benefits of EPR. As expected, EPR produces the best trade-off between permissible clock cycle, required register count, and inevitable combinational area costs. Using EPR area and/or computation time can be saved. In some cases only EPR and no other sequential optimization (Retiming, Peripheral Retiming) was able to master the large complexity of the optimization task. Moreover, using other control parameters, EPR is able to produce faster designs - may be by generating additional pipeline stages or by combining EPR with Retiming. E.g. using one further pipeline stage for *exam1*, the results are 2.81 / 17 / 19 (Clock / #REGs / CAC). Combining EPR with Retiming, the results for *exam2* are 6.96 / 64 / 184.

## 5 CONCLUSION

### *Summary*

We have presented a new approach to optimize sequential circuits. EPR combines sequential and combinational techniques to speed up a circuit at gate level considering area penalties. Working iteratively and heuristically, EPR is applicable to large circuits. Experimental results have shown the practicability of our approach.

### *Future Research*

EPR may be improved by extending it into several directions. Incorporating further performance optimization techniques (e.g. repowering and buffering) may lead to better combinational optimization results. Considering false paths may enhance the efficiency of our algorithm significantly. Also, the iterative manner of EPR allows for looking at don't-care conditions which may result in an improved combinational optimization potential. Moreover, since high-level synthesis is getting more and more practicable, EPR could be applied to simplify timing optimization of extended pipelining structures resulting from high-level synthesis.

## 6 REFERENCES

Fehlauer, E., Rülke, St. and Franke, G. (1993) Design Improvement by Combining Sequential and Combinational Techniques. Proceedings of the Third PATMOS Workshop, 21 - 29.

Leiserson, C., Rose, F. and Saxe, J. (1983) Optimizing Synchronous Circuits by Retiming. Proceedings of the Third Caltech Conference on VLSI, 87-116.

Martin, H.-G. (1993) Retiming by Combination of Relocation and Clock Delay Adjustment. Proceedings of the EURO-DAC, 384-389.

DeMicheli, G. (1994) Synthesis and Optimization of Digital Circuits. McGraw-Hill, New York.

Malik, S. et.al (1991) Retiming and Resynthesis: Optimizing Sequential Networks with Combinational Techniques. IEEE Transactions on CAD, 74-91.

Münzner, A. and Hemme, G. (1991) Converting Combinational Circuits into Pipelined Data Path. Proceedings of the ICCAD, 368-371.

---

This work was supported in part by BMFT (01 M 2897 C) and JESSI (AC8)

# 7 BIOGRAPHY

Dr.-Ing. Erhard Fehlauer

Diplom-Ingenieur (Information Technology) from Technological University of Dresden (TUD) in 1984; 1984-1988: Assistant Professor with TUD; Ph.D. Degree (Information Technology) from TUD in 1988; Present: Research Assistant with FhG IIS / EAS Dresden; Current research interests: high level and logic synthesis, esp retiming

Dr.-Ing. Steffen Rülke

Diplom-Ingenieur (Information Technology) from Technological University of Dresden (TUD) in 1981; 1981-1991: Research Assistant with Academy of Science; Ph.D. Degree (Applied Informatics) from Academy of Science in 1990; Present: Research Assistant with FhG IIS / EAS Dresden; Current research interests: high level and logic synthesis, esp retiming

Dr.-Ing.habil. Günter Franke

Diplom-Ingenieur (Electronical Engineering) from Technological University of Dresden (TUD) in 1968; 1969-1991: Research Assistant with Academy of Science; Ph.D. Degree (Control Engineering) from TUD in 1972; Present: Research Assistant with FhG IIS / EAS Dresden; Current research interests: high level and logic synthesis, HW/SW-codesign

PART SIX

# Controllers

# 16

# ROM-based Multi Thread Controller

*H.A. Hilderink and J.A.G. Jess*
*Eindhoven University of Technology*
*Eindhoven University of Technology, Department of Electrical Engineering,*
*Design Automation Section, P.O. Box 513, 5600 MB Eindhoven,*
*The Netherlands. email:* H.A.Hilderink@ele.tue.nl

**Abstract**

A new controller architecture is presented which can handle multiple threads efficiently without state explosions. The architecture is based upon a ROM, where the address decoders are replaced by latches on each row, which makes it possible activating more rows (states) at the same time. The extra logic for controlling the control flow is minimal.

**Keywords**

Controller Generation, Controller Architectures

## 1 INTRODUCTION

The current generation DSP chips is able to handle large sequences of arithmetic operations in an efficient way. Nowadays the specifications become more and more complex, resulting in small data introduction intervals (the time between the current and the new data to be offered to the system: *dii*) for delivering data, like for video applications. At medium throughput applications the *dii* is larger than the time needed for the calculation of a result based on data, but at high throughput applications the wanted dii becomes smaller. This results that serialization of the algorithm becomes too slow and that other techniques must be found, like:

**concurrency** where two (or more) independent tasks are executed at the same time on different resources;
**pipelining** where the process is executed before its previous execution has finished.

A controller is fully specified by the results of a High Level Synthesis (HLS) system. Starting from a behavioral description operations are placed in time (scheduling) and each operation is assigned to a module which can perform the operation (module binding). The controller is specified as a token flow graph (similar to that of [Eijn93]), where nodes correspond with states and edges with state transitions. Each state is associated with a control vector indicating the operations the modules must perform. A state is active when it contains a token, which will be passed to the next state(s) at the end of the clock cycle.

For the implementation of the controller, different architectures can be used. A first consideration is the partitioning of the controller. For each set of resources a sub-controller is defined, while all these sub-controllers are mastered by a global controller. Each sub-controller can be implemented on the most appropriate architecture. The next consideration is the choice of the implementation. There are different forms like a standard cell implementation, a Programmable Logic Array (PLA) or a ROM-based implementation. But all these implementations appear to be inefficient when the control flow contains a large grain of concurrency.

At the CATHEDRAL-II system [Zege90], the controller is implemented on a ROM-based architecture, which is accompanied by a branch and status finite state machine for the calculation of new (conditional) branch addresses, and an incrementer. The control output signals are directly provided by the ROM. At [Rane93] the controller specification is first partitioned, based upon sub-routine callings. Then the most suitable architecture is chosen for the implementation, which is based upon extended finite state models. The same method is also followed at [Gerb92]. There the implementation is based upon a ROM-based architecture, where control logic is added for the providing of the addresses by special structures according special constructions at the control specification, like conditional statements, loops and sub-routine callings. A modular interconnection of finite state machines architecture is used at [Ku91], where it is possible to implement multi thread control flows on separate finite state machines. The resulting set of communicating finite state machines are implemented on standard cells.

For all these implementations, except the last, holds that the concurrent specification is implemented on a single threaded architecture, i.e. an architecture where only one state can be active at the same time. This limitation often results in state explosions and consequently large implementations.

This paper presents a new controller architecture which can handle concurrency and pipelining efficiently. The architecture is based upon a ROM-structure, where the conventional ROM structure is slightly modified for permitting the activation of more than one row at a time. First the impact of concurrency and loop-pipelining is treated, followed by the description of the architecture. As result some implementation-templates are shown, illustrating the efficiency of the structure.

## 2 CONTROL STRUCTURES

A controller specification consists of an ordered set of sequences. A sequence is a consecutive list of states, where the states are executed one after the other. State transitions within a sequence are performed synchronous with a clock signal and are unconditional. A sequence can be executed independently of other sequences by using its own resources. Conditional execution of sequences is achieved by conditional state transitions between sequences.

*Concurrency of sequences* Concurrency is defined as the execution of two (or more) sequences independent of each other, where each sequence controls an own set of resources. Take for example the control flow of figure 1, where loops are started along sequence S. State $T_n$ can only execute when all the loops have finished their executions. There are two forms of synchronization: first is when the number of iterations of each sequence is known beforehand, where synchronization is achieved by waiting for the slowest sequence (*implicit synchronization*); and second is when the number of iterations of one (or more) sequences is data dependent, requiring special actions which keep track whether all sequences are finished or not (*explicit synchronization*).

*Pipelining of sequences* Pipelining of sequences is defined when the sequence is restarted before its previous execution has been finished. This is only possible when the resources needed for the new invocation are released before the start of a new invocation, and that at every state of the sequence this criterion holds. Moreover it is needed that data, needed at the new invocation, has become available from the current invocation (or before). The data introduction interval (*dii*) is defined as the maximum time needed for the release of a particular resource controlled by states of the sequence.

If each state controls only one set of resources which is not controlled at other states, the dii is one cycle, like in figure 2(a). Here a loop with 6 states is pipelined by restarting the loop each cycle. Looking at the control of the

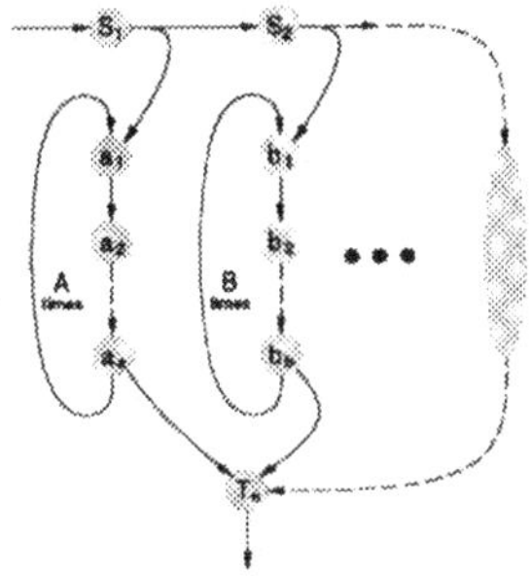

**Figure 1** Set of concurrent sequences, where sequence `S` starts other sequences (loops) `a, b, ...`

loop, we see that its execution can be divided into three parts: first the prologue sequence, where the pipeline is filled, second the steady state sequence, and last the epilogue sequence, where the pipeline is being emptied. For the implementation on a single thread control flow this results in extra states for both the prologue and epilogue, but the body of the steady state sequence contains less states as the body of the original loop.

# 3 MULTI THREAD ROM BASED CONTROLLER ARCHITECTURE

Conventional controller models, based on a single ROM architecture, are characterized by the implementation of a single thread control flow, which is caused by the limitations of the conventional ROM. Although the ROM is efficient for large controller implementations [Gerb92], its structure is not suitable for a multi thread controller. Looking more in detail at the ROM structure, it is evident that the address decoders on its inputs (which selects one row at a time, according the provided address) forms the bottleneck. When these address decoders are replaced by latches, each row can be controlled separately (figure 2(b)).

The control signals (output signals of the controller), are generated by the the columns of the ROM. Each column (set of columns) controls one resource. During the synthesis of the behavioral description, it is guaranteed by the scheduler and allocator that a specific resource is controlled by only one state at a time. This results that for the controller implementation each column can be activated only by one row at each cycle.

A row is activated when a signal ("1") is clocked into the corresponding latch. At the next cycle, the signal can be shifted into the next latch, activating the next row. In this way, a simple sequence of states can be executed.

At a conditional statement, the signal must be passed to one of the possible latches, according the test signal (from the datapath). Only that latch is selected for which the test condition evaluates true. Thus the evaluated test signal for each alternative must be "anded" with the token for each conditional latch (figure 3). After termination of the conditional sequence, the token must be passed to the next latch. This results that both outputs of the conditional sequences (*then* and *else* body) must be "orred" for the next latch. Thus only two *and* gates, one *or* gate (and one invertor) are needed for the implementation of *if-then-else* control flows.

For a loop implementation the same number of gates are required.

Concurrency is achieved when the token is fed not to only one latch, but to more latches, each starting an independent sequence. The next state following the concurrent sequences can only be executed when all concurrent sequences are finished (figure 4b). Synchronization becomes *implicit* by taking the token from the slowest chain,

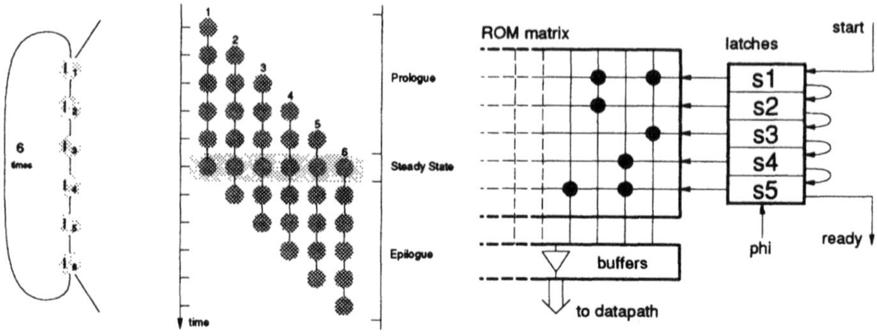


**Figure 2** a: Example of a pipelined loop with 6 states, b: Proto Multi Thread ROM based controller

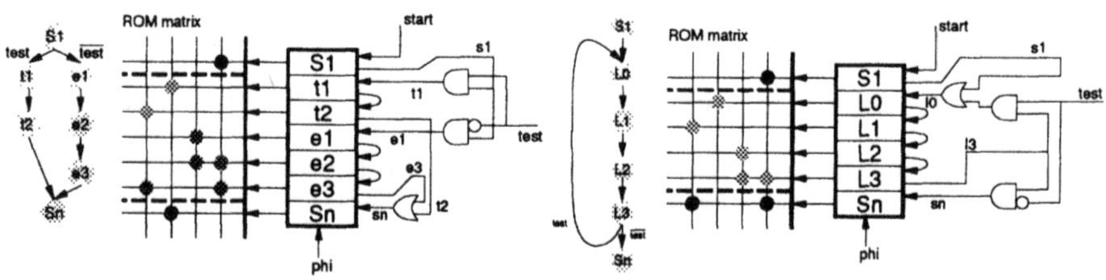


**Figure 3** a: If-then-else implementation, b: Loop implementation

ignoring the other tokens (figure 4a). Synchronization becomes *explicit* when a state machine is needed which keeps track of sequences which are finished and sequences which are not. Only when all sequences are finished, a token is passed to the next latch. Take for example figure 4c, where explicit synchronization is implemented. If the concurrent sequences are started, the FSM is set to state $q_0$. When one of the concurrent sequences ends, the status of the FSM changes (to state $q_a$ if sequence b is ended or to state $q_b$ if sequence a is ended). If both (all) sequences end at the same time, the final state $q_n$ is reached, which issues a signal to the next state $S_n$.

## Pipelining

Pipelining of a sequence is started when a token is inserted while the previous token still has not leaved the sequence yet. This results that more than one states are active at the same time, resulting in two (or more) active rows within

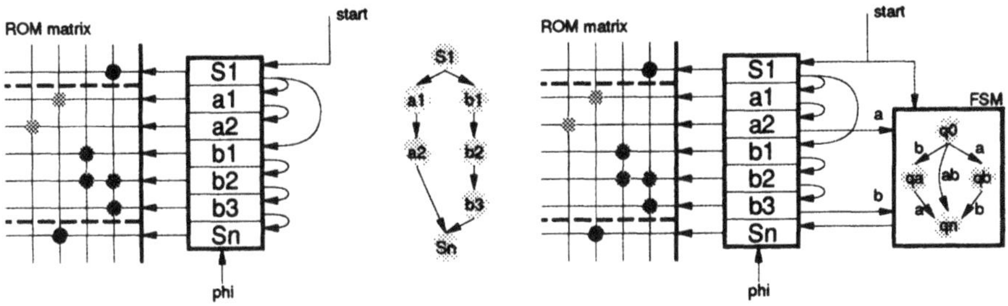


**Figure 4** a: Implementation of concurrent sequences with implicit synchronization, b: A concurrent sequence construction, c: Implementation of concurrent sequences with explicit synchronization

the ROM. Take for example the implementation of a simple sequence of figure 2: a new token can be inserted at least three cycles after the previous token was inserted (the *dii* of the sequence is therefore 3 cycles). This does not mean that the next token must be inserted directly if possible, but it might also be delayed some cycles. This results in extra freedom for defining the pipelines while the implementation remains unaltered.

Also pipelining can be used within special constructs (i.e. *if-then-else* constructs, *loops* or *concurrency*), but then special attention must be paid to for maintaining the correct sequence of the tokens. Consider for example the *if-then-else* construction of figure 3 where the sequences of the alternatives do not have the same number of states. When a first token enters the construction and goes the longest way (*else-branch*), while a second token enters the construction a cycle later and goes the shortest way (*then-branch*), it can occur that both tokens arrive at the end state at the same moment. This non-determinism can be prohibited by either implementing a kind of synchronization within the construction, or by defining the *dii* of the construction equal to the number of states of the longest sequence.

## Partitioning

The most efficient way for partitioning a controller is to define sub-controllers which control only a limited set of resources (probably two or more resources), and that each resource is controlled by only one controller. This results that the outputs of the controllers never have to be combined to deliver the final control word for the resource. Problems arise when a concurrent set of sequences is followed by another concurrent set of sequences, where the resource partition each sequence controls can differ (figure 5). So is resource *B* controlled together with resource *C* by sequence `b`, but is the resource together with resource *A* controlled by sequence `p`. This makes efficient partitioning difficult, but not entirely impossible.

# 4 EVALUATION

This adaptation of the ROM structure and the use of it as a controller has its impact on the total size of the implementation of the controller, and also on the delay and power consumption. This section compares the differences between the Multi Thread ROM controller implementation and a conventional ROM based controller implementation.

| State | Resources *A* | *B* | *C* |
|---|---|---|---|
| S1 | × | × | × |
| a | × | | |
| b | | × | × |
| S2 | × | × | × |
| p | × | × | |
| q | | | × |
| S3 | × | × | × |

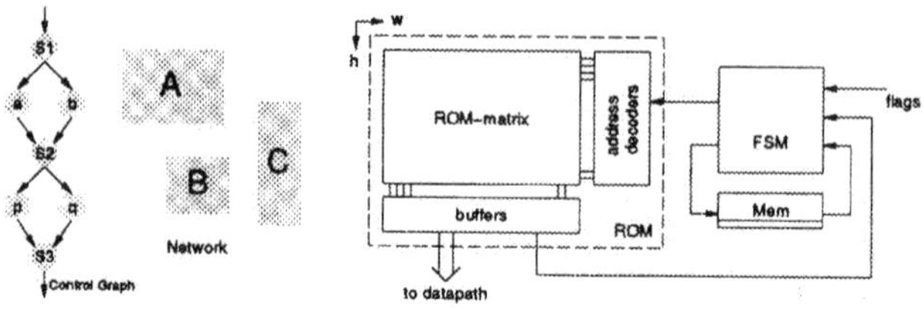


**Figure 5** a: Resource occupation by different concurrent statements, b: Area occupation of a conventional ROM

*Area* A conventional ROM based controller consists of the following parts (figure 5):

- The ROM:
  Matrix: area depends on the number of columns and rows;
  Buffers: area depends on the number of columns;
  Address Decoders: area depends on the number of rows and the number of address lines;
- Finite State Machine: area depends roughly on the number of address lines for which it has to determine a value for each state transition, and on the number of states.

The Finite State Machine provides the next address of the ROM which depends on the output of the ROM of the current address, combined with the status signals from the data path. The more complex the state transitions are, the larger the area will be.

Comparing this conventional architecture with a Multi Thread ROM based solution, the following differences are evident:

- The address decoders are replaced with latches: the size of a single latch is about 4 times larger than an address decoder for 8 bits (The sizes are compared within the Compass Design System where the area of an address decoder, part of its generic ROM structure, is compared with the area of a single transparent latch: this is nevertheless a rough estimation).
- The area for the additional gates is slightly less than compared with the area needed for the Finite State Machine at the conventional solution.
- A minor difference is the size of the matrix: this is smaller caused by a smaller number of states (rows) and the fact that no state information is obtained from the matrix, which reduces the number of columns.

The area needed for the latches and the additional gates must be weighted against the area needed for the address decoders and the Finite State Machine. Especially when a concurrent controller description results in an "exploded" single thread controller description, the Multi Thread solution can be smaller. Also when flexible pipelining is wanted, this solution can be smaller than conventional designs.

*Delay* The difference between both architectures lies within the size of the FSM, whose critical path is for the Multi Thread ROM solution considerably shorter. This improves the overall speed off the controller and probably does not require extra pipelines within the controller.

*Power* A CMOS gate consumes power only when its output changes its value: switching. At the conventional ROM implementation this happens mostly at the address decoders, where for each state transition a new address must be decoded, and at the Finite State Machine where a new address must be determined. At the Multi Thread ROM structure power is consumed when a token is shifted into a latch (and when it is shifted out of the latch), while the rest of the latches remain idle, and at the small number of gates when the transition is a conditional one.

The power consumption for the output buffers of the ROM matrix remain the same for both implementations. The Multi Thread ROM solution consumes therefore less power.

# 5 EXAMPLES

For proving the efficiency of the structure, examples have to be provided, but these are still under construction. This means that no comparison can be made yet between this multi thread solution and other conventional controller implementations. Also the lack of large examples, especially those with a large grain of concurrency, makes a comparison difficult. The controller generator is written as an application on the NEAT system [Heij93], using C++. As backbone the Compass Design System will be used, whereat the Multi Thread ROM module must be generated.

# 6 CONCLUSION

Although the Multi Thread ROM based controller has not yet been implemented, its preliminary results are encouraging. The advantages of the structure are its flexibility and relatively small area and power consumption also when the controller specification becomes more complex. The mapping of the control graph onto the network architecture is a straight forward job and does not require much time, this in contrary with the conventional state assignment procedures.

Remaining work will be the mapping of the controller on hardware and proving its efficiency, but also a more detailed study about partitioning and communicating controllers must be performed.

# REFERENCES


[Eijn93] J.T.J. VAN EIJNHOVEN, J. JESS, AND J.P. BRAGE. Behavioral Specifiation for Synthesis. In F. CATTHOOR AND L. SVENSSON, editors, *Application–Driven Architecture Synthesis*, chapter 2, pages 23–45. Kluwer Academic Publishers, 1993.

[Gerb92] L. GERBAUX AND G. SAUCIER. Automatic Synthesis of Large Moore Sequencers. *Integration, the VLSI journal*, 13:259–281, September 1992.

[Heij93] M.J.M. HEIJLIGERS, H.M.A.M. ARTS, J.T.J. VAN EIJNDHOVEN, H.A. HILDERINK, J.A.G. JESS, W.J.M. PHILIPSEN, AND A.H. TIMMER. The New Eindhoven Architectural synthesis Toolbox. In *Workshop on Circuits, Systems and Signal Processing*, pages 71–75, Houthalen (B), March 1993. ProRisc/IEEE Benelux.

[Ku91] D. KU AND G. DE MICHELI. Optimal Synthesis of Control Logic from Behavioral Specifications. *Integration, the VLSI journal*, 10:271–298, October 1991.

[Rane93] K. RANERUP, L. PHILIPSON, J. MADSEN, O. OLESEN, AND G. JANSSEN. Controller Synthesis and Verification. In *Application–Driven Architecture Synthesis*, chapter 10, pages 211–230. Kluwer Academic Publishers, 1993.

[Zege90] J. ZEGERS, P. SIX, F. RABAEY, AND H. DE MAN. CGE: Automatic Generation of Controllers in the CATHEDRAL-II Silicon Compiler. In *Proceedings of the European Conference on Design Automation*, pages 617–621, Glasgow, February 1990. IEEE Computer Society Press.

# 17

# State Assignment Selection for FPGAs and CPLDs

*Hichem BELHADJ, Aziz FORTAS and Gabrièle SAUCIER*
*Laboratoire CSI-Institut National Polytechnique de Grenoble*
*46, Avenue Félix Viallet 38031 Grenoble Cedex, FRANCE*
*Tel. (33) 76 57 48 20, Fax. (33)76 50 34 21, Hichem.Belhadj@imag.fr*

**Abstract**

This paper deals with the selection of state assignment options for finite state machines implemented in various targets. This choice depends on the intrinsic controller complexity, the partitioning strategy, the area/delay constraints and the target technologies. Industrial controllers as well as a sub-set of MCNC benchmarks have been synthesized and analyzed after place and route to define the selection strategy.

**Keywords**

Finite state machine synthesis, State assignment, FPGAs and CPLDs.

## INTRODUCTION

The optimization of finite state machine synthesis depends primarily on the state assignment and secondly on the minimization of the next state and output logic. The problem has been extensively addressed in the past. The optimization criteria being either the number of product terms or the number of literals of the next state or output logic. An important parameter is the number of memory elements used to store the internal state. In some cases, a minimal number is an important goal ; in some others a larger number of memory elements is quite acceptable or suitable. All these options have to be analyzed with regards to target technology and in the past a lot of attention has been given to PLA implementation [(De Micheli, 1983), (Villa, 1986), (Paulin, 1989), (Amman, 1988)] and to standard cells or gate arrays [(Lin, 1989), (Devadas, 1987)] and only a few works have considered the programmable devices [(Golson, 1993), (Boriello, 1993)]. This paper considers 3 types of state assignments : the well known one hot encoding, the compact encoding aiming at reducing both the number of product terms and the literal count of the next state and output functions, and the linear encodings tending to identify paths in the finite state machine state groups and to encode the states of these paths with successive linear code elements.

The goal is to analyze which state assignment is the "best" for a given constraint (area or speed), for a controller of a given complexity, and for a given target technology. The target technologies are namely FPGAs and PLDs, and experiments have been done on technologies where the prediction of the critical path is easy and accurate namely Actel, Xilinx, QuickLogic, Crospoint, MACH, PAL and for comparison for standard cell library.

This paper is organised as follows. Section 1 presents the different encoding technics used in the experimentation. The complexity of a finite state machine is introduced in section 2. The significant differences between several technological targets and their effect on the selection of the state assignment is detailed in section 3. Section 4 presents the experimentation scripts and analyzes the obtained results. It also presents the state assignment selection strategy.

# I. THE DIFFERENT STATE ASSIGNMENTS

## I.1. One Hot encoding

This technique uses one register for each state, i.e for each state only one of n internal variables has the value 1, where n is the number of states.

## I.2. Compact state assignment

The compact state assignment uses the minimal number of memory elements and tends to minimize both the number of product terms and the literal count in the next state and output functions. It uses a sophisticated and powerful method tending to reduce the complexity of these equations. This method is the extension of the partition pair theory of Stearn and Hartmanis (Hartmanis, 1966) and defines sub-sets of states to be placed on faces of the hypercube to reach this reduced dependency. In a second step, an efficient face embedding technique yields the final state assignment (Duff, 1991). Extensive practical experiments on about a hundred of FSMs have shown after place and route significant benefits in both area and critical path over JEDI (Newton, 1989) and NOVA (Villa, 1989).

## I. 3. Path encoding

This state assignment is suitable when the state graph exhibits long paths. In this case, a pre-processing phase partitions the state graph into paths and branching arcs and then applies linear codes on the paths.

### I.3.1 Partitioning strategy

Some basic definitions need to be introduced before the partitioning strategy can be explained.

* The minterm associated with the binary code ($c_n \ldots c_1\, c_0$) assigned to a node of the control flow-graph is $\tilde{y}_n \ldots \tilde{y}_0$, where $\tilde{y}_i = y_i$ ( $\overline{y_i}$ ) if $c_i$=1 (0). For instance, if the code of the node S is "1010" then the minterm associated with S is $\overline{y0}\ y1\ \overline{y2}\ y3$.

* The *Boolean expression produced by an arc* ***a*** , denoted by $SF_a$, is the product of the minterm associated with the origin node multiplied by the input firing condition, or input predicate, labelling the arc **a**. An example of this is to assume that an arc **a** issued from the state S is labelled by the input predicate (e + f). Then the Boolean expression produced by this transition (or arc) is $SF_{\mathbf{a}} = (e+f) \cdot (\overline{y0}\ y1\ \overline{y2}\ y3)$.

* The next state equation is defined as $Y_{t+1} = \sum_{a \in A} SF_a(y_t, \text{Input})$ where A is the set of arcs.

Let A denote the set of arcs of the state graph. If $SF_a$ is the Boolean expression produced by an arc a, and $L(SF_a)$ represents the literal count of F; then a partitioning strategy, consisting of splitting A into sub-sets A1, …, An, is cost effective if $\sum_{a \in A} L(SF_a) < L(\sum_i SF_{A_i})$.

It has been stated in [BGLS93], that one of the best strategies consists of spliting the set of arcs A into three sub-sets A1, A2 and A3, where

- A1 is the set of arcs whose origin node has only one successor node and whose destination node is the destination of only one arc.
- A2 is the set of arcs whose origin node has several successor nodes.
- A3 is the set of arcs whose the destination nodes have several predecessor nodes.

In the following only path encoding is addressed. In a path, each arc maps bijectively a present state code to a next state code. A path encoding consists in defining a path function fp such that, for each couple (present state $S_i$, next state $S_{i+1}$), $Code(S_{i+1})=fp(Code(S_i))$.
The linear code assignment consists of assigning the nodes of the sub set A1 to linear codes and the arcs to faces of the hypercube; this leads to the study of the cubical compactness .

### I.3. 2 Cubical compactness of a path encoding

The path encoding is said to be **cubical compact** if by fixing a predefined code to the initial state $S_0$, the following property can be obtained on the hypercube[1]:
The codes ($S_0$, $\delta(S_0)$, ..., $\delta^3(S_0)$) define a 2-cube in the hypercube,
The codes ($S_0$, $\delta(S_0)$,..., $\delta^7(S_0)$) define a 3-cube in the hypercube, and more generally,
The codes ($S_0$, $\delta(S_0)$, ..., $\delta^{2^n-1}(S_0)$) define an n-cube in the hypercube.
The cubical compactness plays an important role in the global state assignment and has therefore to be analyzed for the main codes which can be chosen on a path.

### I. 3. 3. Study of the different path encodings

Two commonly used path encodings will be analyzed with respect to the complexity of the logic they produce, estimated by the literal count, and to the cubical compactness property.

#### A. Radix-2 or sequential code

The radix-2 code is one of the most commonly used encodings. The path is encoded by assigning to the states $s_0$, ..., $s_{n-1}$ sequential binary codes. The code for the state $s_j$ is therefore equal to the code for the state $s_{j-1}$ plus one. If ($y_1$, ..., $y_k$) represents the current state then the next state ($Y_1$, ... $Y_k$) is obtained by means of the following formula :

$$Y_1 = \overline{y_1} \quad \text{and} \quad Y_i = y_i \cdot \sum_{j=1}^{i-1} \overline{y_j} + \overline{y_i} \cdot \prod_{j=1}^{i-1} y_j \qquad \text{for } i = 2, \ldots, k$$

#### B. Gray code

Let $c_m \ldots c_1 c_0$ denote a code in the Gray code, and let $b_m \ldots b_1 b_0$ designate the corresponding binary number, where the subscripts n and 0 denote the most significant and the least significant digits respectively. Then the ith digit $c_i$ can be obtained form the corresponding binary number as follows: $c_i = b_i \oplus b_{i+1}$ $0 \leq i < m$; (m= log2 n)]) and $c_m = b_m$, where $\oplus$ denotes the modulo-2 sum. If ($y_1$, ..., $y_k$) represents the current state then the next state ($Y_1$, ... $Y_k$) is obtained by means of the following formula :

$$Y_{1,k} = \overline{(y_k \oplus \ldots \oplus y_2)} \;;\; Y_{i,k} = y_i \oplus ( \overline{(y_k \oplus \ldots \oplus y_i)} \cdot y_{i-1} \cdot \prod_{j=1}^{i-2} \overline{y_j}) \quad \text{for } i = 2, \ldots, k-1, \text{ and}$$

---

[1]The hypercube is represented as the lattice of the parts of N elements. Each node is labelled by its relative coordinates denoted by $\varepsilon$. $\varepsilon$ is a set of integers ($e_1$, ..., $e_n$). It may be understood as follows: $e_i \in \varepsilon$ if and only if the ith bit of the code of that node is equal to 1. Figure 1 shows three representations of a 3-cube using binary codes, equivalent decimal codes and the lattice of the parts of N elements. A cube is defined by a pair [$\varepsilon_{min}$, $\varepsilon_{max}$] with $\varepsilon_{max} \supseteq \varepsilon_{min}$. For instance [2,123] is a cube of the hypercube of dimension 2. $\varepsilon_{max} = \varepsilon_{min} + \varepsilon_c$; $\varepsilon_c$ is called the characteristic set of indices of a cube, and $|\varepsilon_c|$ gives the dimension of the cube. [2, 123] has as characteristic indices {13}.

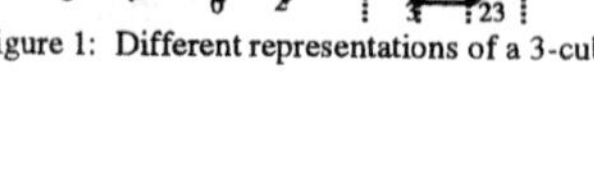

Figure 1: Different representations of a 3-cube

$$Y_{k,k} = y_k \oplus ((y_k \oplus y_{k-1}) \cdot \prod_{j=1}^{i-2} \overline{y_j})$$

Figure 2 illustrates the cubical compactness of the sequential and the Gray codes, and Figure 3 shows the complexity of the derived next-state functions, in terms of literal count, for paths of different lengths. On the other hand, it is easy to verify that the one hot encoding does not verify the cubical compactness property since $(S_0, \delta(S_0), \ldots, \delta^{2^n-1}(S_0))$ defines a $(2^n-1)$-cube.

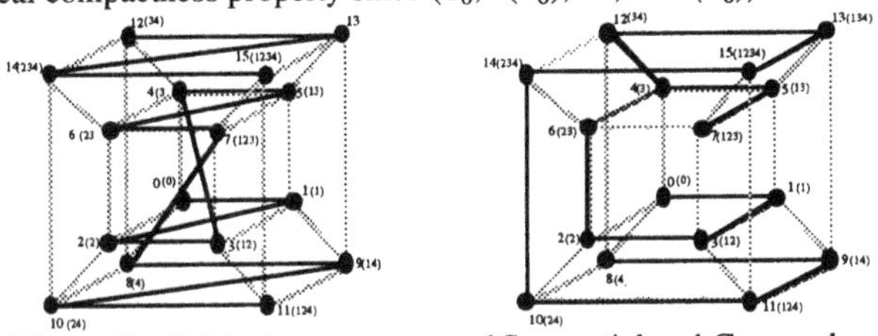

**Figure 2** : Cubical compactness of Sequential and Gray codes.

| Path Lengths | 8 | 16 | 32 |
|---|---|---|---|
| Sequential | 11 | 17 | 23 |
| Gray | 12 | 26 | 42 |
| Optimized | 13 | 30 | 45 |

**Figure 3** : Complexity of next state function (literal count) for several path lengths

One can state that Radix-2 code is to be used in the case of a controllers with large paths.

## II. FINITE STATE MACHINE COMPLEXITY

Commonly the complexity of controllers is related to the number of states and the number of inputs and/or outputs. Other important parameters have a critical influence on the efficiency of the state assignment. One of them is the complexity, in terms of literal counts, of the predicates on transitions between states. This parameter is correlated with the outdegree of states, i.e the number of transitions origin from the same state. Another topological characteristic of a state machine is the existence of paths. Paths and their lengths and their number of occurrences are very important if a partitioning strategy is adopted particularly in the case of controllers with more than one hundred states (Belhadj, 1993). A particular attention is given to this characteristic. In this paper, two sets of controllers are considered. The first one, called small sequencers, are state machines with a reduced number of states . The second set includes controllers with a number of states greater than 50.

## III. THE TARGET TECHNOLOGIES

The experimental study concerns the synthesis of a large set of industrial sequencers using the pervious state assignments on the following technologies : Xilinx, Actel, Crosspoint, QuickLogic and standard cells. Mapping on Max and MACH PLDs is in progress and the whole results will be included in the final version of this paper.

The choice of this set of technologies for mapping is motivated by the significant variety of their characteristics. Some significant differences and their influences are discussed.

i) The FPGAs have a larger number of registers than CPLDs ; but have finer logic granularity

ii) The fanin and fanout of basic modules are typically small for FPGAs than for CPLDs

iii) The FPGA architecture requires incremental extra routing resources to send external and internal signals to each of its loads; whereas in CPLD all the bank of registers within a block are accessible to its inputs without need of extra resources. Some CPLDs architecture has *shared and parallel expanders* that provide additional product terms directly to a macro cell (MAX7000 from ALTERA). The use of these expanders ensures that the logic is synthesized with the fewest logic resources to obtain the fastest possible speed.

The standard cell and PLAs technologies are not addressed here since the most important state assignments were specifically developed for these technologies. The crosspoint technology is closed to standard cells and the results should be closed or similar.

A first consequence is that the FPGAs require next state and output functions that fan out as little as possible. Another consequence is that it is necessary to reduce the number of state variables for implementation on CPLDs, since the reduction of state variables increases the combinational logic parts of the state machine, and allows a higher use of their logic blocks .

According to these remarks, we could predict that the optimized state assignment, Gray and sequential codes are more appropriate than the one hot code for implementations on CPLDs. The one hot encoding should be suitable for a mapping on FPGAs due to the availability of large number of registers and the fine granularity of their architectures.

The optimization criterion (speed, area, trade offs) has a considerable influence. It is expected that the one hot encoding leads to the best critical paths if the delay is the criterion fixed by the user. But this statement must be taken with care, especially when the number of states of the controller is reduced (less than 15). To explain this, let us assume that N, the number of states, is reduced and that it will be implemented on Xilinx. Then N CLBs are needed for the one hot code and [log2 (N)] CLBs are necessary for the other encoding techniques. Even if the major parts of logic blocks of these N CLBs are not used the critical path increases due to extra-delays introduced by intra-CLBs connections (see i) and ii)).
If the optimisation criterion is area, the choice of the encoding technique becomes more complex. It considers all the parameters introduced.

## IV EXPERIMENTAL RESULTS

All the benchmarks have been run first using the ASYL area option then the speed option with the compact, random, sequential, one hot and Gray codes for implementation on standardcell, FPGAs (Xilinx 4000, Actel (ACT2), QuickLogic, Crosspoint) and MACH. The given area (or number of modules) results are obtained using the area option and the criticalpath results are obtained using the speed option.

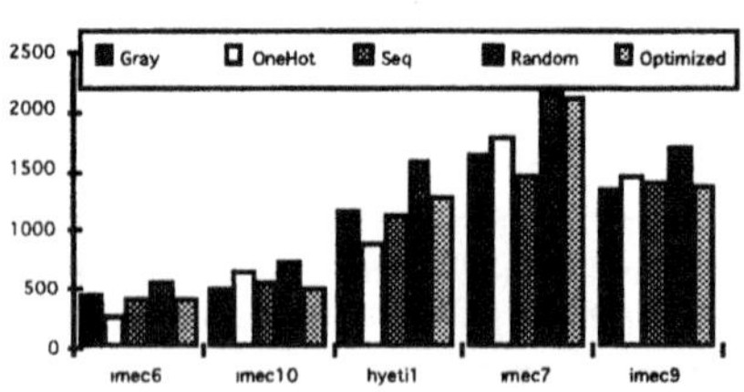


**Figure 4** : Area measures for ACT2.

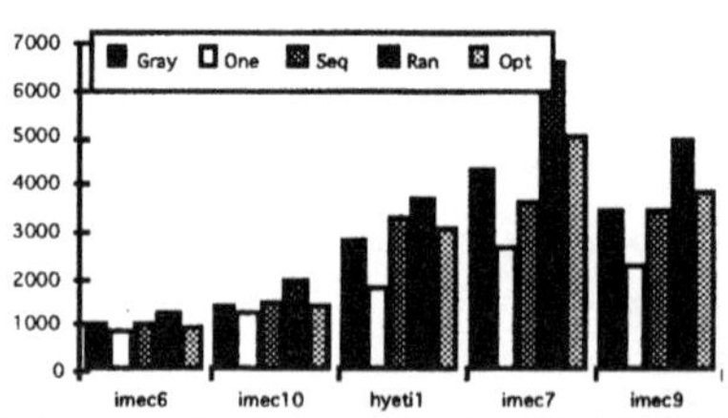


**Figure 5**: Area results for standard. cells (VSC370).

Figures 4 and 5 give the area measures for the complex controllers implemented on respectively Actel (ACT2) and standard cells (VSC370 library). Finally, Figures 6 and 7 show the area obtained for AMD PAL22V10 and MACH.

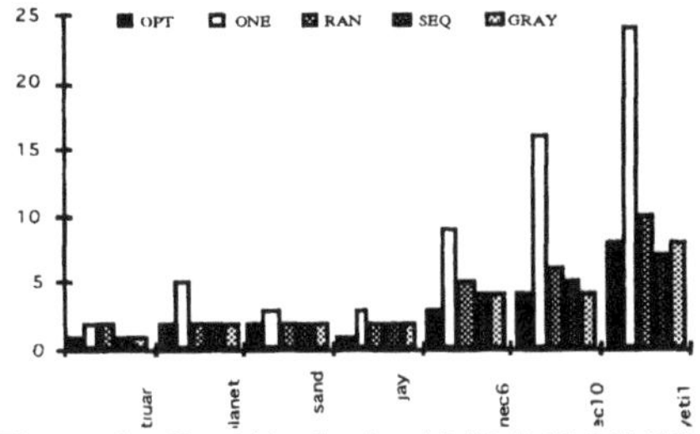


**Figure 6**: Nb of blocks for AMD PAL22V10.

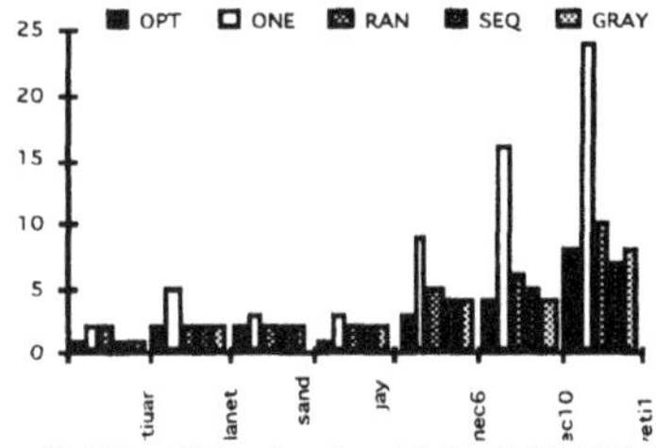


**Figure 7**: Nb of blocks for AMD MACH 130.

## Results analysis for the area option

A quick analysis of the area results shows that the compact encoding leads to the best results for the small sequencers (less than 50 states), with an exception for those that exhibit long paths, for which better results are obtained using linear codes, and this is true for all the targets. For complex sequencers (controllers with more than 50 states), the one hot encoding, as expected, leads to the best results for implementation on standard cells and FPGAs with some exceptions for controllers containing large amount of paths. But as we stated the one hot encoding is not suitable for PLDs. For these technologies, the state assignments with the minimum number of internal variables are more suitable. The compact, sequential and Gray options give approximatively the same results depending on the complexity of the predicates on the transitions and the existence or not of long paths. Table II summarizes these conclusions and draws a rought version of the state assignment selection strategy when the performance criterion is the area.

| | 2<#states<15 | 15<#states<50 | 15 < #states < 50 | #states > 50 ** | #states > 50 |
|---|---|---|---|---|---|
| | | Long Paths | Short Paths | Long Paths | Short Paths |
| Std Cells | Compact | Seq/Gray | Compact | OneHot/Comp. | One Hot |
| Xilinx | Compact | Seq/One Hot | Comp./One Hot | One Hot/Seq | One Hot |
| Actel | Compact | Seq/Gray | Compact | Seq/OneHot | One Hot |
| QuickLogic | Compact | Seq/Gray | Compact | Seq/OneHot | One Hot |
| Crosspoint | Compact | Gray/Seq | Compact | One Hot/Seq | One Hot |
| PAL | Compact | Seq/Gray | Compact | Seq/Gray | Compact |
| MACH | Compact | Compact/Seq | Compact | Seq/Gray | Compact |

**Table II**: State assignment selection strategy for area criterion.

## Results analysis for the speed option

The analysis of the delays obtained with the ASYL speed option confirms the expectation that the one hot encoding is globally suitable for all the targets with some exception where the radix-2 (sequential) option is better due to the existence of long paths. However, this is not the case for very small controller (with less than 15 states), for which the state assignement options with the minimum number of state variables lead to more valuable critical paths. Thus the experimental results validate the statements informally introduced in the previous sections. The Table III summarizes these remarks and draws the state assignment selection strategy when the delay is the performance criteria.

| | 2<#states<15 | 15<#states<50 | 15<#states<50 | #states > 50 | #states > 50 |
|---|---|---|---|---|---|
| | | Long Paths | Short Paths | Long Paths | Short Paths |
| Std Cells | Seq./Comp. | Seq./One Hot | One Hot | One Hot/Seq. | One Hot |
| Xilinx | Seq./Comp. | One Hot | One Hot | One Hot | One Hot |
| Actel | Seq./Comp. | One Hot | One Hot | One Hot/Seq. | One Hot |
| QuickLogic | Seq/Comp. | One Hot | One Hot | One Hot/Seq. | One Hot |
| Crosspoint | Seq./Comp. | One Hot | One Hot | One Hot/Seq. | One Hot |
| PAL | Seq./Comp. | Seq./One Hot | One Hot | Seq./One Hot | One Hot |
| MACH | Seq./Comp. | Seq./One Hot | One Hot | Seq./One Hot | One Hot |

**Table III**: State assignment selection strategy for delay criterion.

## CONCLUSION

This paper introduces a strategy for the selection of a state assignment based on a set of parameters that includes the intrinsic complexity of the considered controller, the technological target and the performance criterion fixed by the designer.

Future works will explore several improvements of the optimized state assignment in order to make different versions; each one is more or less specific to a category of controller and a technological target given a fixed performace criterion. An other direction of investigation concerns the comparison between different implementations, within a partitioning strategy, using a ROM or PLAs and logic part (for multiple output branching primitives) implemented on standard cells, on FPGAs or Complex PLDs. The results of these extensionswill be used as basic start point for efficient synthesis of communicating FSMs (Saucier, 1992).

## REFERENCES

R. Amman et al.. (1988) PLA Based Finite State Machines Using Johnson Counters as State Memories. ICCD, 267-270.

H. Belhadj et al. (1992) Specification and Synthesis of Communicating Finite State Machines. Synthesis For Control Dominated Circuits, Saucier and Trilhe Ed. North Holland.

S. Devadas. (1989) General Decomposition of Sequential Machines: Relationship to State Assignment. 26th DAC, 333-338.

S. Devadas et al. (1987) MUSTANG: State Assignment of FSM for Optimal Multi-Level Implementations. ICCAD, 16-19.

S. Devadas and A. R. Newton (1988) Decomposition and Factorisation of Sequential Finite State Machines. ICCAD, pp. 148-151.

G. de Micheli, et al. (1983) Comp. Aided Synthesis of PLA-based FSMs. ICCAD, 154-156.

C. Duff. (1991) Codage d'automates et théorie des cubes intersectants, I.N.P. Grenoble.

S. Golson. (1992) One Hot state machine design for FPGAs. Int'l Workshop on Logic Synthesis.

L. Gerbaux, G. Saucier. (1992) Automatic Synthesis of Large Moore Sequencers. EDAC.

J. Hartmanis and R.E. Stearns. (1966) Algebraic Structure Theory for Sequential Machines. Prentice-Hall, Englewood Cliffs, N. J.

S. Hassoun and G. Boriello. (1993) Improving Finite State Assignment for Two-Level PLDs. International Workshop on Logic Synthesis, Califonia.

Bill Lin and A. R. Newton. (1989) Synthesis for Multiple-Level Logic from Symbolic High Level Description Languages. VLSI'89 Conference, 187-196, Germany.

P. Paulin. (1989) Horizontal Partitioning of PLA-based Finite State Machines. DAC, 333-338.

T. Villa, et al.. (1986) NOVA : state Assignment of Finite State Machines for Optimal Two Level Logic Implementations.26th DAC, Las Vegas.

18

# Control optimization and hardware translation of Esterel programs

*François-Xavier Fornari*
*INRIA*
*INRIA, BP 93, 06902 Sophia-Antipolis, FRANCE. Telephone: (33) 93 95 74 93, Fax: (33) 93 95 74 88. e-mail:* `fornari@sophia.inria.fr`

**Abstract**

This paper describes an original application of hardware optimization techniques for ESTEREL programs. These techniques are tuned to the target technology which may be software or hardware. The control portion of an ESTEREL program consists of a set of sorted boolean equations and a set of boolean memory cells. It computes the next state of the program and drives data handling actions. The presented techniques heavily reduce the size of the control while preserving the sequential constraints between the actions. For hardware implementations, we describe a translation into HARDWAREC. This final synthesis step is done using the Stanford OLYMPUS synthesis system.

**Keywords**

Synchronous languages, ESTEREL, digital circuits, sequential optimizations, high level synthesis.

## 1 INTRODUCTION

Designing and implementing controllers is known to be a hard task especially when concurrent components are used. One way to reduce the development time, to increase the reliability, and to be technology independent is to use a high-level programming language. In this paper we present work on the translation of **full** ESTEREL (Berry and Gonthier 1992) language into digital circuits, extending previous work by Berry implementing the PURE ESTEREL synchronization subset of ESTEREL (Berry 1992). The ESTEREL language was designed for modular description of control. It fully combines sequencing, parallelism, instantaneous broadcast, exceptions and watchdogs. It has a well-defined mathematical semantics that relies on the **perfect synchrony hypothesis**, which states that the response of a program to a stimulus from its environment is instantaneous. This model is known in the hardware domain as the zero-delay model.

ESTEREL was primarily designed for control intensive software applications. Several tools have been built for the ESTEREL language: a compiler that produces software implementations of encoded finite state machines, graphical simulators, a symbolic debugger, and automata verification tools. Since programs can be compiled into FSMs, the existing FSM techniques for sequential hardware synthesis can be used. Unfortunately, optimal state encoding is hard to achieve and current techniques do not scale with the size of programs.

```
loop
  [
    await A
  ||
    await B
  ];
  emit O
each R
```

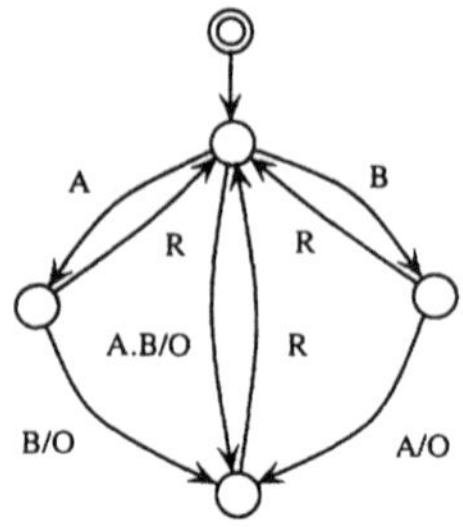


**Figure 1** An ESTEREL program and its FSM representation

In (Berry 1992) Berry proposed a structural hardware translation that follows the structure of the behavioral description and translates the synchronization kernel of the program into a sequential circuit. The state encoding is done implicitly. This work was limited to the PURE ESTEREL language, which is the synchronization kernel of ESTEREL where data handling is forbidden. Pure programs are translated into hardware and the circuits obtained are heavily optimized (Touati and Berry 1993).

The problem we solve here is the translation of the **full** language into hardware including data handling. We rely on the existing ESTEREL v4 compiler that transforms an ESTEREL program into a control part and a separate set of data handling actions. The control part is expressed as a set of sorted boolean equations that drive the actions. Optimizing the control becomes not trivial since one must respect sequentiality constraints between actions. The optimization step is done with the SIS program (Sentovich *et al.*) and hierarchical optimization is available. Once the control is optimized the final translation to hardware is done using the Stanford OLYMPUS (De Micheli *et al.*) synthesis system to recombine the control and the data path.

Section 2 presents the language and gives a flavor of the structural translation to hardware. Section 3 presents the control optimizations: control part extraction, specific optimizations and control-data recombination. The final hardware translation is presented in Section 4. Section 5 summarizes and offers directions for future development.

# 2 THE ESTEREL LANGUAGE

## 2.1 The language

ESTEREL has three main features: it assumes a perfect synchrony hypothesis, it permits communications through instantaneous broadcast signals, and it has a deterministic parallelism. Computations are triggered by input events. Sequencing, concurrency, looping and presence of signals perform instantaneous control transmission. Instantaneous preemption are done by `watching` and `trap` statements. Non instantaneous statements are various forms of explicit delay. The language and its semantics are presented in (Berry and Gonthier 1992).

Figure 1 shows a program that waits for the signals `A` and `B` in parallel. The program emits signal `O` once both signals are received and stops. This behavior is restarted at each occurrence of signal `R`.

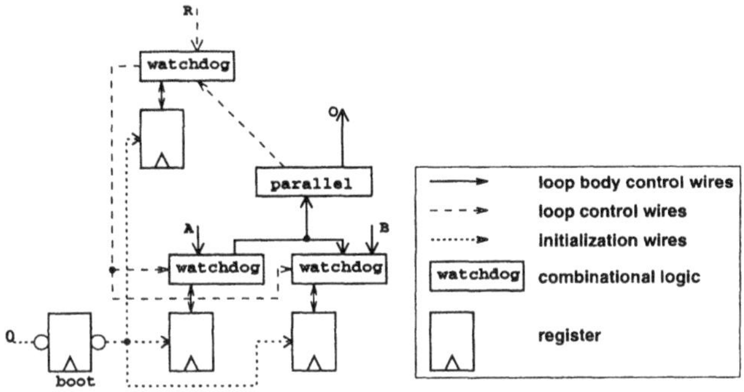


**Figure 2** Structural translation of the `loop` ... `each` program

## 2.2 The structural translation

The state of an ESTEREL program is fully determined by its **distributed program counter**, i.e the points where the program is waiting for an input event. The structural translation proposed by Berry implements each of these halt points by a register. For instance, the program given in Figure 1 contains three halt points. Figure 2 shows its structural translation in a simplified way. A register is set to "1" when the corresponding halt point is **active**. The `watchdog` boxes distribute control according to the presence of the associated signal and the associated register value. The difficulty is to generate the correct combinational logic that implements instantaneous constructs (Berry 1991). Notice also a fourth register, named `boot`, that is used for the initialization step. Its initial value is "1", and then "0" thereafter.

An ESTEREL program is compiled into an internal description composed of two parts. The first part describes the control of the program as a set of sorted boolean equations and registers. It is generated using the structural translation. The second part is the data path and describes actions such as tests of signals, calls to procedures and test evaluations. Given the current inputs, the control computes the next state and enables the appropriate data computations.

# 3 CONTROL OPTIMIZATION

Since the control is produced by direct structural translation, it has poor speed performance and is usually too large. But it may be easily reduced and improved. Control optimization is done in three steps. The first step is the separation of the control part and the data manipulation part. During this operation, dependencies between actions must be found and recorded. The second step is the optimization process done with the SIS program. The last step is the merging of the optimized control part with the data manipulation part.

## 3.1 Extraction step

The aim of the extraction step is to translate the control part into the input format of a sequential logic optimizer. The control part is already separated from the data manipulation part in our internal format (Section 2.2). As we said, it is expressed as a set of sorted boolean equations and registers, and drives the data manipulation part through input and output signals. Inputs correspond to the inputs of the program and outputs enable the execution of actions. Test instructions generate input/output pairs of signals. The control description is behavioral: equations are evaluated following the sorted order. In particular, sequentiality constraints between actions are implicitly coded by the equation order.

To optimize the control, we consider it as a sequential circuit. The main problem is that we do not want to include the sequentiality constraints into the description of this sequential circuit. There are two reasons. First, it is not possible to express directly these constraints using the optimizer format. Second, we could code them as micro-steps, but it would add more logic and registers and it would change the behavior of our control. Therefore, we keep track of the sequentiality constraints and we record them. This done by analysis of the control path and data dependencies. The control part is then translated into the SIS input format. It is described as a single global circuit, or as a hierarchy of sub-circuits. This decomposition will be detailed later.

Another circuit may be possibly produced at this stage. It is a combinational circuit that describes the execution environment of the program. One can specify explicit relations between inputs of an ESTEREL program in the source code. There relations give signals that are mutually exclusive, and signals that imply other signals. The combinational circuit represents these relations and is used by the optimizer to define sequential don't care conditions.

## 3.2 Optimization step

The ESTEREL compiler generates circuits with some particularities:

- redundant registers,
- interdependent registers,
- equivalent nets.

Two nets are said equivalent if they express the same function with respect to reachable states and external don't care conditions. The following example illustrates the generation of redundant registers:

```
do
     every LowerLeft do emit StopwatchStartStop end every
||
     every Upperleft do emit StopwatchLap end every
watching LowerRight % exit stopwatch mode
```

This example is extracted from the wristwatch example (Berry 1991). The loop instruction `every S do` *stat* `end every` terminates and starts *stat* afresh whenever `S` occurs. The structural translation generates one register per `every` instruction. The two registers produced are always equal since the two instructions are always active together. Thus, they are redundant and one can suppress one of them. Furthermore, there is a third register which is the boot register. The example actually has only two states: the inactive state when the program is not yet started and one active state. The optimized register can be derived from the boot register by a simple combinational logic circuit, so finally only one register is needed.

**Table 1** Hardware optimization results

| Circuit | Standard | | | Optimized | | | num. of states |
|---|---|---|---|---|---|---|---|
| | lit(fac) | regs | levels | lit(fac) | regs | levels | |
| abc | 227 | 13 | 9 | 157 | 6 | 3 | 9 |
| arbiter12 | 959 | 50 | 52 | 224 | 12 | 3 | 14 |
| controller | 578 | 67 | 10 | 51 | 34 | 2 | 372 |
| lights | 153 | 8 | 7 | 86 | 8 | 2 | 9 |
| game | 184 | 7 | 17 | 185 | 5 | 2 | 7 |
| tcint | 1032 | 82 | 62 | 491 | 62 | 3 | 287 |
| vme_control | 116 | 14 | 5 | 26 | 7 | 1 | 44 |
| watch | 775 | 35 | 25 | 312 | 10 | 3 | 42 |
| Gain | 2.71 | 1.97 | 7.13 | | | | |

It is not necessary to rewrite programs to generate less registers because the optimizations used automatically remove extra registers. They are based on the algorithms described in (Touati and Berry 1993). These algorithms exploit the state reachability of a circuit and are able to drastically reduce the number of registers while maintaining proper behavior.

Table 1 shows the results obtained with our standard benchmarks. *lit(fac)* is the number of literals in the factored form. *Gain* is the geometric mean of *Standard* over *optimized*. These examples show that our optimizations are well-suited to ESTEREL circuits, as shown by the *Gain* row.

These circuits have been improved using SIS. Practically, specialized scripts that group SIS commands have been developed to do specific optimizations, such as redundancy removal or depth reduction. These scripts have been then combined to form a general script that gives those results.

It should be noted that the control of an ESTEREL program is technology-independent at this level of abstraction. Therefore, we have developed other scripts to optimize the control for software implementation. The control equations are directly written in C logical expressions. The goal is to reduce the size of the object code and to improve the execution speed. This is done by simplifying the logic and factorizing the expressions as much as possible. Our experiments have shown that optimized code is 40% smaller and 50% faster than unoptimized code.

The decomposition into sub-circuits (Section 3.1) allows the optimization of larger circuits using **hierarchical optimization** (Fornari 1995). The decomposition is done during the extraction step. It exploits the module call tree of an ESTEREL program. A sub-circuit contains the part of the global control corresponding to a module call. The decomposition is used to do specific optimizations on particular sub-circuits. It may also help for circuits that are too large for optimizations based on state reachability. Strong optimizations are done for the leaves of the sub-circuit hierarchy, while lighter optimizations are applied to the whole hierarchy.

### 3.3 Insertion step

The insertion step recombines the optimized control and the data manipulation. It is the opposite problem of the extraction step. Here we start from a structural description and we end with a behavioral one. This conversion is done by a topological sorting. The difficulty comes from the fact that the optimizer has no knowledge about the sequentiality constraints. We have shown that the new structure of the circuit after optimization is always compatible with the sequentiality constraints (Fornari 1995). Thus, topological sorting is always possible. The final conclusion is that we can always optimize ESTEREL program control using the extraction-insertion technique.

## 4 HARDWARE GENERATION

ESTEREL programs are of two kinds: programs that only use the synchronization kernel (also called PURE ESTEREL) and programs that also use the data flow instructions. PURE ESTEREL programs can be directly translated into gates. This is possible because the control is the circuit, since it only uses pure signals and do not manipulate data. Full ESTEREL programs require the implementation of the implicit sequentiality between actions.

### 4.1 Pure Esterel Case

As PURE ESTEREL programs are directly translated into gates, optimizations are performed on the final circuits. We avoid the extraction-insertion steps, but we still use the same scripts developed for control optimization in case of hardware implementation. Experimental results (Touati and Berry) show that the quality of the circuits is close to hand-crafted circuits.

PURE ESTEREL allows the use of static counters to count the occurrences of a signal or the number of iterations in a bounded loop. Programs with such counters are also translated into circuits. The final circuit is then composed of the control part and several fast counters. Finally, all circuits generated from PURE ESTEREL programs execute an ESTEREL transition within one clock cycle.

### 4.2 Full Esterel Case

We consider our description of a circuit as a behavioral description. So, we need a synthesis system that it is able to handle this level of abstraction. For our experiments we used the OLYMPUS synthesis system (De Micheli *et al.*) developed by De Micheli, Ku, Mailhot and Truong at Stanford University. It is a set of tools for ASIC synthesis from the high-level hardware description language HARDWAREC. We use the fact that the language allows the user to express several forms of parallelism and that the system generates an optimal control.

Thus, we translate an ESTEREL circuit into a HARDWAREC program. The user must provide the HARDWAREC implementation of the functions and procedures that were declared external in the ESTEREL program. The circuit is then synthesized in a technology-independent fashion: operations on data take an unknown number of clock cycles to execute. To ensure the correspondence between an ESTEREL transition and an execution of the circuit, we define a synchronization signal to start a transition. The circuit waits until this signal is set to "1" before it starts proceeding any other input event. The synchronization signal need not to be maintained during the computation. A special output signal synthesized by the OLYMPUS system defines the end of the transition. Using these techniques, we have implemented a hardware version of the Reflex Game (Boussinot 1991) for a Xilinx XC3090 circuit. The design was down-loaded onto the PeRLe-1 board (Bertin *et al.* 1989).

PURE ESTEREL programs may be translated into circuits in the same fashion. There is no need for synchronization signals in this case. We still guarantee that an ESTEREL transition is performed within one clock cycle.

## 5 CONCLUSION

ESTEREL programs can be translated into hardware, using structural translation following by specific optimizations. These optimization techniques have been successfully extended to software compilation of ESTEREL programs. The ability to produce either software or hardware allows verification of a design before it is fabricated. One can use the full ESTEREL environment to do simulation, analysis and optimizations.

We now produce output in BLIF (Berkeley Logical Interchange Format), Lustre (Halbwachs *et al.* 1991) and VHDL formats. We are currently working on the improvement of the optimizations in terms of efficiency and speed.

## REFERENCES

Berry, G. (1991) Programming a digital watch in Esterel v3_2. Rapport de recherche 08/91, Centre de Mathématiques Appliquées, Ecole des Mines de Paris, Sophia-Antipolis.

Berry, G. (1992) A hardware implementation of pure Esterel. *Sadhana, Academy Proceedings in Engineering Sciences, Indian Academy of Sciences*, 17(1):95–130. Rapport Centre de Mathématiques Appliquées de l'Ecole des Mines de Paris, numéro 06/91.

Berry, G. and Gonthier, G. (1992) The Esterel synchronous programming language: Design, semantics, implementation. *Science Of Computer Programming*, 19(2):87–152.

Bertin, P. Roncin, D. and Vuillemin J. 1989 Introduction to programmable active Memories. In E. Swartzlander Jr, J. McCanny, J. McWhirter, editor, *Systolic Array Processors*, 300–9. Prenctice Hall.

Boussinot (1991) Programming a reflex game in Esterel v3_2. Rapport de recherche 07/91, Centre de Mathématiques Appliquées, Ecole des Mines de Paris, Sophia-Antipolis.

Fornari, F-X. (1995) *Optimisation du contrôle et implantation en circuits de programmes* ESTEREL. Thèse d'informatique, Ecole des Mines de Paris.

Halbwachs, N., Caspi, P. and Pilaud, D. (1991) The synchronous dataflow program ming language lustre. *Another Look at Real Time Programming, Proceedings of the IEEE, Special Issue*.

De Micheli, Ku, Mailhot, and Truong The Olympus Synthesis System for Digital Design.

Sentovich, E. M. *et al.* (1992) SIS: a system for sequential circuit synthesis. memorandum UCB/ERL M92/41, University of California, Berkeley.

Touati, H. and Berry, G. (1993) Optimized controller synthesis using esterel. In *Proc. International Workshop on Logic Synthesis IWLS'93*, Lake Tahoe.

19

# SEC state assignment selection: consequences on the area and reliability of fault-tolerant controllers

*R. Rochet, R. Leveugle*
*Institut National Polytechnique de Grenoble / CSI*
*46, Avenue Félix Viallet - 38031 Grenoble Cedex - FRANCE*
*Phone: (33) 76.57.46.86   Fax: (33) 76.50.34.21*
*E-mail: Raphael.Rochet@imag.fr*

**Abstract**

This paper addresses the synthesis of fault-tolerant controller architectures based on the use of Single Error Correcting codes. The impact of two state assignment procedures, implemented in a specific synthesis tool (ASYL-SdF), is studied in terms of area and reliability. The first procedure uses a specific unseparable SEC code, while the second is based on the use of the Hamming code. This second state assignment often gives better results in terms of area and dependability, but the first state assignment remains more interesting for small controllers.

**Keywords**

State assignment, Single Error Correcting codes, synthesis efficiency, area, dependability

## 1 INTRODUCTION

When integrated circuits are dedicated to critical applications, it is necessary to provide fault-tolerance capabilities in the major circuit blocks, in particular in the Finite State Machines (FSMs) which are implemented to control the operations performed by the circuit. For these applications, fault tolerance must often avoid any break in the service delivery. In that case, a classical solution is the Triple Modular Redundancy approach (TMR: triplication and voting). An alternative approach, initially proposed by Armstrong (1961), provides fault-tolerance in the sequencing part (next-state logic and state register) by using error correcting codes when assigning binary codes to the FSM states. This approach, studied by several authors, was expected to lead to lower hardware overheads and higher reliability than TMR (Armstrong 1961, Meyer 1971, ...). However, few practical evidences were available.

Applying this approach in the case of integrated circuits and providing the associated CAD tools was considered only recently, with the synthesis of the Single Independent Decoder (SID)

architecture (Leveugle 1993 - FTCS). This first version of the ASYL-SdF tool used a state assignment procedure based on a specific Single Error Correcting (SEC) code, presented in (Leveugle 1993 - DAC). The analysis of the results showed that the induced area overheads were essentially due to the decoding logic (Rochet 1993). A new state assignment and synthesis flow, aiming at decreasing the size of the decoding logic, was thus developed. After a brief presentation of the basic architecture in section 2, the two state assignment procedures are presented in section 3, and the results are discussed in section 4 in terms of silicon area and dependability properties.

# 2 SID ARCHITECTURE

## 2.1 Preliminaries

A finite state machine is defined as a quintuple (I,O,S,δ,ω) where I, O and S are respectively the sets of inputs, outputs and states. $\delta: I \times S \rightarrow S$ is the state transition function. $\omega: S \rightarrow O$ or $\omega: I \times S \rightarrow O$ (Moore or Mealy-type machine) is the output function. Generally, these machines are described by their state transition graph and implemented using a register to store the current state code, and combinational logic for the next-state and output computation (Figure 1). Such an implementation without fault-tolerant capabilities is named "simplex".

The classical synthesis of a FSM involves three main phases: state assignment, Boolean equation generation, according to the binary state codes and to the δ and ω functions, and logic structure generation, including logic minimization, factorization and technology mapping. The state assignment is often performed using the minimal number of bits, which will be denoted by k ($k = \lceil \log_2(|S|) \rceil$ where |S| is the number of states in the FSM).

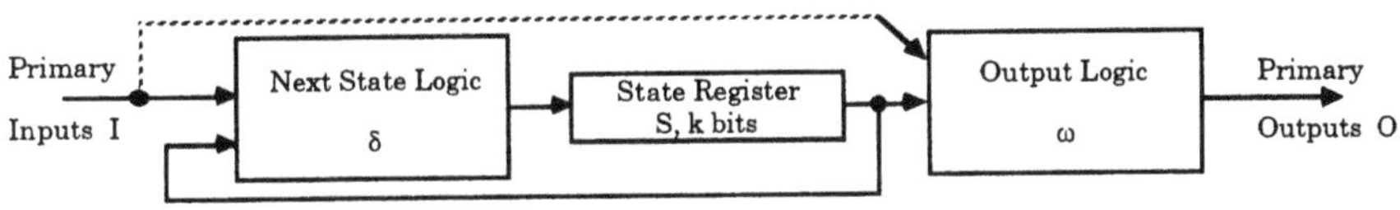


**Figure 1** Simplex Finite State Machine.

## 2.2 SID architecture

The SID architecture is based on the concept presented by Armstrong (1961).With this concept, to achieve single fault tolerance in the sequencing logic (next-state logic and state register), a SEC code with a minimum Hamming distance equal to 3 must be used during the state assignment. Then, to guarantee the tolerance of a fault in the next-state logic, this logic is minimized locally and the factorization and mapping phases avoid any logic shared between two next-state functions. It can be shown that this is not really costly for this architecture, especially from the area point of view (Rochet 1993). Finally, the fault tolerance is achieved by an independent logic block connected on the outputs of the state register (Figure 2), so that an erroneous state code is corrected before it is used to compute the next-state and output functions. It must be noticed that a fault in this decoder block is not necessarily tolerated since it may directly induce errors on the outputs.

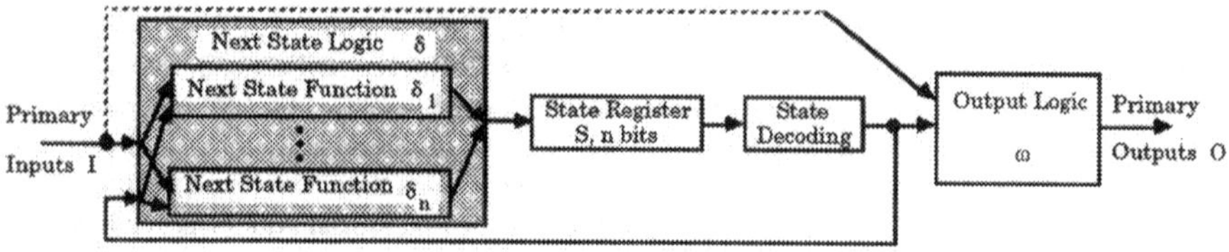


**Figure 2** SID Architecture (Single Independent Decoder).

# 3 STATE ASSIGNMENT PROCEDURES

## 3.1 Classical and distance 3 state assignments

In classical optimized state assignment procedures two steps are clearly identified. The first one is the recognition, in the controller specification, of situations which will lead to minimization ; to each identified situation is associated an "adjacency constraint" on a group of states. The second one is the embedding in the hypercube, or binary code assignment, trying to satisfy as many adjacency constraints as possible (distance minimization).

A distance 3 state assignment begins with the same first step. The second step can be performed following two different approaches. The first approach, named "Global state assignment", is similar to the classical optimized assignment and tries to satisfy adjacency constraints for all state code bits, taking into account the minimum distance 3 constraint. The second approach, named "Hamming state assignment", consists in adding check bits to the optimized distance 1 state code, and then optimizes only k out of the n next-state functions.

## 3.2 Global state assignment and related synthesis flow

When considering distance 3 state assignment, the adjacency constraints cannot be handled in the same way as in the classical state assignments, which prevents us to take advantage of the existing algorithms. However, the basic principle of distance minimization can be extended to the synthesis of single fault-tolerant FSMs (Leveugle 1993 - DAC). This led to the definition of a procedure based on a specific SEC code with the following properties:

P1: The minimal distance between two code words is 3.

P2: The total number n of bits necessary for any code word is the minimal known value for a SEC code, with respect to the total number of generated code words.

P3: Two successive code words are at distance 3.

P4: The sum of distances between 4 or 8 successive code words is minimum if the last word is, in the code sequence, in a position corresponding to a multiple of 4 or 8 respectively.

The embedding phase of the Global state assignment first orders the state list according to the adjacency constraints and to the distance properties of the code list, so that the distances between the codes of the states in a given adjacency constraint are minimized. Then, the *i*th binary word of the SEC code is assigned to the *i*th state of the ordered state list. In order to use the properties P3 and P4 at best, in a first pass, the Global state assignment searches for the constraint inclusions (inclusion of constraints is relatively easy to process by means of permutations). The result is a reduced list of constraints within which the states are partially ordered. The relative position of the different constraints, in the final ordered list, is then determined in a second pass according to the intersections between constraints.

For each assigned state code, the list of codes at distance 1 is built. These codes correspond to the erroneous state codes to tolerate and the decoding equations are generated so that each of these codes is mapped to the correct state code on the output of the decoding logic.

Then, the output and next-state equations are computed. In order to avoid shared logic among next-state functions, an independent local minimization, factorization and technology mapping is performed for each function. As faults are not necessarily tolerated in the output and decoding logic, each of these blocks can be globally minimized and implemented with shared logic.

As the code generated by the global state assignment procedure is not separable, the decoding block is generated with n inputs and n outputs (n = k+r, r being the number of redundant bits). With a separable code, only k outputs would be necessary for the decoder block (see section 3.3). Due to the minimization simplifications, the next-state and output equations generally use only a subset of the current state code bits to compute the next-state code and primary outputs. So a procedure is implemented to eliminate the unnecessary decoding functions. This allows us to optimize the decoding logic, but the resulting number of functions is often greater than k.

## 3.3 Hamming state assignment and related synthesis flow

The analysis of the results obtained with the Global state assignment demonstrated that the most critical block, in terms of implementation penalties, was the decoding logic (Rochet 1993). The second state assignment procedure we implemented aims therefore at simplifying this logic, potentially at the cost of a complexity increase of the other blocks. A Hamming SEC code was selected for the state assignment. This code is known to lead to small decoder implementations. Furthermore, using this separable code, the decoder structure can be precisely controlled during the synthesis, and the minimum number k of outputs can be easily guaranteed.

Once the decoding logic has been optimized, it is then necessary to avoid a too large overhead in the next-state and output logic. To reach this goal, the state assignment is performed in two phases. First, a distance 1 state assignment is performed with the same optimization procedure as used for the simplex implementation, e.g. an adjacency constraint recognition followed by an optimized embedding (Saucier 1990). More optimization constraints can be satisfied during this distance 1 state assignment, compared with a global distance 3 state assignment. This allows us to achieve a better optimization of the output functions and of k out of the n next-state functions, compared to the global state assignment. Secondly, using the generator Hamming code matrix (Hamming 1950), redundant code bits are added to the words of the compact code in order to obtain the distance 3 Hamming SEC code.

With this approach, the functions related to the check bits are therefore not taken into account during the state assignment optimization. However, after state assignment, decoding equations can be generated using Hamming code properties. This leads to very small decoders, with mainly two stages. The first one computes a s-bit syndrome ($s = \lceil \log_2(n) \rceil$). Each syndrome bit is the result of an Exclusive-OR of some state code bits. In the second stage, the syndrome value indicates which state code bit is erroneous, i.e. the bit to inverse. As only information bits are corrected, the number of outputs of the decoder block is always k.

After the decoder generation, the output and next-state equations are computed, minimized and factorized as described in the previous section. The n next-state equations compute the next-state distance 1 code plus the r redundant bits according to the current corrected state code (k-bit wide). The output equations compute the primary outputs according to the corrected state code.

# 4 DISCUSSION OF IMPLEMENTATION RESULTS

About 100 FSM benchmarks have been implemented using ASYL-SdF, including examples with 4 to 257 states, and 10 to 3598 transitions in the initial specification. These examples have been implemented using a CMOS 1.2μ standard cell library from VLSI Technology (vsc370) and with the area-oriented logic synthesis optimizations in ASYL. The areas after placement and routing were obtained using the COMPASS tools.

## 4.1 Area Results

Due to the Hamming code properties, the decoder block is very often smaller when the SID architecture is synthesized using the Hamming state assignment procedure rather than the Global one. This is illustrated in Figure 3, which shows the decoder size (for all implemented examples) according to the state number of the FSM and the state assignment procedure used. A very interesting characteristic of the Hamming decoding logic is to be almost not sensitive to the state number. So, even if the smallest decoders have similar sizes for the two state assignment procedures, the gain due to the Hamming assignment increases quickly with the number of states. This is explained by the regularity of the Hamming decoder structure, and the number of decoder outputs, which is guaranteed equal to k.

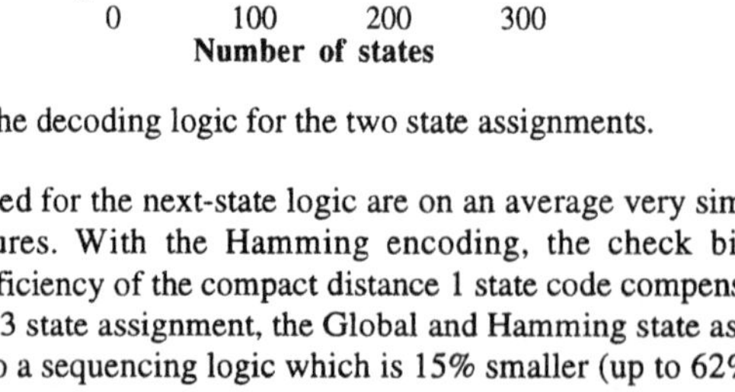


**Figure 3** Area of the decoding logic for the two state assignments.

The results obtained for the next-state logic are on an average very similar for the two state assignment procedures. With the Hamming encoding, the check bit equations are not optimized, but the efficiency of the compact distance 1 state code compensates. Compared with the random distance 3 state assignment, the Global and Hamming state assignment procedures lead on an average to a sequencing logic which is 15% smaller (up to 62% smaller for Cpt100 with the Global encoding). Results are given for some representative examples in Table 1.

Due to the optimized distance 1 code used for the k information bits, the output logic is very well optimized with the Hamming state assignment. So the average of the gain due to this state assignment procedure is greater than 20% compared with the random distance 3 encoding. The Global state assignment cannot satisfy adjacency constraints as well as the classical encoding, due to the minimum distance 3 constraint. So its average gain is only equal to 5%.

**Table 1** Sequencing logic area according to the state assignment procedure

| Name | Random (μm2) | Global gain | Hamming gain | Name | Random (μm2) | Global gain | Hamming gain |
|---|---|---|---|---|---|---|---|
| cpt100 | 974 804 | 62.9% | 29.3% | kirkman | 112 065 | 16.4% | 21.62% |
| dk17 | 89 175 | 15.2% | 18.1% | s1a | 774 752 | 25.3% | 27.1% |
| ex7 | 143 349 | 23.7% | 16.1% | sr8a-slave | 211 433 | -4.6% | -0.6% |
| imec1 | 1 981 778 | 43.3% | 36.3% | tav | 4 309 | -41.6% | -49.1% |
| keyb | 712 141 | 22.6% | 11.3% | zeegers | 2 052 919 | 9.36% | 13.6% |

In fact, even when the Global state assignment gives good results for the sequencing block, the FSM implemented with the Hamming state encoding is generally smaller, due to the output and decoding logic overheads. This is well illustrated by the ex7, imec1 and keyb examples in Table 2. So the average gain of the Hamming state assignment compared with the Global one is finally 25%. Actually, the Global state assignment remains interesting for few particular examples such as cpt100 and for the small FSMs where the Hamming decoder structure does not lead to the best implementation (Figure 3).

**Table 2** Gains of the Hamming state assignment compared with the Global state assignment

| Name | Whole logic | Seq. logic | Out. logic | Dec. logic | Name | Whole logic | Seq. logic | Out. logic | Dec. logic |
|---|---|---|---|---|---|---|---|---|---|
| cpt100 | -5.7% | -90.6% | 30.54% | 78.0% | kirkman | 40.9% | 6.3% | 31.5% | 71.4% |
| ex7 | 25.5% | -10.1% | -25.1% | 68.5% | s1a | 22.7% | 2.4% | N.R. | 71.4% |
| dk17 | 10.0% | 3.3% | 13.9% | 6.4% | sr8a-slave | 27.0% | 3.8% | 2.1% | 78.3% |
| imec1 | 31.0% | -12.3% | 44.15% | 82.7% | tav | 3.43% | -5.3% | 11.0% | 21.6% |
| keyb | 8.45% | -14.6% | 22.1% | 72.0% | zeegers | 17.7% | 4.7% | 13.8% | 77.9% |

## 4.2 Dependability results

Since the FSMs are part of an integrated circuit, they are non-repairable systems and their reliability can essentially be characterized by the Mean Time to First Failure (MTFF), computed starting from the initial fault-free state, and the Mission Time (MT), representing the expected operation time for a given reliability level.

**Table 3** Mission Time and MTFF (both $10^3$ hours) results.

| Name | Global | | Hamming | | Name | Global | | Hamming | |
|---|---|---|---|---|---|---|---|---|---|
| | MT | MTFF | MT | MTFF | | MT | MTFF | MT | MTFF |
| cpt100 | 24 620 | 86 021 | 37 815 | 98 609 | kirkman | 27 192 | 111 061 | 55 048 | 203 736 |
| dk16 | 27 215 | 86 611 | 52 596 | 149 143 | log | 35 940 | 131 967 | 68 243 | 208 062 |
| imec10 | 3 766 | 15 696 | 5 834 | 23 642 | scf | 9 493 | 30 718 | 16 127 | 45 738 |
| jay | 13 711 | 46 356 | 33 949 | 97 537 | zeegers | 5 411 | 18 870 | 7 158 | 23 621 |

MT and MTFF have been obtained with a Markov model, using the MIL-HDBK-217F procedure to evaluate the expected failure rates (US Defense 1991) and the tool SURF-2, from LAAS-CNRS (Beounes 1993) to solve the model. The results in Table 3 illustrate the impact of the state assignment on the Mission Time (computed for a reliability equal to 0.8) and on the MTFF. These results show the interest to optimize the logic of the critical blocks of the SID architecture (i.e. the output and the decoder logic).

## 5 CONCLUSION

The synthesis efficiency of fault-tolerant FSMs based on SEC codes is very dependent on the efficiency of the state assignment procedure. Two state assignment procedures have been described and compared. The results show that the state assignment based on the Hamming code leads in general to a noticeable area improvement compared to the Global optimization procedure, which is however a good solution for small FSMs. In terms of reliability, the Hamming state assignment also proved its efficiency.

## ACKNOWLEDGEMENTS

This work has been partially supported by the European JESSI AC8 project.

## REFERENCES

Armstrong D. B., 1961, "A general method of applying error correction to synchronous digital systems", The Bell System Technical Journal, vol. 40, no. 2, March, pp. 577-593.

Beounes C. et al., 1993, "SURF-2: a program for dependability evaluation of complex hardware and software systems", 23rd FTCS, pp. 668-673.

Hamming R. W., 1950, "Error Detecting and Correcting Codes", The Bell System Technical Journal, April, pp. 147-160.

Leveugle R., 1993, "Optimized state assignment of single fault tolerant FSMs based on SEC codes", 30th DAC, pp. 14-18.

Leveugle R., Rochet R., Saucier G., Martinez L., Pitot C., 1993, "A synthesis tool for fault-tolerant finite state machines", 23rd FTCS, pp. 502-511.

Meyer J. F., 1971, "Fault-tolerant sequential machines", IEEE trans. on Comp., vol. C-20, no. 10, October, pp. 1167-1177.

Rochet R., Leveugle R., Saucier G., 1993, "Analysis and comparison of fault tolerant FSM architectures based on SEC codes", "The IEEE International Workshop on Defect and Fault Tolerance in VLSI Systems, Venice, Italy, October 27-29, 1993", F. Lombardi, M. Sami, Y. Savaria, R. Stefanelli editors, IEEE Computer Society Press, Los Alamitos, California, 1993, pp. 9-16.

Saucier G., Duff C., Poirot F., 1990, "State assignment of controllers for optimal area implementation", EDAC, pp. 547-551.

United States Department of Defense, 1991, Military Standardization Handbook: Reliability Prediction of Electronic Equipment, MIL-HDBK-217F.

# 20

# On Multi-Cycle False Paths in Sequential Circuits

*Zafar Hasan and Maciej J. Ciesielski*
*University of Massachusetts*
*Department of Electrical & Computer Engineering*
*Amherst, MA 01003, USA*
*Telephone: (413)-545-0401. Fax: (413)545-0401.*
*email: ciesiel@ecs.umass.edu*

**Abstract**

In this paper we analyze the problem of multi-cycle false paths in sequential circuits and their impact on circuit performance. We present a study on the identification and removal of multi-cycle false paths from sequential machines specified at the functional and behavioral levels and show that multi-cycle false paths are not a property of the gate-level implementation but a property of the behavior of the machine and its state encoding. Based on behavioral level analysis, we derive the necessary and sufficient condition for the encoding of an FSM to obtain a false path free implementation and present a systematic encoding methodology to obtain a sequential false-path-free circuit. The experimental results show significant clock cycle reduction in sequential circuits encoded with this method.

**Keywords**

Sequential circuits, FSM, false paths, encoding, state assignment.

## 1 INTRODUCTION

The problem of unsensitizable, or false, paths in logic circuits has been studied extensively in recent years, e.g. (McGeer, 1991), (Devadas, 1991), (Chen, 1992), (Saldanha, 1992). False path in the circuit is a path that is never exercised during the operation of the circuit due to the circuit functionality and/or delay values of the circuit components. We shall differentiate between false paths in combinational circuits and false paths in sequential circuits. False paths in *combinational circuits* are due to the particular implementation or structure of the circuit. In this case the path is physically present in the structure but it cannot be sensitized by an input vector. The presence of false paths obstructs the analysis of circuit performance, since the actual delay in the combinational circuit is determined by its critical path, i.e., longest true path in the circuit.

False path in a *sequential circuit* have a different nature: it is a *multi-cycle path*, obtained by unrolling the sequential circuit over a number of clock cycles, as shown in Fig. 3. A sequential circuit may contain a *multi-cycle false path* even if there are no (single-cycle) false paths in the combinational logic block of the circuit. This is due to uncontrollability of the state registers between the consecutive time frames. We shal also refer to multi-cycle false paths as $MFP's$ or *sequential false paths*.

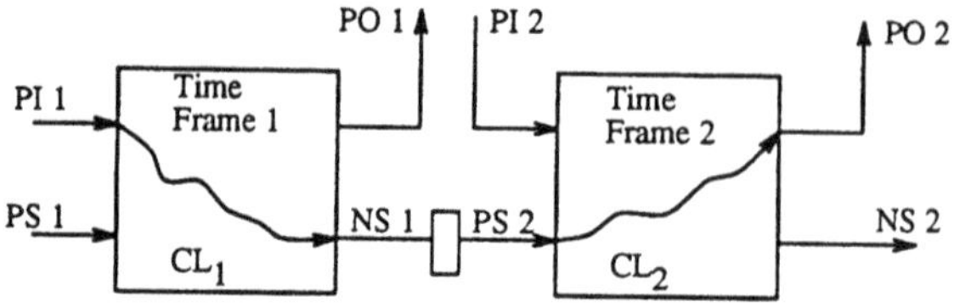


**Figure 1** Unrolling a sequential circuit over 2 clock cycles

Sequential false paths have also been investigated in literature to some extent, but their removal from a circuit were only considered in the context of restructuring the circuits on a gate level (Ashar, 1992), (Lam, 1994). Even though the presence of sequential false paths does not obstruct the estimation of the circuit delay, it has been demonstrated that the these paths may significantly affect the circuit performance. Removing multi-cycle false paths from the sequential circuit often results in circuit restructuring that may enable additional retiming and in effect can lead to performance improvement (Ashar, 1992). The presence of sequential false paths also reduces the testability of sequential circuits. It is therefore desirable to identify and remove multi-cycle false paths from sequential circuits to help improve their performance and testability.

We have demonstrated in our earlier work (Hasan, 1993) that the presence of MFPs in a sequential circuit can be readily detected on *behavioral* and *functional* levels, where only the behavior of the machine (e.g., state transition table) or its logic is available. In this paper we review these observations and present the encoding of Finite State Machines (FSMs) to eliminate the MFPs. In particular, we present the necessary and sufficient conditions for the encoding of an FSM such that it does not contain any multi-cycle false paths.

## 2 PRELIMINARIES

In this section we briefly define some of the basic terms used throughout the paper.

A sequential circuit is composed of simple gates, latches/registers and leads with the primary inputs and primary outputs defining its boundary. Let $\pi = (f_0, g_1, f_1, \ldots, f_{m-1}, g_m$ be a path. Lead $f_i$, $0 \leq i \leq m-1$, is defined as an **on-input** of $\pi$. Other leads which are inputs to gate $g_{i+1}$ are defined as **side inputs** of $f_i$. A **controlling value** at an input to a gate is the value that determines the output value of the gate independent of the other inputs. The **non-controlling value** to a gate is the complementary value of the controlling value. An **input vector** $v$ is a vector of logic values 0 or 1 at all the primary inputs. A path is **statically sensitizable** if there exists an input vector which sets all side inputs to the path to non-controlling values. A path is considered to be **statically false path** if it is not statically sensitizable. In this paper we only consider statically false paths.

A **variable** is a symbol representing a single coordinate of the Boolean space (e.g., $\overline{a}$). A **literal** is a variable or its negation (e.g., $a$ or $\overline{a}$). A **cube** is a set $C$ of literals such that $x \in C$ implies $\overline{x} \notin C$. A cube represents the conjunction of its literals. An **expression** is a set of cubes. For example, $\{a,\ b\overline{c}\}$ is an expression consisting of two cubes: $a$ and

$b\bar{c}$. An expression represents the disjunction of its cubes. The **support**, $sup(f)$, of an expression $f$ is defined as follows: $sup(f) = \{x \mid cube\ C \in f \implies x \in C\ or\ \bar{x} \in C\}$. For example, $sup(ab + \bar{a}\bar{c}) = \{a, b, c\}$. Two expressions $f$ and $g$ have **disjoint support** if $sup(f) \cap sup(g) = \emptyset$.

Given a binary encoding $\{z_{n-1} \ldots z_0\}$ of the states of a machine, each bit of the code induces a **2-block partition** of the states, such that one block contains symbols (states) with code 1, and the other block contains symbols with code 0. This 2-block partition is referred to as a **dichotomy**, and denoted as $D_i = (l_i : r_i)$ (Yang, 1991). The dichotomy $(l_i : r_i)$ implies that for all the symbols $z_i$ in $l_i$, $z_i = 1$ (or 0), while for all the symbols $z_i$ in $r_i$, $z_i = 0$ (or 1). That is, each dichotomy defines a one-bit encoding, associated with bit $z_i$, that allows to distinguish between the symbols in $l_i$ from those in $r_i$.

# 3 IDENTIFICATION OF MULTI-CYCLE FALSE PATHS

In this section we shall demonstrate that multi-cycle false paths are a property of the sequential behavior or function of the circuit and not a property of the particular structure (gate level implementation). This is in contrast with the false paths observed in combinational circuits. Let us briefly review the following problems addressed in our earlier work (Hasan, 1993):

**P1:** *Given a sequential circuit on functional level, represented by its state transition information and state encoding, identify and remove multi-cycle false paths.*

**P2:** *Given a sequential circuit on behavioral level, represented by its state transition information without the state encoding, identify all multi-cycle false paths.*

We shall explain the method of identifying multi-cycle false paths using an example sequential machine shown in Figure 2. The behavior of the machine is defined by its state transition graph (STG), Figure 2(a). The encoding, shown in Figure 2(b), together with the STG defines the circuit on the functional level.

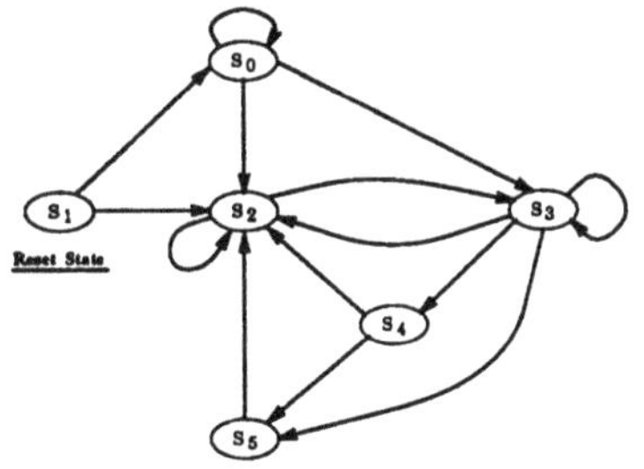


(a)

| State | Code | | |
|---|---|---|---|
| | $r_1$ | $r_2$ | $r_3$ |
| $S_0$ | 1 | 0 | 0 |
| $S_1$ | 0 | 0 | 0 |
| $S_2$ | 1 | 1 | 0 |
| $S_3$ | 1 | 1 | 1 |
| $S_4$ | 0 | 1 | 1 |
| $S_5$ | 0 | 1 | 0 |

(b)

**Figure 2** STG with an encoding

Figure 3 represents the gate-level implementation of the machine with this state encoding. This gate-level circuit was shown to have multi-cycle false paths terminating on output $o_2$ (Ashar, 1992). If restructured to remove the MFPs, it results in a circuit shown in Figure 4(b) (Ashar, 1992), to be analyzed later. Here we show how to obtain a false-path free (FPF) circuit *directly* from high-level (behavioral) description.

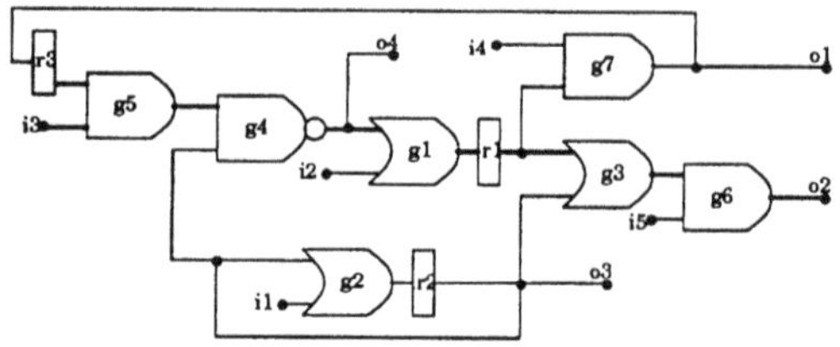


**Figure 3** Circuit with multi-cycle false path: $(i_3, g_5, g_4, g_1, r_1, g_3, g_6, o_2)$

## 3.1 False Paths on Functional Level

The presence of static multi-cycle false paths on the functional level can be detected by analyzing the *functional* output and next state logic equations of the circuit. The equations can be derived in terms of the Timed Boolean variables (Lam, 1994), representing primary inputs $i_i[t]$ and the state variables $r_i[t-1]$ as a function of the time frame $[t]$. For example, output $o_i[t]$ expressed in terms of state variable $r_j[t-k]$ gives the output expression for $o_i$ in $k$ clock cycles. The following are the *one-cycle* next state and output equations for the example machine in Figure 2.

$$o_1[t] = i_4[t] r_1[t-1]; \quad (4)$$

$$r_1[t] = i_2[t] + \overline{i_3[t]\, r_2[t-1]\, r_3[t-1]}; \quad (1) \qquad o_2[t] = i_5[t](r_1[t-1] + r_2[t-1]); \quad (5)$$

$$r_2[t] = i_1[t] + r_2[t-1]; \quad (2) \qquad o_3[t] = r_2[t-1]; \quad (6)$$

$$r_3[t] = i_4[t] r_1[t-1]; \quad (3) \qquad o_4[t] = \overline{i_3[t]\, r_2[t-1]\, r_3[t-1]}; \quad (7)$$

(8)

Consider the Boolean expression for output $o_2$ and its logic dependence on the primary inputs. In one clock cycle $o_2[t]$ depends only on $i_5[t]$. It also depends on the register variables $r_1[t-1]$ and $r_2[t-1]$, which in turn depend on $i_1[t-1], i_2[t-1]$ and $i_3[t-1]$. Thus we can say that there are physical two-cycle paths $i_1 \rightarrow r_2 \rightarrow o_2$, $i_2 \rightarrow r_1 \rightarrow o_2$, and $i_3 \rightarrow r_1 \rightarrow o_2$. Now we will show that these paths are false.

Let us substitute the variables $r_1[t-1]$ and $r_2[t-1]$ in the expression for $o_2[t]$ by their corresponding one-cycle expressions (1 & 2) to obtain the *two-cycle expression* for $o_2[t]$.

$$o_2[t] = i_5[t]((i_2[t-1] + \overline{i_3[t-1]\, r_2[t-2]\, r_3[t-2]}) + (i_1[t-1] + r_2[t-2])) \quad (9)$$

Consider variable $i_3$, which appears in the two-cycle expression for $o_2$. Hence, there is a two-cycle physical path from $i_3$ to $o_2$. However, by applying De Morgan's Law, we obtain

$$o_2[t] = i_5[t]((i_2[t-1] + \overline{i_3[t-1]} + \overline{r_2[t-2]} + \overline{r_3[t-2]}) + (i_1[t-1] + r_2[t-2])) \quad (10)$$

Since $\overline{r_2[t-2]} + r_2[t-2] = 1$, we can simplify the above equation to $o_2[t] = i_5[t]$. Hence, output $o_2$ is logically independent of $i_3$. Clearly such a path cannot be sensitized, since there is no assignment of side inputs (variables in the expression different than $i_3$) that will make the output sensitive to $i_3$; hence the two-cycle path ($i_3 \rightarrow o_2$) is declared false. In fact there are also two-cycle false paths from $i_1$ and $i_2$ to $o_2$, since these input variables also disappear from the two-cycle equation for $o_2$ after simplification.

We can now extend our approach to multi-cycle false paths over an arbitrary number of clock cycles.

**Definition 1** *Let $f^k(o_i)$ be a k-cycle expression for output $o_i[t]$ obtained by substituting the current inputs $i_i[t]$ and state variables $r_j[t-1]$ by their corresponding $(k-1)$ cycle expressions.*

**Definition 2** *Let $f^k_{min}(o_i)$ be a function logically equivalent to $f^k(o_i)$ but with fewer input variables. This function can be obtained by simplifying $f^k(o_i)$ using known boolean simplification methods We assume that both $f^k(o_i)$ and $f^k_{min}(o_i)$ do not contain any single-cycle false paths.*

**Lemma 1** *The path $x_j \rightarrow o_i$ is a k-cycle false path if and only if $x_j \in sup(f^k(o_i))$ and there exists a simplification $f^k_{min}(o_i)$ of $f^k(o_i)$ such that $x_j \notin sup(f^k_{min}(o_i))$. Such a simplification corresponds to an implementation of a circuit with the k-cycle false-paths being removed.*

*Proof.* **Sufficiency:** Let $x_j \in sup(f^k(o_i))$. This implies that there exists a physical path $x_j \rightarrow o_i$. Whether this path is also a logical path or not depends on the possible simplification of $f^k(o_i)$. If the resulting $f^k_{min}(o_i)$ can be made independent of $x_j$, i.e., if $x_j \notin sup(f^k_{min}(o_i))$, then there is no logical path from $x_j$ to $o_i$. That is, $o_i$ is independent of the variable $x_j$ and the path cannot be sensitized. Hence it is false.
**Necessary condition:** Let $x_j \rightarrow o_i$ be a false path. We need to show that this implies that $x_j \in sup(f^k(o_i))$, and there exists $f^k_{min}(o_i)$ such that $x_j \notin sup(f^k_{min}(o_i))$. We shall prove both by contradiction: 1) Assume that $x_j \notin sup(f^k(o_i))$. This means that there is no physical path from $x_j$ to $o_i$, in which case the false path is not defined ($x_j$ must be one of the on-inputs of the path). This brings us to a contradiction; for a false path to exist, $x_j$ must belong to $sup(f^k(o_i))$.
2) Assume that there is no simplified expression $f^k_{min}(o_i)$ which is free of $x_j$. The presence of $x_j$ in any of the expressions for $o_i$ indicates that there is always a logical path between $x_j$ and $o_i$. The variables appearing in these expressions constitute the side inputs that can be assigned respective noncontrolling values and used to sensitize the path. Hence the path is sensitizable. This contradicts our initial assumption that the path is false. Therefore, for a false path to exist there must exist a simplification such that $x_j \notin sup(f^k_{min}(o_i))$.
□

## 3.2 False Paths on Behavioral Level

Here we consider problem **P2** stated earlier. We show that a behavioral specification (STG, STT, state charts, etc) may contain multi-cycle false paths. Our goal is to identify these false paths.

Let us examine the STG of our example machine, Figure 2(a), without assuming any encoding. We can obtain the next state equations for the machine by representing the state transitions in the form of *symbolic expressions*, as shown below. The variables in these expressions are the input variables $i_i$, and the symbolic state variables $S_i$, expressed as a function of previous time frame. Clearly, these expressions are independent of the encoding and the implementation of the circuit.

$$
\begin{aligned}
S_0[t] &= \overline{i_1[t]}(S_0[t-1] + S_1[t-1]); && (11)\\
S_1[t] &= \textit{Initial state}; && (12)\\
S_2[t] &= i_1[t]\overline{i_4[t]}S_0[t-1] + i_1[t]S_1[t-1] + \overline{i_4[t]}S_2[t-1] + S_5[t-1] + (\overline{i_3[t]}\ \overline{i_4[t]} && (13)\\
&\quad + i_2[t]\overline{i_4[t]})S_3[t-1] + (\overline{i_3[t]} + i_2[t])S_4[t-1]; &&\\
S_3[t] &= i_1[t]i_4[t]S_0[t-1] + i_4[t]S_2[t-1] + (\overline{i_3[t]}i_4[t] + i_2[t]i_4[t])S_3[t-1]; && (14)\\
S_4[t] &= \overline{i_2[t]}i_3[t]i_4[t]S_3[t-1]; && (15)\\
S_5[t] &= \overline{i_2[t]}i_3[t]\overline{i_4[t]}S_3[t-1] + \overline{i_2[t]}i_3[t]S_4[t-1]; && (16)
\end{aligned}
$$

Similarly, one can obtain the symbolic output equations for each output $o_i$ of the machine.

$$
\begin{aligned}
o_1[t] &= i_4[t](S_0[t-1] + S_2[t-1] + S_3[t-1]); && (17)\\
o_2[t] &= i_5[t](S_0[t-1] + S_2[t-1] + S_3[t-1] + S_4[t-1] + S_5[t-1]); && (18)\\
o_3[t] &= (S_2[t-1] + S_3[t-1] + S_4[t-1] + S_5[t-1]); && (19)\\
o_4[t] &= \overline{i_3[t]} + (S_0[t-1] + S_1[t-1] + S_2[t-1] + S_5[t-1]); && (20)
\end{aligned}
$$

In general, a $k$-cycle ***symbolic output expression*** $h^k(o_i)$ is obtained by substituting the state symbols by their corresponding ($k$-1) cycle expressions. We similarly define $h^k_{min}(o_i)$ as the ***simplified k-cycle symbolic expression*** for $o_i$. A single time-frame substitution of the state equations in our example results in the 2-cycle output expressions $h^2(o_i)$ and $h^2_{min}(o_i)$.

As before, to determine if there is a multi-cycle false path terminating at a given output, we need to compare the support of the original symbolic output expressions, $h^k(o_i)$, with the symbolic expression after the simplification, $h^k_{min}(o_i)$. This time the absence of certain variables in the support after simplification does not necessarily indicate a false path, but a *potential false path.* Whether the path is false or not, depends on the final encoding of the machine states.

**Definition 3** *A* **potential false path (PFP)** *is a path which, under certain encoding, may appear as a false path in the circuit.*

**Lemma 2** *If $x_j \in sup(h^k(o_i))$ and there exists a simplification $h^k_{min}(o_i)$ such that $x_j \notin sup(h^k_{min}(o_i))$, then path $x_j \rightarrow o_i$ is a potential k-cycle false path.*

*Proof.* Let us assume one-hot encoding for the state variables of the machine. Replacing the state symbols by the the state variables, corresponding to the one-hot code, yields the functional equations similar to the symbolic equations. Then by lemma 3.1 each PFP is a false path. □

## 4 FALSE-PATH-FREE ENCODING

In this section we describe a method to eliminate the detected potential false paths by means of state encoding. Specifically, we consider the following problem.

**Table 1** Properties of paths on behavioral and functional level

| Type of Path | Behavioral Level | Functional Level |
|---|---|---|
| False Path | | $x_j \in sup(f^k(o_i)),\ x_j \notin sup(f^k_{min}(o_i))$ |
| Potential FP | $x_j \in sup(h^k(o_i)),\ x_j \notin sup(h^k_{min}(o_i))$ | |
| True Path | $x_j \in sup(h^k(o_i)), x_j \in sup(h^k_{min}(o_i))$ | $x_j \in sup(f^k(o_i)), x_j \in sup(f^k_{min}(o_i))$ |
| No Path | $x_j \notin sup(h^k(o_i)), x_j \notin sup(h^k_{min}(o_i))$ | $x_j \notin sup(f^k(o_i)), x_j \notin sup(f^k_{min}(o_i))$ |

**P3:** *Given a high-level specification of an FSM, find an encoding of the states which does not introduce false paths in the circuit.*

We refer to such an encoding as a *false-path free*, or *FPF*, encoding. We shall show that such an encoding always exists and that under certain conditions the corresponding path does not appear as a physical path in the final implementation; otherwise the path remains in the circuit as a false path. We shall also derive the encoding conditions under which PFP is removed from the circuit.

A potential false path that does not appear as a physical path in the final implementation is called a *no-path (N)*. Such a path has the trivial property that it exists neither in the original nor in the simplified functional output expression. On the other hand, a path $x_j \rightarrow o_i$ is a *true path (T)* if it exists both in the original and the simplified $k$-cycle functional output expression. Table 1 summarizes the properties of different types of paths in the context of our approach.

We formulate the FPF encoding problem as a constrained encoding, which can be solved using *dichotomy covering* technique, (Yang, 1991), (Ciesielski, 1991). We shall illustrate our approach with the machine example used earlier, Figure 2(a). It can be shown that the symbolic output equations of the circuit can be represented in the following sum of symbolic product form

$$o_j = \sum_j \prod_l i_l \sum_k S_k \tag{21}$$

Here $i_l$ is a primary input variable, and $\sum_k S_k$ is a conjunction of the state symbols. Each symbolic product term $P_j = \prod_l i_l \sum_k S_k$, induces a set of constraints, called *false-path dichotomies* or FPDs.

**Definition 4** *Consider output $o_j$ with a potential false path. Each sum term $\alpha_j = \sum_k S_k$ in the symbolic output expression for $o_j$ defines a dichotomy, $FD_j$, called* **false path dichotomy**. *$FD_j$ distinguishes the states in the sum term $\alpha_j$ from the remaining states of the machine: $FD_j = ((S_k \mid S_k \in \alpha_j) : (S_k \mid S_k \notin \alpha_j))\ \forall S_k \in S$.*

Consider a 1-cycle equation from our circuit example: $o_1[t] = i_4[t](S_0[t-1] + S_2[t-1] + S_3[t-1])$. This expression contains one symbolic product term, $P_1 = i_4(S_0 + S_2 + S_3)$, and thus contains one false-path dichotomy, $FD_1 = (S_0, S_2, S_3\ :\ S_1, S_4, S_5)$. We can generalize this to the equations for arbitrary number of time-frames and define the $k$-*cycle* false path dichotomies in a similar fashion. We refer to the set of dichotomies obtained from the $k$-cycle equations as the *dichotomy-set* $DS_i^k$.

The following 1-cycle dichotomies are obtained from the 1-cycle output expressions with potential false paths for our example machine. In this case each dichotomy-set $DS_j$ contains only one false-path dichotomy $FD_j$.

$$\begin{aligned} o_1[t] &= i_4[t](S_0[t-1] + S_2[t-1] + S_3[t-1]); \\ &\Longrightarrow DS_1^1 = \{(S_0, S_2, S_3 \ : \ S_1, S_4, S_5)\} \\ o_2[t] &= i_5[t](S_0[t-1] + S_2[t-1] + S_3[t-1] + S_4[t-1] + S_5[t-1]); \\ &\Longrightarrow DS_2^1 = \{(S_0, S_2, S_3, S_4, S_5 \ : \ S_1)\} \\ o_3[t] &= S_2[t-1] + S_3[t-1] + S_4[t-1] + S_5[t-1]; \\ &\Longrightarrow DS_3^1 = \{(S_2, S_3, S_4, S_5 \ : \ S_0, S_1)\} \\ o_4[t] &= \overline{i_3[t]} + (S_0[t-1] + S_1[t-1] + S_2[t-1] + S_5[t-1]); \\ &\Longrightarrow DS_4^1 = \{(S_0, S_1, S_2, S_5 \ : \ S_3, S_4)\} \end{aligned} \tag{22}$$

## 4.1 Necessary and Sufficient Condition

In general, if dichotomy $FD_j$ is satisfied by some bit $r_j$ of the state code, then each sum term $\alpha_j$ will reduce to a single state variable $r_j$ in its true or complemented form, and the expression for $o_i$ will become a simple sum of products expression, $\sum_{j \in DS_i} r_j \prod_l i_l$. By our initial assumption the 1-cycle expressions for state variables $r_j$ are irredundant, i.e., do not contain one-cycle false paths. Therefore, the substitution of the state variables to obtain next time-frame expression will never eliminate any of the inputs from the associated product $\prod_l i_l$, and thus will not introduce redundancies, hence false paths.

Let us illustrate this sufficiency condition by examining the output equation $o_1[t] = i_4[t](S_0[t-1] + S_2[t-1] + S_3[t-1])$ of our machine. After substituting the symbolic equations (11-16) for the states we obtain the following 2-cycle equation

$$\begin{aligned} o_1[t] = \; & i_4[t](S_0[t-2] + S_1[t-2] + S_2[t-2] \\ & + (i_2[t-1] + \overline{i_3[t-1]})(S_3[t-2] + S_4[t-2]) + S_5[t-2]) \end{aligned} \tag{23}$$

Comparison of the original and simplified 2-cycle expression reveals that $i_1$ is not present in the simplified expression, hence there is a potential false path $i_1 \rightarrow o_1$. We therefore derive the false path dichotomy $FD_1 = (S_0, S_2, S_3 \ : \ S_1, S_4, S_5)$ associated with this potential false path.

**Sufficiency condition:** Let dichotomy $FD_1$ be satisfied by bit $r_1$ of the code, as in Figure 2(b). That is, $r_1 = 1$ for the states $S_0, S_2, S_3$ on the left block of the dichotomy, and $r_1 = 0$ for the states $S_1, S_4, S_5$ on the right block. The codes (001, 101) of the states that do not appear in the original specification can be used as don't cares. By adding the don't care states with $r_1 = 1$ in their code (in this case the state 101) to the sum term $\alpha_1 = S_0 + S_2 + S_3$, the expression for $\alpha_1$ simplifies to a single state variable $r_1$ (or $\overline{r_1}$: reversing the bit polarity does not change the encoding or its property). As a result, the one-cycle *functional output equation*, $o_1[t] = i_4[t]r_1[t-1]$ is independent of $i_1$[t-1]. Furthermore, the expression for $r_1$ does not contain the variable $i_1$ because $r_1$ represents the function $(S_0[t] + S_2[t] + S_3[t])$ which is independent of $i_1$, as observed at the behavioral level. That is, under this encoding there is no *physical* two-cycle path $i_1 \rightarrow o_1$ in the final implementation.

The satisfaction of the false-path dichotomies is also necessary to obtain a false-path-free implementation of the circuit. We illustrate the necessary condition with the same example.

**Necessity:** If $FD_1$ is not satisfied, then there will be no single product term which can represent the sum term $(S_0 + S_2 + S_3)$ uniquely. Therefore at least two product terms are required to represent this term, with the hamming distance $\geq 2$ (if it were less than 2, the two product terms would have been combined). These two product terms will therefore differ in at least 2 state variables, as only state variables are used to encode the state. Since each of the state symbols $(S_0, S_2, S_3)$ independently depends on $i_1$, the functional sum term for $(S_0 + S_2 + S_3)$ and the resulting functional output expression for $o_1$ will contain $i_1$ in its support. Recall that the simplified equation for $o_i$ (both functional and behavioral) does not depend on $i_1$, hence, there is false path $i_1 \rightarrow o_1$.

The necessary and sufficient condition for FPF encoding can now be summarized by the following theorem; its proof follows directly from the discussion above.

**Theorem 1** *Given an output $o_i$ of a sequential machine and the corresponding k-cycle false path dichotomy set $DS_i^k$. There exists an implementation of the machine which does not contain any (k+1)-cycle false paths terminating at $o_i$ if and only if the encoding of the states of the machine satisfies all dichotomies in $DS_i^k$. Furthermore, such an implementation is directly defined by the encoding and the corresponding functional equations.*

## 4.2 Encoding as a Dichotomy Covering Problem

The removal of false paths on behavioral level can now be formulated as a dichotomy covering problem (Yang, 1991), (Ciesielski, 1991). For each output with a potential false path we generate the corresponding set of dichotomies and include it in the final constraint set for the state code of the FSM. For a false path to be removed, the corresponding dichotomy must be covered by at least one of the encoding bits. Any constrained encoding method, such as (DeMicheli, 1985), or (Yang, 1991), capable of satisfying all of these constraints will result in a code, and subsequently in a circuit, which does not contain any multi-cycle false paths.

Table 2 shows the transformation of 2-cycle potential false paths ($P$) into false paths ($P$) and no-paths ($N$) for the encoding in Figure 2(b). Notice that this encoding satisfies false path dichotomies for outputs $o_1, o_3, o_4$ but not for $o_2$. Consequently, the resulting expression for $o_2$ contains redundant variables and thus has multi-cycle false paths. All other outputs are free of two-cycle false paths.

Figure 4(a) shows the gate-level implementation of our example machine using the encoding which satisfies all two-cycle false path dichotomies. This circuit uses 4 registers and can be operated at a clock period of 2 gate delays without additional retiming (the original circuit required 3 gate delays). The circuit is smaller and has fewer registers than the circuit in (Ashar, 1992) obtained after gate-level false path removal and retiming.

**Table 2** Transformation of potential false paths into False Paths or No-Paths

| | $o_1$ | $o_2$ | $o_3$ | $o_4$ |
|---|---|---|---|---|
| $i_1$ | P | P | T | P |
| $i_2$ | T | P | P | P |
| $i_3$ | T | P | P | T |
| $i_4$ | T | P | P | T |
| $i_5$ | N | T | N | N |

(a)

encoding $\Longrightarrow$

$S_0 = 100$ $S_1 = 000$
$S_2 = 110$ $S_3 = 111$
$S_4 = 011$ $S_5 = 010$

| | $o_1$ | $o_2$ | $o_3$ | $o_4$ |
|---|---|---|---|---|
| $i_1$ | N | F | T | N |
| $i_2$ | T | F | N | N |
| $i_3$ | T | F | N | T |
| $i_4$ | T | N | N | T |
| $i_5$ | N | T | N | N |

(b)

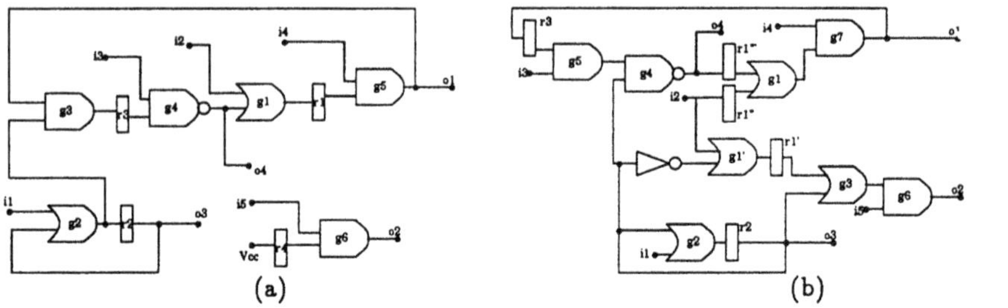

**Figure 4** (a) Circuit with FPF encoding (b) Final circuit of (Ashar, 1992) after retiming

# 5 RESULTS AND CONCLUSION

We have implemented our FPF encoding technique as a C program. It identifies potential false paths in a given FSM and generates the necessary FPDs. Table 3 shows the results of our experiments performed on a set of 66 different benchmark examples. The 25 circuits shown in the table were found to have 2-cycle potential false paths. The FPF encoding was obtained using DIET (Yang, 1991), a constraint-encoding program, which accepts a user-defined set of encoding constraints in form of dichotomies. The results were compared to the encoding obtained with multi-level encoding program JEDI (Lin, 1989) (with 3 different encoding options), and to that with standard DIET (without FPDs). The encoded circuits were optimized using the `script.delay` script, mapped to the *lib2* SIS library and retimed. The following 5 examples: *bbsse, donfile, ex6, shiftreg,* and *sse*, did not generate any additional FPD's; Hence, the delays for FPF circuits for these examples were identical to those obtained with original DIET. In the table, the column labeled *Original Delay* refers to the delay values for the encoding without considering false path constraints. The column labeled *Delay with FPF code* gives the delay value for FPF encoding obtained using DIET.

In summary, we have presented a study on the identification and removal of multi-cycle false paths from a sequential circuit specified at the behavioral level. We derived the necessary and sufficient condition to remove the multi-cycle false paths from the circuit and presented a systematic encoding methodology to obtain a false-path-free circuit based on the dichotomy covering approach. It can be argued that removing *all* multi-cycle false

Table 3 Experimental results

| Name | Original Delay (ns) | | | | Delay with FPF code (ns) | % improvement | | | |
|---|---|---|---|---|---|---|---|---|---|
| | with DIET | with JEDI $-o$ | $-i$ | $-c$ | | w.r.to DIET | w.r.to JEDI $-o$ | $-i$ | $-c$ |
| bbsse | 9.34 | 9.72 | 11.37 | 9.34 | 9.34 | 0 | 4 | 18 | 0 |
| donfile | 3.14 | 11.25 | 11.51 | 8.90 | 3.14 | 0 | 72 | 73 | 65 |
| ex1 | 7.25 | 13.24 | 12.95 | 13.17 | 7.03 | 3 | 47 | 46 | 47 |
| ex6 | 9.26 | 10.22 | 9.81 | 10.61 | 9.26 | 0 | 9 | 6 | 13 |
| false* | 6.60 | 3.92 | 4.29 | 3.57 | 3.43 | 48 | 13 | 20 | 4 |
| mark1 | 11.82 | 8.91 | 11.08 | 8.74 | 8.74 | 26 | 2 | 21 | 0 |
| mc | 6.13 | 5.61 | 5.88 | 5.37 | 5.37 | 12 | 4 | 9 | 0 |
| planet | 13.62 | 13.95 | 17.75 | 13.85 | 12.82 | 6 | 8 | 28 | 7 |
| rie | 12.28 | 11.54 | 12.36 | 12.91 | 11.53 | 6 | 0 | 7 | 11 |
| scf | 12.31 | 15.79 | 16.98 | 14.65 | 11.01 | 11 | 30 | 35 | 25 |
| shiftreg | 4.93 | 5.07 | 4.93 | 4.93 | 4.93 | 0 | 3 | 0 | 0 |
| sse | 9.34 | 9.72 | 11.37 | 9.34 | 9.34 | 0 | 4 | 22 | 0 |
| s1488 | 16.96 | 13.63 | 16.68 | 15.05 | 13.63 | 20 | 0 | 18 | 9 |
| s1494 | 14.04 | 17.54 | 13.72 | 16.23 | 13.72 | 2 | 22 | 0 | 15 |
| s386 | 10.82 | 10.46 | 10.11 | 9.85 | 9.85 | 9 | 6 | 3 | 0 |
| s510 | 13.76 | 12.93 | 15.01 | 12.18 | 12.18 | 12 | 6 | 19 | 0 |
| s820 | 13.28 | 15.34 | 14.60 | 14.30 | 13.20 | 1 | 14 | 10 | 8 |
| s832 | 13.65 | 13.32 | 16.65 | 13.00 | 12.88 | 6 | 3 | 23 | 1 |
| poptest | 15.34 | 22.80 | 21.12 | 15.08 | 14.10 | 8 | 38 | 58 | 7 |

*example shown in the paper

paths (instead of aiming at *critical* false paths along longest paths) is not a good strategy, as it may lead to implementations with longer state codes than necessary. However, our initial experience indicates that such an encoding results in a circuit with reduced delay and often with simpler structure and thus a smaller area (see the example in the paper). Although the effect of false path free encoding on retiming is not yet fully understood, our results seem to suggest that removal of multi-cycle false paths may result in circuits that either do not require additional retiming or allow for retiming that was not possible in the original circuit, thus resulting in improved performance.

## REFERENCES

Ashar, P., S. Dey, and S. Malik. Exploiting multi-cycle false paths in the performance optimization of sequential circuits. In *Proceedings of the International Conference on Computer-Aided Design*, pages 510–517, November 1992.

Chen, H.C, D.H. Du, and S. W. Cheng. Circuit enhancement by eliminating long false paths. In *Proceedings of the Design Automation Conference*, pages 249–252, June 1992.

Ciesielski, M, J.J. Shen, and M. Davio. A unified approach to input-output encoding for FSM state assignment. In *Proceedings of the Design Automation Conference*, June

1991.

DeMicheli, G., R.K. Brayton, and A. Sangiovanni-Vincentelli. Optimal state assignment of finite state machines. *IEEE Transactions on Computer-Aided Design*, pages 269–285, July 1985.

Devadas, S., K. Keutzer, and S. Malik. Delay computation in combinational circuits: Theory and Algorithms. In *Proceedings of the International Conference on Computer-Aided Design*, November 1991.

Hasan, Z. and M. J. Ciesielski. On multi-cycle false paths. In *Proceedings of the IFIP Workshop on Logic and Architecture Synthesis*, December 1993.

Lam, W.K., R.K. Brayton, and A.L. Sangiovanni-Vincentelli. Exact minimum cycle time for finite state machines. In *Proceedings of the Design Automation Conference*, June 1994.

Lin, B. and A.R. Newton. Synthesis of multiple level logic from symbolic high-level description languages. In *Proceedings of IFIP International Conference on Very Large Scale Integration*, pages 187–196, August 1989.

McGeer, P.C. and R. K. Brayton. *Integrating Functional and Temporal Domains in Logic Design*. Kluwer Academic Publishers, 1991.

Saldanha, A., R.K. Brayton, and A.L. Sangiovanni-Vincentelli. Circuit structure relations to redundancy and delay: The KMS algorithm revisited. In *Proceedings of the Design Automation Conference*, pages 245–248, June 1992.

Yang S. and M. J. Ciesielski. Optimum and suboptimum algorithms for input encoding and its relationship to logic minimization. *IEEE Transactions on Computer-Aided Design*, pages 4–12, January 1991.

## 6 BIOGRAPHY

**Zafar Hasan** received his B.Tech degree in Electrical Engineering from Banaras Hindu University, Varanasi, India in 1989. He received his M.S. and Ph.D. degrees from the department of Electrical and Computer Engineering, University of Massachusetts at Amherst in 1992 and 1995, respectively. He will be joining shortly as a CAD Engineer at Intel's Hillsboro, OR location. His research interests include Logic Synthesis, CAD for VLSI Design, Testing and Computer Architecture. He is a member of the IEEE.

**Maciej Ciesielski** received his M.S. in Electrical Engineering from Warsaw Technical University, Poland, in 1974, and Ph.D. in Electrical Engineering from the University of Rochester in 1983. From 1983 to 1986 he was a Senior Member of Technical Staff at GTE Laboratories, Waltham, MA, working o a silicon compilation project. Currently he is Associate Professor in the Department of Electrical and Computer Engineering at the University of Massachusetts, Amherst. His research interests include: CAD for VLSI systems, high-level and logic synthesis, performance optimization of IC's, and mathematical optimization methods. He has served on the technical program committees for ICCAD, ICCD, IFIP/IWLAS, and IWLS. He is a member of the IEEE.

PART SEVEN

# Control Part and Operative Part

21

# Low Power VLSI Design Method for Data Path and Controllers

*Kunihiro ASADA, Makoto IKEDA and Junichi AKITA*
*University of Tokyo*
*Department of Electronic Engineering, Faculty of Engineering,*
*University of Tokyo, 7-3-1 Hongo, Bunkyo-ku, Tokyo, 113, JAPAN.*
*Telephone: +81-3-3812-2111 ext 6771. Fax: +81-3-3818-5706.*
*email: {asada,ikeda,akita}@silicon.t.u-tokyo.ac.jp*

**Abstract**

We propose design methods for power reduction of data path and controllers in VLSIs.

For data path, we propose a partitioned-bus architecture for reducing power consumption in bus lines with taking account of geometric information of function units. Applying this method to ASIC examples, we demonstrate power reduction of $1/5 \sim 1/12$ in bus lines compared with the conventional bus architecture.

For controllers, we propose a state code assignment method orienting reduction of power in finite state circuits. Simulation results show that finite state circuits implemented with our state code assignment method have an advantage of 15% over those with the conventional methods in power consumption, with a small disadvantage of 8% in hardware cost.

**Keywords**

low power, high-level design, partitioned-bus architecture, state code assignment, finite state circuits.

## 1 INTRODUCTION

Low power design of VLSIs is an important theme. A lot of studies for reducing power consumption of VLSIs have been reported with various approaches[M. Kakumu 1993]. Dynamic power consumption due to charging / discharging load capacitance is dominant in CMOS circuits, which is described as $CV_{dd}V_s\overline{f}$, where $C$ is the total load capacitance, $V_{dd}$ is supply voltage, $V_s$ is signal swing, and $\overline{f}$ is an effective signal frequency. In data path, it becomes necessary to reduce power consumption in long global signal lines such as bus lines. From a view point of circuit design to solve this problem, reduction methods of signal swing $V_s$ have been studied [Y. Nakagome et al. 1992]. In controllers, as the effective signal frequency $\overline{f}$ is depending on both the system clock and circuit topologies, power can be reduced by optimizing logic circuits implementation for a given logic function. In this paper, we use two approaches to reduce $\overline{f}$; a partitioned-bus architecture using high-level VLSI design method and state code assignment method for circuit implementation optimization in finite state machines.

In section 2, we propose a partitioned-bus architecture and a high-level VLSI design with taking account of geometric information to reduce power consumption in bus lines. In section 3, a method of state code assignment for synchronous CMOS finite state machines is described, which is effective for reducing $\overline{f}$ at outputs of flip-flops without reducing a system clock frequency.

# 2 A POWER REDUCTION METHOD USING PARTITIONED-BUS ARCHITECTURE IN HIGH-LEVEL VLSI DESIGN

Conventionally, the high-level synthesis aims to optimize number of control steps or/and function units under constraints on number of control steps or/and function units[P. G. Paulin et al. 1989]. Although there is an attempt to reduce power consumption in a high-level synthesis, HYPER-LP[A. C. Chandrakasan et al. 1992], it uses a typical CMOS gate characteristic that the delay depends on supply voltage. HYPER-LP achieves power consumption reduction by lowering supply voltage, with keeping system performance from speed degradation by optimizing system critical paths in high-level synthesis. There is also an attempt that takes account of geometric information of function units in high-level synthesis[H. J. Jang and B. M. Pangrle 1993]. It mainly aims to produce designs with short interconnection length.

The target of this section is to reduce power consumption in bus lines in high-level design. For this purpose, we assume a 1-dimensional placement of function units to estimate real interconnection length. We minimize power consumption in bus lines, linking scheduling, allocation and placement together. We also propose a partitioned-bus architecture, which restricts bus driving range so as to minimize power. In section 2.1, we describe the partitioned-bus architecture. In section 2.2, we show that a high-level design, based on the partitioned-bus, is effective in reducing power consumption in bus lines, using a small design example with an exhaustive search method. In section 2.3, we show a cluster growing method for practical applications, that can derive a power-minimal solution within allowable CPU time.

## 2.1 Partitioned-bus architecture

In the conventional bus architecture, power is wasting to drive wire sections irrelevant to data transmission as shown in Fig. 1. On the other hand, in a case of point-to-point connection, increase in area for interconnections and multiplexors come into problem.

A partitioed bus architecture was proposed before[C. Ewering 1990], which aimed, however, to yield a space efficient interconnection structure by means of conditionally interconnecting multiple buses in ASIC chips, and was not adequate for power reduction. The present partitioned-bus architecture, which consists of bus lines and repeaters with control inputs to partition bus lines as shown in Fig. 2, restricts bus driving range to minimal so as to minimize power. The repeaters are controlled so that bus partitions relevant to

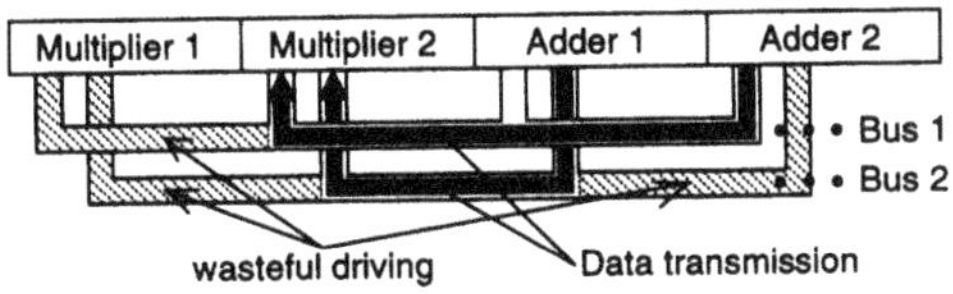


**Figure 1** Model of the conventional bus architecture.

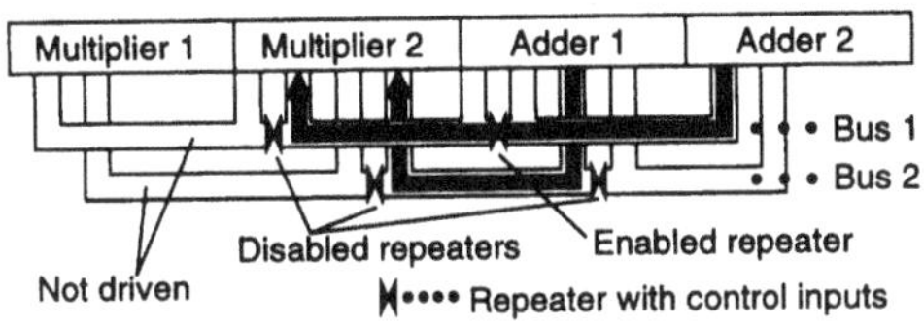


**Figure 2** Proposed partitioned-bus architecture.

the data transmission are selectively driven, and the rest partitions are kept in the same status. This method has following advantages:

1. There is no wasteful driving of bus lines because bus lines are separated by repeaters with conditional drivability.
2. Every partition can be used for independent data transmission simultaneously, so far as it is isolated from the others by disabled repeaters.

Disadvantages are additional repeaters and delay requested for bus separation and control signal generation for them.

The effects of this partitioned-bus lines are shown in the next section, along with optimization in high-level design.

## 2.2 Power reduction in partitioned-bus architecture.

In this section, we show that the partitioned-bus in high-level design is effective in reducing power consumption in bus lines using an example and an exhaustive design optimization.

A process of the exhaustive design optimization is carried out in three domains in design space as follows:

1. Search all the possible schedulings under restrictions of number of function units between ASAP(As Soon As Possible) and ALAP(As Late As Possible) for a given control steps. (domain of schedulings)
2. Search all the possible allocations of operations to function units for each scheduling given in step 1. (domain of allocations)
3. Search all the possible placements of function units in the 1-dimensional layout for each allocation given in step 2. (domain of placements)

**Table 1** Area and power consumption of function units and bus lines.

| module | area[$mm^2$] | | power[$pJ$] |
|---|---|---|---|
| adder | 0.18 | $0.8 \times 0.2$ | 34.0 |
| multiplier | 12.92 | $2.0 \times 6.3$ | 1,080 |
| register | 0.018 | $0.8 \times 0.02$ | 28.9 |
| bus lines(32bit) | — | | $1.28 \times 10^{-2} \times L$ |

L: bus length[$\mu m$]

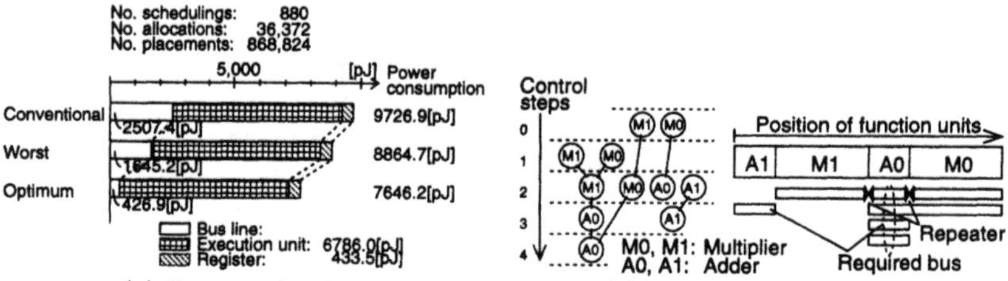


(a) Power reduction. (b) An optimum solution.

**Figure 3** Results of the exhaustive optimization for the differential equation solver.

Total length of interconnection can be derived from geometric information of the placement. Using table 1, we calculate power consumption in bus lines and function units.

We use a differential equation solver [G. Borriello and W Detjens 1988] as an example. The number of adders, multipliers and control steps are restricted up to 2, 2 and 5, respectively. The results of the exhaustive optimization are shown in Fig. 3(a). Figure 3(b) shows one of the optimum solutions of scheduling, allocation and placement, which minimizes power consumption in bus lines.

Assuming that all function units consist of CMOS static circuits, function units and registers consumes power only when they are accessed. So, the total power consumption of function units and registers are constant (6786.0, 433.5[$pJ$], respectively, as shown in Fig. 3(a)) in full cycles of the differential equation solver. Therefore, the partitioned-bus architecture achieves 21% reduction in total power consumption (7646.2[$pJ$]) compared with the conventional bus architecture (9726.9[$pJ$]) in this example. Even if using the partitioned-bus architecture, the worst case assignment of scheduling, allocation and placement causes 16% increase of the total power consumption (8864.7[$pJ$]) compared with the optimum case. These results indicate that an appropriate scheduling is necessary to reduce power consumption in the partitioned-bus architecture.

## 2.3 Heuristic design method for optimal power design

In this section, we show a heuristic method based on a cluster growing method that makes power consumption minimal within a practical CPU time.

To obtain optimal scheduling, allocation and placement, there can be two main methods. One is an iterative method that improves initial solution toward optimum, and another

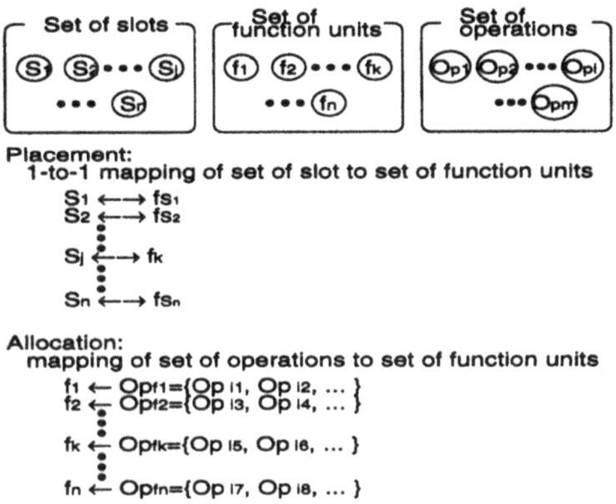


**Figure 4** The relationship of placement and allocation with the set of slots, function units and operations.

is a constructive method to obtain the optimal without iteration. In this paper we use a constructive method by a cluster growing method that successively assigns each operation to a control step, a function unit, and a placement, so that increase of power consumption in bus lines is minimal, under given restrictions on number of control steps and function units. We process each operation in the reverse order of operation's freedom between ASAP and ALAP. The freedom is defined as a distance between control steps in ASAP and ALAP scheduling.

Let the number of the total function units be $n$. We need $n$-slots in 1-dimensional layout for accommodating $n$-function units. Let the slots be $S_1, \cdots S_n$. The target of the heuristic procedure is to assign the function units, $f_1, \cdots f_n$, to the slots, $S_1, \cdots S_n$, as a 1-to-1 mapping. At the same time we need to assign the operations, $Op_1, \cdots Op_m$, to the function units, so that each operation is done by a suitable function unit under the control step constraint. The relationship of these mappings is shown in Fig. 4. Thus the procedure is described as follows:

1. For each operation, $Op_i$, in order of less freedom, iterate step 2~6.
2. Find an optimal slot, $S_j$, for $Op_i$ in terms of minimal power increase.
3. If no function unit is assigned to the slot, $S_j$, yet, and there is a function unit, $f_k$, compatible with $Op_i$, which is not yet assigned to any slot, assign $f_k$ to $S_j$ and assign $Op_i$ to $f_k$.
4. If a function unit, $f_k$, compatible with $Op_i$ is already assigned to $S_j$, assign $Op_i$ to $f_k$ so as to satisfy the given control step constraint.
5. If a function unit, incompatible with $Op_i$ is already assigned to $S_j$, and there is a function unit, $f_k$, that is not yet assigned to any slot, assign $Op_i$ to $f_k$ and assign $f_k$ to any unused slot.
6. Otherwise, try all the possible assignments of $Op_i$ to function units compatible with $Op_i$. Find an optimal assignment in terms of increase of power.

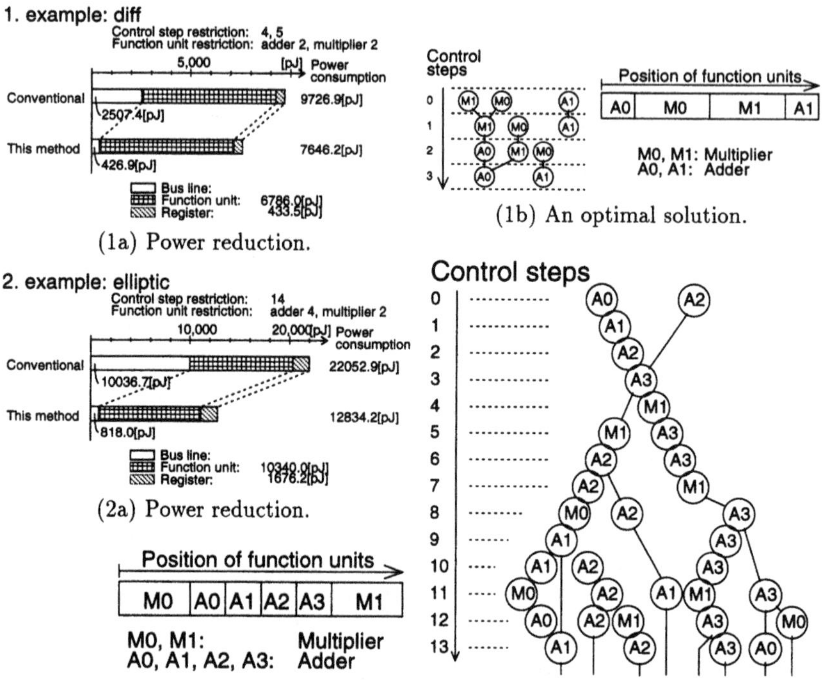

(2b) An optimal solution.

**Figure 5** Results of the heuristic design method.

Note that we always carry out a rearrangement of the slot assignment of function unit, $f_k$, when a new operation is assigned to $f_k$ in the step 3~6, so that the power increase become minimal. The rearrangement is not an exhaustive trial but that the slot for $f_k$ is checked from $S_1$ to $S_n$ successively without changing the order of the other function units.

Figure 5(1a)(2a) shows results of this heuristic method. In this figure, "diff" is the differential equation solver already shown, and "elliptic" is a fifth-order digital elliptical wave filter [G. Borriello and W Detjens 1988]for more substantial behavioral description that contains 43 operations. In the case of diff, one of the optimum assignments of scheduling, allocation and placement is obtained from this method as shown in Fig. 5(1b). CPU time was about 8.5 sec. on Sparc Station 10. In the case of elliptic, the solution of this method is shown in Fig. 5(2b). Although we can't always obtain the optimum solution using this method for such a large design as elliptic, the power consumption in bus lines (818.0[$pJ$]) can be reduced to 1/12 compared with conventional bus architecture(10036.7[$pJ$]) in this example.

# 3 OPTIMUM STATE ASSIGNMENT FOR CMOS IMPLEMENTATION OF LOW POWER FINITE STATE MACHINE

In this section a new method of state code assignment for synchronous CMOS finite state machines is described. Concept of the method and model of power consumption is described in section 3.1 along with brief description of probability calculation required for the model in section 3.2. In section 3.3 a heuristic algorithm for the power-minimum state code assignment is presented. The effect of power reduction will be demonstrated in section 3.4 by examples.

## 3.1 Concept and Model of Power Consumption

In CMOS circuits power consumption $P_{\text{total}}$ is composed of static power $P_{\text{static}}$ and dynamic power $P_{\text{dynamic}}$ as $P_{\text{total}} = P_{\text{static}} + P_{\text{dynamic}}$. In synchronous CMOS circuits, the dynamic power $P_{\text{dynamic}}$ is further decomposed into power at clock lines $P_{\text{clock}}$ and the others $P_{\text{logic}}$ as $P_{\text{dynamic}} = P_{\text{clock}} + P_{\text{logic}}$. Reduction of $P_{\text{clock}}$ is difficult to achieve without reducing system clock frequency. $P_{\text{logic}}$, however, can be reduced by using a suitable circuit topology without reducing system clock frequency.

For reducing power consumption in synchronous finite state circuits, there are at least two possible ways; state code optimization and combinational logic optimization. This paper focuses on the former approach.

We assume that probabilistic characteristics of the primary input are given and fixed. The target of the state code assignment is reduction of average signal frequency at the state input, which is directly related to the power consumption at the state input lines. Furthermore it would be indirectly related to the power consumption in the combinational logic part, because reduction of average input signal frequency is expected to result in power reduction in whole combinational logic.

We defined, therefore, a cost function for the power consumption of finite state circuits as follows:

$$\overline{d} = \sum_{i=1}^{n}\sum_{j=1}^{n} d_{i,j} \cdot p_{i,j}. \tag{1}$$

Here, $i$ and $j$ stand for state. $d_{i,j}$ is the Hamming distance between state codes corresponding state $i$ and $j$. $p_{i,j}$ is an event probability of state transition from state $j$ to $i$, and $n$ is the number of states. Target of our study is minimization of $\overline{d}$ for a given state transition diagram and probabilistic characteristics of the primary input. Note that $d_{i,j}$ is equal to the number of flip-flops which toggle their output at the transition from $j$ to $i$. Derivation of $p_{i,j}$ is described in the following section.

## 3.2 Transition Probability

In finite state machines, the transition probability between each pair of states can be derived from the probabilistic characteristics of the primary input signals and state transition diagram. Here we introduce a conditional transition probability matrix, $\boldsymbol{A} = \{a_{i,j}\}$, where each element $a_{i,j}$ is the conditional probability of transition from state $j$ to $i$ under the condition that the current state is $j$.

The conditional transition probability, $a_{i,j}$ can be derived from probabilistic characteristics of the primary input. Usually a branch condition in a state transition diagram is described as a logic function of the primary input. Assuming that probability of logic "1" for each primary input is given, probability that the logic function becomes "1" is calculated using the conventional relation between the boolean algebra and the set theory[Gary D. Hachtel et al. 1994].

Let $\boldsymbol{q}$ be a vector, whose element $q_j$ is a probability that the current state is $j$. Assuming that $\boldsymbol{q}$ is in a stationary state, we have the equation $\boldsymbol{q} = \boldsymbol{A}\boldsymbol{q}, \quad \sum q_j = 1$. [M. S. Bartlett 1966] We can calculate $\boldsymbol{q}$ by solving this equation, and obtain an unique solution.

With both the conditional transition probability $a_{i,j}$ and occupation probability $q_j$, the event transition probability from state $j$ to $i$, $p_{i,j}$ is described as $q_j \cdot a_{i,j}$. With these probabilities, we can calculate the cost function, $\overline{d}$, as an average Hamming distance between each pair of states from eqn.1.

## 3.3 Optimal State Assignment Algorithm

We are going to propose a heuristic algorithm to assign an optimal binary code to each state that minimize the cost function.

An exhaustive assignment algorithm is trivial to undergo all the possible code assignments for finding the lowest cost function This method is not practical because calculation time is proportional to $n!$. In our study, this method is used in case of no more than $n = 8$, for evaluating the quality of the heuristic algorithm.

For a larger $n$, we developed a heuristic assignment algorithm as follows.

1. Make a bidirectional transition probability matrix, $\tilde{\boldsymbol{P}} = \{\tilde{p_{i,j}}\}$, so that

$$\tilde{p_{i,j}} = \begin{cases} p_{i,j} + p_{j,i}, & \text{for } i < j \\ 0, & \text{for } i \geq j \end{cases} \tag{2}$$

2. Sort the non-zero element, $\tilde{p_{i,j}}$ in decreasing order.
3. For the largest element, $\tilde{p_{i,j}}$, assign the code "0" to $i$ and "1" to $j$. ( Note that $i$ and $j$ are exchangeable.)
4. Iterate the following procedures 5 $\sim$ 7 for each $\tilde{p_{i,j}}$ until all the elements have been processed.
5. When both the states, $i$ and $j$, have been already assigned codes, simply skip $\tilde{p_{i,j}}$
6. When exactly one of the states, $i$ and $j$, has been already assigned a code, assign the nearest code in Hamming distance to the other state. If there are more than two nearest codes, choose the best one such that increase of the cost function calculated for the partially assigned state machine is minimal. In other words, the cost function $\tilde{d}$ for the partially assigned state machine is defined as

$$\tilde{d} = \sum_{i=1}^{n}\sum_{j=1}^{n} \tilde{d_{i,j}} \cdot p_{i,j}, \quad \tilde{d_{i,j}} = \begin{cases} d_{i,j}, & \text{for the case that both } i \text{ and } j \text{ have been assigned codes,} \\ 0, & \text{otherwise.} \end{cases} \tag{3}$$

7. When both the state, $i$ and $j$ have been not yet assigned codes, suspend $\tilde{p_{i,j}}$ and go forward to the next. Each time the procedure 6. has been done, the suspended $\tilde{p_{i,j}}$ list is checked and processed as in 6. when the condition is satisfied.

## 3.4 Results and Discussions

We have carried out the assignment algorithm for several examples to evaluate the power reduction effect.

### *Evaluation of the heuristic algorithm*

We have carried out state assignment for 8-state finite state machines. For this size of finite state machines, we could also carry out the exhaustive searching algorithm in practical time. We generated both the state transition diagrams and state transition probabilities randomly for 8-state finite state machines of 2000 cases, and compared the cost with the optimum assignment by the exhaustive searching. The mean of the cost function $\overline{d}$ with exhaustive assignment, heuristic assignment and random assignment was 1.19, 1.24 and 1.54, respectively.

From these results, we can accomplish about 4% larger cost with heuristic assignment than optimum assignment. The calculation time of heuristic assignment is measured in proportion to $O(n^3)$, though that of exhaustive assignment is expected in proportional to $O(n!)$.

### *Gain of the optimum assignment.*

Reduction of the cost function, $\overline{d}$, is the gain obtained by the optimum assignment. The gain is dependent on characteristics of the transition matrix, $\boldsymbol{P}$. When $\boldsymbol{P}$ is uniform, namely all the elements have uniform values, the gain is apparently small. Contrarily when $\boldsymbol{P}$ is non-uniform, the gain becomes bigger by assigning codes so that the Hamming distance of state two codes becomes smaller for a pair of states with lager transition probability. For evaluating the gain for various finite state machines, we have systematically generated state transition diagrams and transition matrices with characteristic parameters $(n_b, n)$. Here $n_b$ is an average number of outgoing edged from a state. $(1, n)$ means a ring transition diagram, while $(n, n)$ means a perfect transition diagram with edged for all the pair of states. Table 2 summarizes the gain for various state machines. Here the gain is defined as a cost reduction ratio between the optimum assignment obtained by the heuristic algorithm and an average cost of 100 random assignments. Gain is defined as follows:

$$\text{Gain} = \frac{\overline{d_{\text{av}}} - \overline{d_{\text{opt}}}}{\overline{d_{\text{av}}}} \times 100 \quad [\%], \tag{4}$$

where $\overline{d_{\text{av}}}$ is the average cost and $\overline{d_{\text{opt}}}$ is the cost for the optimum assignment.

From Table 2, we can conclude that the gain becomes larger for the larger $n$ and smaller $n_b$.

### *Relation between the cost function and real power consumption*

In the above discussions, we used the cost function $\overline{d}$ as a measure of $P_{\text{logic}}$. Synchronous circuits always consume some power $P_{\text{clock}}$ at clock lines even when $\overline{d} = 0$ ($P_{\text{logic}} = 0$).

**Table 2** Gain (reduction of cost) by optimum assignment.

| $n_b$ | $n$ | bits | Gain |
|---|---|---|---|
| 2 | 8 | 3 | 25.3% |
| | 64 | 6 | 45.7% |
| 8 | 8 | 3 | 5.7% |
| | 64 | 6 | 15.6% |

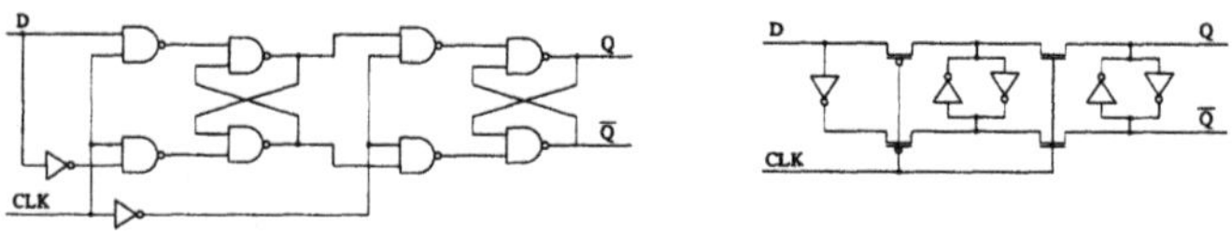


(a) (b)

**Figure 6** Two type of D flip-flops.

Here we describe a relation between the gain of the cost and real power reduction in flip-flops. (We do not here discuss the power reduction in whole combinational logic because it depends on the circuit topology of the combinational logic.)

Let $P_t$ and $P_n$ be power consumption of flip-flops for cases of toggling and non-toggling operation, respectively. Ratio of $P_n/P_t$ depends on a circuit configuration of flip-flops. Two types of flip-flops are shown in Fig.6 A typical ratio of $P_n/P_t$ is 0.35 for Fig.6(a), while it is 0.15 for Fig.6(b). Apparently the real power reduction is larger for flip-flops with smaller ratio. If the ratio is ideally zero, the real power reduction is same as the reduction of cost function $\overline{d}$(gain). Table 3 shows several examples of real power reduction for the two flip-flops as well as the ideal one.

**Table 3** Power reduction dependence on flip-flop characteristics.

| | | | | Power reduction ratio | |
|---|---|---|---|---|---|
| $n_b$ | $n$ | bits | Gain | $P_n/P_t = 0.15$ | $P_n/P_t = 0.35$ |
| 2 | 8 | 3 | 25.3% | 18.8% | 12.2% |
| | 64 | 6 | 45.7% | 33.8% | 22.0% |
| 8 | 8 | 3 | 5.7% | 4.2% | 2.7% |
| | 64 | 6 | 15.6% | 11.6% | 7.5% |

**Table 4** Comparison of mean transient of flip-flop per cycle for FSM with two code assignment: power-optimal assignment(Pow.-opt.) and circuit-optimal assignment(Cir.-opt.).

| example | Tran. per cycle | |
|---|---|---|
| | Pow.-opt. | Cir.-opt. |
| bbsse | 0.77 | 1.41 |
| beecount | 0.47 | 0.48 |
| cse | 0.23 | 0.35 |
| dk15 | 0.85 | 0.99 |
| donfile | 1.44 | 1.79 |
| exl | 1.14 | 2.13 |
| kirk | 0.98 | 1.54 |
| planet | 1.17 | 2.99 |
| tbk | 1.08 | 1.33 |

### *Comparison with conventional assignment*

In a practical circuit design the state codes are not randomly assigned. They would be assigned, for example, to minimize the size of the implemented circuit of combinational logic[S. Devadas et al. 1988].(Here we call this method as "circuit-optimal assignment.")

We carried out Monte-Carlo simulation for some FSMs in MCNC89 benchmarks with both methods of state code assignment; the power-optimal assignment with our method and the circuit-optimal assignment with MUSTANG.

We assigned state code with both methods at first, and synthesized combinational logic circuits with misII, a multiple-level logic optimization system developed by UCB[R. Brayton et al. 1987]. Next we estimated load capacitance $C_L$ of each flip-flop, and assumed that each flip-flop consumes energy in proportion to its $C_L$ only when its output changes. (In other words, we assumed that $P_n/P_t = 0$.) Finally we counted transition of flip-flop's outputs for randomly given primary inputs of 10,000 cycles.

Results of the above Monte-Carlo simulation for some of 31 MCNC89 benchmarks are shown in Table 4. The power-optimal assignment gives about 30% less transition than the circuit-optimal assignment in average. This result implies the possibility of reducing practical power consumption of FSM with power-optimal state code assignment.

## 4 CONCLUSIONS

In this paper, we proposed a high-level design method with taking account of geometric information in bus lines and a optimization method of state code assignment to reduce power consumption in data path and controllers in CMOS VLSIs.

For data path, we proposed a partitioned-bus architecture to restrict bus driving range to minimal. Applying the present method, we demonstrated $1/5 \sim 1/12$ and $20\% \sim 40\%$ power consumption reduction in bus lines and total chip, respectively, compared with the conventional bus architecture. From the results, it is shown effective for low power VLSI design to take into account of placement in high-level design. As a future study, it is desirable to establish a 2-dimensional layout model for this approach.

For controllers, simulation results show that finite state circuits implemented with our state code assignment method have an advantage of 15% over those with the conventional methods in power consumption, with a small disadvantage of 8% in hardware cost. In practical circuit design, we need to consider the optimization of combinational logic circuits depending on the code assignment, though the reduction of average frequency of state input is expected to cause the reduction of power consumed in combinational logic circuits as well.

## REFERENCES

P. G. Paulin, et al., "Force-Directed Scheduling for the Behavioral Synthesis of ASIC's." *IEEE T-CAD*, pp. 661-679, vol. 8, No. 6, 1989.

A. C. Chandrakasan, et al., "HYPER-LP: A System for Power Minimization Using Architectural Transformations," *ICCAD-92, Dig. of Tech. Papers*, pp. 300–303 (1992).

H.J Jang and B.M. Pangrle, "A Grid-Based Approach for Connectivity Binding with Geometric Costs," *ICCAD-93, Dig. of Tech. Papers*, pp. 94–99 (1993). G. Borriello and W. Detjens, "High-Level Synthesis: Current Status and Future Directions," *25th DAC, Proc. 1988*, pp. 477–482, (1988).

Y. Nakagome et al., "Sub-1-V Swing Bus Architecture for Future Low-Power ULSIs," *Symp. on VLSI Cir., Dig. of Tech. Papers*, pp. 82–82 (1992).

C. Ewering, "Automatic High Level Synthesis of Partitioned Buses," *ICCAD-90,* Dig. of Tech. Papers, pp. 304–307, Nov. 1990.

M. Kakumu, "Process and Device Technologies of CMOS Device for Low-Voltage Operation," *IEICE Trans. Electron*, Vol. E76-C, No. 5, pp. 672–680, May 1993.

S.Devadas et al., "MUSTANG: State Assignment for Finite State Machines Targeting Multilevel Logic Implementations," *IEEE Trans. of Computer-Aided Design*, Vol. 7, No. 12, Dec. 1988.

K. Roy and S. Prasad, "SYCLOP: Synthesis of CMOS Logic for Low Power Applications," *Proc. of ICCAD'92*, pp. 464–467, 1992.

Gary D. Hachtel et al., "Symbolic Algorithms to Calculate Steady-State Probabilities of a Finite State Machine," *Proceedings of* EDAC-ETC Euro ASIC '94, pp.214–218, Mar. 1994.

M.S.Bartlett, "An Introduction to Stochastic Processes — With Special Reference to Methods and Applications," Cambridge University Press, 1966.

R.Brayton et al., "MIS: A Multiple-Level Logic Optimization System," *IEEE Trans. of Computer-Aided Design*, pp.1062–1081, Nov., 1987.

22

# Behavioral synthesis : control schemes in question

*B. Rouzeyre, D. Dupont*
*L.I.R.M.M., UMR CNRS C9928, Université Montpellier II*
*161 rue Ada, 34392 Montpellier Cedex 5, FRANCE*
*Tel : (33) 67148525, Fax: (33)67148500, e-mail :rouzeyre@lirmm.fr*

**Abstract**

In many behavioral synthesis systems, a one–clock scheme is assumed for both controller and datapath implementation An alternative solution to this purely synchronous control scheme is presented here. It is based on tuning the clock period according to data–path delays in order to suppress idle times due to synchronism. This scheme is mainly attractive when datapath operations have heterogeneous delays. An implementation for such a control scheme is proposed. The silicon area overhead is very small while it allows a good performance enhancement.

**Keywords**

Behavioral synthesis, clocking, performance

## 1 INTRODUCTION

Since a decade, an emerging design technique referred as High Level Synthesis (HLS) or as behavioral synthesis has been developed in CAD industry and in many research centers. It is intended to automate the earliest steps of the ASIC's design task. It consists in generating automatically the architecture of a circuit at the register transfer level from its behavioral specification, generally given as an input algorithm (see Mc Farland 1988 or IEEE 1990) for a general presentation). Generically, the architecture consists of a data–path and of a controller in charge of sequencing the instructions of the input algorithm on the synthesized data–path.

The two main tasks of data path synthesis are scheduling and allocations. The performance of the future architecture is fixed mainly during the scheduling phase of the synthesis: the operations of the data flow graph (DFG) are apportioned among control steps (c–steps) that is, among states of the control machine, according to data dependencies and to design constraints. (e.g. Ku 1990, Paulin 1989, Paulin 1990, Cloutier 1990, Camposano 1991). The allocation phase consists in mapping the variables and operations of the input algorithm onto a minimal amount of hardware.

The underlying control scheme adopted for synthesis strongly influences the circuit performance as well as the scheduling process itself.

Generally a "synchronous" clocking scheme is taken : a Finite State Machine (F.S.M) implements the sequencing of the input algorithm and drives the data–path signals through combina-

tional logic. Depending whether there are data dependent transitions in the behavioral description or not, the lower bound T of the clock period is :
$T = T_{FSM} + T_{comb} + T_{datapath}$ or $T = Max\,(T_{FSM} + T_{comb}, T_{datapath})$.
$T_{FSM}$ is the internal delay of the FSM for delivering the next state, $T_{comb}$ is the delay of the control logic. $T_{datapath}$ is the time requested for completion of data transfers from register to register through functional units (f.u.*s*). Due to the synchronism of the controller, $T_{datapath}$ must be taken equal to the longest delay in the data–path, i.e. :

$$T_{datapath} = \max_{s \in \{c\_steps\}} d(s) \qquad \text{with } d(s) = \max_{\text{data-transfers during } s} T_{data\text{-}transfer}$$

where $T_{data\text{-}transfer}$ denotes the time requested for a data transfer from registers to register through operations.

Thus, in both cases, the period of the controller is fixed by the *slowest* operation (i.e. the slowest data–transfer on the data–path) over all c–steps. Consequently, unnecessary slack time intervals are added to the other c–steps. This may result in poor performance if the delays of operations are heterogeneous. For instance, if a ($\times$ : 80 ns) is performed in a c–step, 50 ns are wasted at every c–step involving only (+ : 30 ns). Even for the same type of operation, this heterogeneity can be due to the f.u. type implementing the operations or simply due to the data bitwidth involved. For instance, in experiments given in Chaiyakul 1992, on the same example, d(s) varies from 63ns to 252ns as the data bitwidth ranges from 8 to 32 bits.

With regard to the whole performance of the circuit, the total idle time due to such a clocking scheme is:

$T_{idle} = \sum_{s \in \{c\_steps\}} F_s \cdot (\max_s (d(s)) - d(s))$ where $F_s$ denotes the number of times c–step s is activated. It can be very large in the data operations have heterogeneous delays..

Multi–cycling operations (i.e. making operations overlap c–steps) is a commonly mentioned trick for minimizing these idle times by making slow operations overlap c–steps (this can be done with pipeline or non–pipeline f.u.*s*). But, these idle times can not be totally removed if the delays of the operations cannot be expressed as integral multiples of a common delay. In Narayan 1992, a method for determining, prior to scheduling, the clock period that minimizes slack times is presented.

Table 1 summarizes scheduling results published in numerous papers (e.g. Hwang 91) on the benchmark filter example taken from Dewilde 1985 : multiplications are assumed to overlap two consecutive c–steps and additions are not chained.

**Table 1** Scheduling results for several synthesis parameters

| *Adders* | *Multipliers.* | *2 stage pipeline Mult.* | *# c–steps* | *# c–steps with only +* | *# c–steps with × (and +)* |
|---|---|---|---|---|---|
| 1 | 1 | – | 28 | 8 | 20 |
| 2 | 1 | – | 21 | 5 | 16 |
| 2 | 2 | – | 18 | 10 | 8 |
| 3 | 3 | – | 17 | 9 | 8 |
| 1 | – | 1 | 28 | 17 | 11 |
| 2 | – | 1 | 19 | 8 | 11 |
| 3 | – | 1 | 18 | 7 | 11 |
| 3 | – | 2 | 17 | 9 | 8 |
| 2 | – | 2 | 18 | 11 | 7 |

$T_{idle}$ can be obtained on the following way:
– if d(+) = 1/2 d(×), then Tidle = 0
– if d(+) < 1/2 d(×), then d(s)=1/2 d(×) and Tidle = (1/2 d(×) – d(+)) * column 5
– if d(+) > 1/2 d(×), then d(s)=d(+) and Tidle = (d(+) – 1/2 d(×)) * column 6

# 2 ADAPTATIVE CONTROL

## 2.1 Principle

For non–pipelined designs, a controller architecture allowing to adjust independently the delay of every c–step to the minimum leads to a higher speed since there are no idle times. Such a controller structure is depicted in Figure 1.

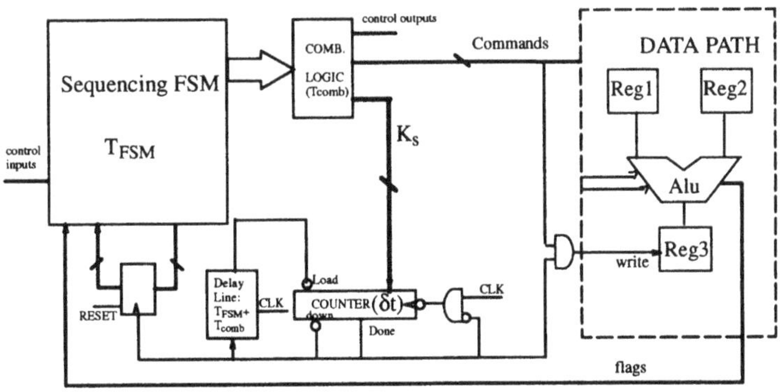


**Figure 1** Controller structure for adjusted clocking

It is mainly composed of a counter in charge of computing the delay d(s) of every c–step s. It is loaded with a parameter $K_s$, depending on s. After $K_s$ clock periods, it delivers a rising edge on its output which is used as the clock for the FSM. Once the outputs of the FSM and the outputs of the combinational logic are stable, i.e. after a $T_{FSM} + T_{comb}$ delay, the counter is loaded with the new value of $K_s$. Due to its simplicity , such a counter can have a very fast clock (typically up to 100Mhz). The $K_s$ coefficients are taken equal to $K_s = \lceil d(s)/Tclk \rceil$ where Tclk is the clock period. The delay line on Figure 1 delivers the load signal of the counter once the combinational block outputs are stable, i.e. after $T_{FSM} + T_{comb}$.

This structure can be simplified for data path dominated designs, i.e. if $T_{FSM} + T_{comb}$ is of the same order as $T_{clk}$. The delay line is removed, $K_s$+1 instead of $K_s$ is sent to the counter, its output signal enables directly the clock of the FSM.

Doing so, the idle time of every c–step is reduced to d(s)–$K_s$.Tclk, instead of being Max(d(s))–d(s) with a "synchronous" control. It must be noticed that, due to the simplicity of the added hardware, Tclk can be very short (see below) involving so a very small total idle time.

With such a clocking scheme, the execution time of the circuit is $\Sigma\ F_s.[T_{FSM} + T_{comb}+d(s)]$ (instead of being $\Sigma\ F_s.[T_{FSM} + T_{comb}+T_{datapath}]$ for synchronous control).

## 2.2 Implementation

Figure 2 gives the mask of such a 4–bit counter with its peripheral logic. It has been designed using the Cadence silicon compiler and the 1.2 μm ES2 module library. Its frequency is 87 Mhz while its silicon area is 0.24 $mm^2$. Let's recall that its frequency is bitwidth independent while its area growths as log(max Ks). On the taken behavioral description example, depending on data–transfers involved at every c–step, d(s) ranges from 15ns to 140 ns. The behavioral description contains 13 c–steps that leads to a controller implementation with $T_{FSM}+T_{COMB} = 40ns$.

**Figure 2** Layout example of a controller allowing "adjusted control"

Thus, on this example, idle times are at most 11ns per control step. Using a "synchronous" control, the clock should be set to 180ns at least. Just considering the fastest operation, the idle time would be at least equal to (140–15)*number of executions.

## 2.3 Considerations for high level synthesis

Such a control scheme has consequences on the synthesis process itself since the delay of control steps is no longer a constant but tuned to the delay of operations :

– the cost function to be considered during the scheduling of operations is no longer the number of c–steps but the sum of the delays of the c–steps. This makes most of the published scheduling algorithms inadequate. So we have developed a new scheduler (see Rouzeyre 1994). The results given in section 3 show that adaptative control minimizes not only idle times but also reduces the number of c–steps and thus $T_{FSM}$.

– the second one, related to the first one, is the importance of the choice of the functional units types for the operations of the input algorithm (module selection and module type binding) since the delay of the operations depends on this choice (and thus d(s)).

– It is clear that the profit of using such a clocking scheme versus "synchronous" control will be all the better that the operations delays are very dissimilar. Thus this control model is very adequate when the design space is very large. This is the case when there are many operations types in the behavioral description or when the synthesis system offers the possibility of implementing several operations of the same type (lets say additions) using different functional units types (lets say ripple–carry and look–ahead carry adders) with different delays in the same design (e.g. Timmer 1993).

Let us illustrate these points on the D.F.G. fragment in Figure 3.a. The different possible execution times of those three functions are given in Figure 3.b in which 0i represents a functional unit able to implement the operation. Figure 3.c depicts the scheduling solution according to synchronous control (cost function : the number of c_step). The clock period is then at least equal to 230 ns or 180 ns depending on the binding of operations to functional units (Figure 3.d, 3.e). Figure 3.f shows the scheduling solution with an adjusted clocking scheme (considering the cost function as the sum of the delay of the c_steps. The binding is given in Figure 3.g. The execution time for this fragment is now equal to (80+50)ns=130ns versus 180ns. Results given in Table 1 show that, conversely to what this example suggests, the use of an adaptative control scheme does not increase the number of c–steps ). These points are addressed and detailed in Rouzeyre 1994 a.

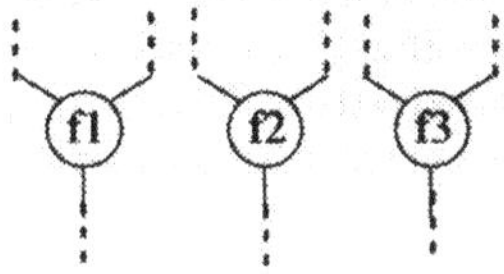


**Figure 3.a Fragment of D.F.G.**

| time(ns) | O1 | O2 | O3 |
|---|---|---|---|
| f1 | 230 | 70 | 50 |
| f2 | 180 | 80 | — |
| f3 | — | — | 70 |

**Figure 3.b Delays and function sets are given for illustration purpose**

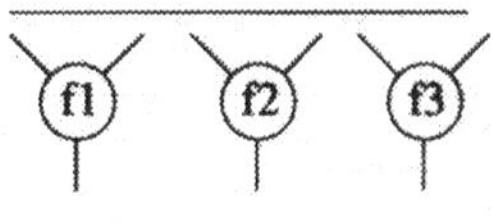


**Figure 3.c Scheduling solution with the number of c_steps as cost function**

| | O1 | O2 | O3 |
|---|---|---|---|
| f1 | 230 | - | - |
| f2 | - | 80 | — |
| f3 | — | — | 70 |

Figure 3.d **T=230ns**

| | O1 | O2 | O3 |
|---|---|---|---|
| f1 | - | 70 | - |
| f2 | 180 | - | — |
| f3 | — | — | 70 |

Figure 3.e **T=180ns**

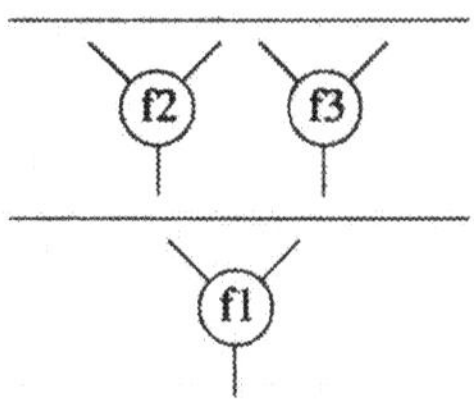


**Figure 3.f Scheduling solution with the sum of the delays of the c_steps as cost function**

| | O1 | O2 | O3 |
|---|---|---|---|
| f2 | - | 80 | — |
| f3 | — | — | 70 |

T1=80ns

| | O1 | O2 | O3 |
|---|---|---|---|
| f1 | - | - | 50 |

T2=50ns

Figure 3.g **T=130ns**

**Figure.3** (Module binding / scheduling / control scheme) impact on circuit performance.

## 3 RESULTS

We have applied this control scheme to the fifth order filter from (e.g. Dewilde 1987). The DFG is composed of 26 additions and 8 multiplications.

First experiment : In order to demonstrate the efficiency of this method with the usual scheduling hypothesis, we have first assumed a synchronous clocking scheme. The delay of the 40ns for the adders and 100ns for the multipliers. Mult–P are two–stage pipeline multipliers with a $2 \times 50$ns delay. Cycle time is taken at 50ns (possible chaining of additions is inhibited). The critical path length is 17 cycles. Table 1 summarizes the results. The scheduling results (number of c–steps) have been obtained from literature (e.g. Hwang1991). The fifth column gives the sum of the slack times Tidle due to the synchronous clocking (under the assumptions on the delays) with Tidle = number of c–steps with only additions multiplied by (50–40)ns.

Second experiment : Then we have assumed an adjusted clocking scheme and the same hypothesis on f.u. delays. Results are reported in the right side of Table 1 using and ad hoc scheduling algorithms. It must be noticed, as mentioned before, that not only the total delay has been reduced but also the number of c–steps. This c–step reduction can not be obtained by using a synchronous control model during scheduling and adjusting the c–step delay afterwards. It is clear that the gain on performance evolves in the same way as the difference between the chosen clock period and the delay of additions.

Third experiment : Table 2 presents results with a richer f.u. library : adders 40 ns, multipliers 100 ns, alus (+ : 50ns, × : 110ns). The sixth column give results obtained from literature with a standard clocking. In the last column, the total delay is computed with a 55 ns clock cycle, with regard to the f.u. delays and assuming that multiplications overlap two c–steps. This shows again that "adjusted control" minimizes the total delay not only by suppressing the idle times but also by minimizing the number of c–steps.

**Table 1** : experiments with synchronous control (clock period 50ns) and adjusted control

| | | | *Synchronous control* | | | *Adjusted control* | |
|---|---|---|---|---|---|---|---|
| *Adders* | *Mult.* | *Mult–P* | *c–steps* | $T_{idle}$ | *total delay* | *c–steps* | *total delay* |
| 1 | ≥1 | – | 28 | 80 ns | 1400 ns | 27 | 1320 ns |
| 2 | ≥2 | – | 18 | 100 ns | 950 ns | 17 | 800 ns |
| ≥3 | ≥3 | | 17 | 90 ns | 850 ns | 16 | 760 ns |
| 1 | – | ≥1 | 28 | 170 ns | 1400 ns | 28 | 1230 ns |
| 2 | – | 1 | 19 | 80 ns | 950 ns | 19 | 870 ns |
| ≥3 | – | ≥2 | 17 | 90 ns | 850 ns | 17 | 760 ns |
| 2 | – | ≥2 | 18 | 110 ns | 900 ns | 18 | 790 ns |

**Table 2**

| | | | *Adjusted control* | | *Synhronous control* | |
|---|---|---|---|---|---|---|
| *Adders* | *Mult.* | *Alu* | *c–steps* | *total delay* | *c–steps* | *total delay* |
| – | – | 2 | 19 | 1190 ns | 23 | 1265 ns |
| 1 | 1 | 1 | 18 | 1000 ns | – | – |
| 1 | – | 2 | 17 | 900 ns | 19 | 1035ns |

## 4 CONCLUSION

We propose a control scheme for high level synthesis systems that allow to minimize the idle time when sequencing data operations. It is mainly valuable for large behavioral description when performance is addressed. This is achieved by using a counter that drives the clock signal of the controller.

The silicon area offset is constant and minor to the controller area itself while the gain on performance is all the better that the heterogeneity of data instructions is high.

## 5 REFERENCES

M.Mc.Farland, A.Parker, R.Camposano (1988) Tutorial on High–Level Sythesis. In Proc. of 25th Design Automation Conference, pp 330–336.

IEEE Design and Test of computers (1990). Special issue on high level synthesis. Vol.7, No.5, Oct. 1990.

D.Ku, G. De Micheli (1990) Relative scheduling under timing constraints. Proc. 27th DAC, pp. 59–64.

P.G. Paulin (1989) Scheduling and binding algorithm for high level synthesis. Proc. 26th DAC, 1989, pp. 1–6.

R. Cloutier, D. Thomas (1990) The combination of scheduling, allocation, and mapping in a single algorithm. Proc. 27th DAC, pp. 71–76.

P.G. Paulin, J.P. Knight (1989) Force–directed scheduling for the behavioiral synthesis of ASIC's. IEEE Trans. on CAD, Vol8, No6, pp. 661–679, June 1989.

R. Camposano (1991) Path–based scheduling for synthesis. Proc. IEEE Transactions on CAD. Vol.10., No.4, January 1991, pp. 85–93.

V. Chaiyakul, A. Wu, D. Gajski (1992) Timing Models for High Level Synthesis. Proc. EuroDac 92, pp. 60–64.

S. Narayan, D.Gajski (1992) System Clock Estimation based on Clock Slack Minimization. Proc. EuroDac 92, pp. 66–71.

C. Hwang, J. Lee, Y. Hsu (1991) A formal approach to the scheduling problem in high level synthesis. Proc. IEEE Transactions on CAD. Vol.10., No.4, April 1991, pp. 464–475.

P. Dewilde, E. Deprettere, R. Nouta (1987) Parallel and pipelined VLSI implementation of signal processing algorithms. In VLSI and Modern Signal Processing. S.Y.Kung,H.J.Whitehouse, T.kailath Editors. Prentice Hall. pp. 257–260.

B. Rouzeyre, D. Dupont, G. Sagnes (1994) Component Selection, Scheduling and Control Schemes for High Level Synthesis" Prop. EDAC–ETC–EUROASIC 94, pp. 482–489.

A. Timmer, M. Heijligers, L. Stok, J. Jess (1993) Module selection and selection using unrestricted libraries. Proc. Euroasic–Edac 93, pp.547–551.

C. Hwang, J. Lee, Y. Hsu (1991) Formal approach to the scheduling problem in high level synthesis. Proc. IEEE Transactions on CAD. Vol.10., No.4, April 1991, pp. 464–475.

## 6 BIOGRAPHY

B. Rouzeyre is Associate Professor at the the ISIM engineering school at the University of Montpellier. He got his PhD degree in 1984 in CAD. He heads the Architecture Group within the Microelectronics Department at LIRMM. Its main research interests are behavioral synthesis of ASICs and testing. He authored and coauthored more than 30 papers on this field.

D. Dupont received PhD degree in computer Aided Design in 1994 from the University of Sciences & Technology, Montpellier II, FRANCE. He have been working since 1990 with the "High Level Synthesis" Department at LIRMM.

PART EIGHT

# Link to Libraries

23

# RAM-based architectural synthesis

*Peter Marwedel, Birger Landwehr*
*Universität Dortmund, Informatik XII*
*D-44221 Dortmund, Germany*
*e-mail: marwedel@ls12.informatik.uni-dortmund.de*

**Abstract**

This paper describes how the TODOS microarchitecture synthesis system* uses information about available library components during the synthesis phases. TODOS stands for "TOp Down Synthesis". TODOS is an extension of the work described in Marwedel (1986). TODOS takes advantage of this information right from the beginning and contains an assignment algorithm considering more component-specific library details than other algorithms. Special care is taken about RAMs as library elements. Possible multiple concurrent accesses are considered in the scheduling and the assignment phases. Possibilities for scheduling reads and writes with common addresses in the same control step are exploited. The assignment algorithm simultaneously generates bindings to ALUs, immediate control fields and memory ports. The paper shows that some control steps do not influence the generated data path. Excluding these control steps from the assigment phase speeds up this phase. An even more important speedup is obtained by using special simplifying rules for the assignment problem at hand.

## 1 INTRODUCTION

Microarchitecture- or high-level-synthesis (HLS) is concerned with the mapping of behavioral descriptions onto register-transfer-level structural descriptions implementing that behavior. The resulting structure is intended to be implemented using components from a technology-specific library.

In contrast to many other HLS systems, our TODOS system favors RAMs over separate registers. The advantages RAMs include the following:

*This research was partially supported by projects BMFT-NT 2850 6 ("SALLY") and ESPRIT-BRA 6855 ("LINK").

- RAM-based designs usually require a simpler interconnection scheme (see e.g. section 8).
- If layout is taken into account, RAM-based designs are even more efficient. The reason is that RAMs usually come with full-custom layout quality. Rauscher et al. (1992) mention the use of RAMs as an important reason for the fact that designs generated by TODOS compete favourably with manual designs.
- Furthermore, RAMs are easier to test than dedicated registers, because they are usually very well-connected to the other components.
- An increasing number of cell libraries now contains multi-port memories. These multi-port memories provide the bandwidth that applications require. Hence, RAM-based architectures are becoming more and more interesting.

A good link between cell libraries and the HLS system is generally considered to be one of the important features that have to be implemented before HLS systems will be commercially successful. This includes a good link to available RAM generators.

For RAM-based synthesis, the way in which the characteristics of available RAMs are taken into account, is very crucial. If these characteristic are taken into account only during the final phases of HLS, there is a large danger of suboptimal designs. For example, the scheduling phase should generate control steps not containing "too many" references to variables. Otherwise, even sophisticated assignment algorithms cannot guarantee any constrained number of memory ports. Furthermore, the assignment of operations to operators and the assignment of variable references to memory ports should be done concurrently. Otherwise, there would be a risc of generating a suboptimal number of interconnections.

In this paper, we will describe how the different tasks have been integrated in our TODOS tool. This paper is structured as follows: Section 2 describes related work. Section 3 presents the information made available in the input to our TODOS algorithm. Section 4 explains our method for variable to storage binding. Section 5 focusses on scheduling and how information about RAMs is used during that phase. Allocation is briefly described in section 6. Section 7 contains the main contribution of this paper. In this section, we describe the assignment phase and the special techniques that we have developed for integrating operator and RAM port assignment. Results and summary are given in sections 8 and 9, respectively.

## 2 RELATED WORK

For a general survey of approaches for architectural synthesis see McFarland (1990). In most approaches, register-based target architectures have been generated. The use of RAMs, despite its well-known advantages, has been described only in few papers.

Some authors (see e.g. Peng (1994)) just consider single-port RAMs. Balakrishnan et al. (1988) describe the assignment of variables to multi-port RAMs. In their algorithm, access requirements for variables are analysed and used in order to map a compatible subset of variables to a RAM. This process is repeated for remaining variables, if such variables exist. The mapping problem is modeled as an IP-problem. In Kim (1993), Kim and Liu map variables (or registers) to RAMs, too. In contrast to Balakrishnan, several RAMs can be considered concurrently. However, the authors assume that both the schedule and the assignment of operations to operators is given. Generating the assignment of operations to operators and of memory accesses to memory ports sequentially may result in non-optimal interconnection patterns. In

Ahmad (1991), Ahmad and Chen present a post processor for simultaneous assignment of operators and memory ports. The assignment problem is modeled as an IP-problem. In this approach, the scheduling does not take information about available RAMs into account.

In all the cases mentioned above, bindings to RAMs and RAM-ports are generated for a fixed given schedule. It is easy to imagine that this potentially results in suboptimal solutions.

Moreover, most existing algorithms use separate phases for binding arithmetic operations to ALUs and for binding variable references to ports. Also, none of the approaches considers interconnections required for generating constants. Suboptimal interconnection schemes may be the result. In this paper, we will describe how the different tasks have been integrated in our TODOS tool.

# 3 INPUT TO TODOS

The input to TODOS consists of the behavior of the system to be designed, the behavior of the library components, and additional user-defined bindings.

The description of the *system behavior* differs from that of other HLS algorithms only by features which are not relevant for the purpose of this paper. The interested reader is therefore requested to refer to other papers on the synthesis from our language MIMOLA (see Marwedel (1986)).

*Example:* As a simple running example of the specified system behavior, we will use the following two assignments:

```
a:= (b - 2 * i) +1;
i:= 0;
```

The description of *library components* contains the functions performed by these components, the codes required for mode control and information about the required wiring of the ports.

Unfortunately, the meaning of the term *port* in the sense of *memory ports* is not consistent with its use in VHDL. In order to access a memory location, a set of (VHDL-) ports (data, address, control and clock ports) is usually required. We will call such a set of VHDL-ports a *meta port* whenever confusion could occur.

For TODOS, we assume that control steps are free of subcycles and that therefore a single clock phase is sufficient. All ports have to remain allocated during each of the control steps. Due to that assumption, we distinguish between four different types of meta ports:

1. *Input meta ports*
   These can be used only to *write* data.
2. *Output meta ports*
   These can be used only to *read* data.
3. *Bidirectional meta ports*
   These can be used to *read* and *write* data, but only one of these operations can be performed in a control step. Since read and write operations are mutually exclusive, they can share the same data port.

4. *Shared i/o meta ports*
   These can be used to *read* and *write* data in the *same* control step.
   Due to the absence of time-multiplexing within control steps, shared i/o meta ports require separate data inputs and output ports. Their address input is not time-multiplexed within a control step. Shared i/o meta ports can be used to read and write during the same control step, provided both operations are for the same address.

*Example:* For the purposes of our running example, let us assume that the libary contains a three-port memory called **ThreePortRAM**. We assume that **ThreePortRAM** has one input (meta) port `p1`, one output (meta) port `p3`, and a shared i/o (meta) port `p2`.

Modules like **ThreePortRAM** denote library components. Such components can be *instantiated.* User-instantiated components will never be deleted by TODOS. Such components therefore form a *minimum set of components* which will always be present in the final generated structure.

Storage components *must* be instantiated by the user. The reason behind this requirement is that there are usually only few choices for RAMs. Choosing RAMs early in the design process results in more information about the library being available to the scheduler. This, in turn, results in a controlled number of accesses to storage components per control step. If required, designs can be repeated with different RAMs. Note that TODOS allows the user to instantiate several RAMs and that the approach of TODOS is not limited to the single RAM that will be used in the example.

A subset of the locations of instances of RAMs can be *designated as locations for user-defined variables.* This is an example of the *user-defined bindings* between hardware resources and the input behavior.

Now that we have described the input to TODOS, we are going to describe the synthesis process with emphasis on handling of RAMs.

# 4 VARIABLE TO STORAGE BINDING

In order to guarantee that the final structure does not require more accesses to variables than possible with the given RAMs, TODOS performs the variable to storage binding for user-defined variables quite early.

*Example:* Assuming that `reg` has been declared as an instance of **ThreeportRAM** and that `reg` is the designated memory for variables, TODOS would transform the variables of our running example into:

```
reg[0]:= (reg[1] - 2 * reg[2]) + 1;   -- a => reg[0]; b => reg[1];
reg[2]:= 0;                           -- i => reg[2]
```

# 5 SCHEDULING

The idea underlying scheduling in TODOS is that an even distribution of operations over control steps like the one generated by force-directed scheduling (Paulin (1987)) is not a primary design goal. The primary goal is to create the fastest possible design under a given set of design constraints. Therefore, it is important not to generate control steps for which the constraints are not met. The constraints should be made as tight as possible. Stok (1991) comes to the same conclusion:

> The best way to seems to develop algorithms, that can deal with both strong hardware bounds and strong time constraints. They can produce good results taking advantage of the reduction in search space provided by these hard bounds.

This requirement is fulfilled by the TODOS ASAP scheduler, which relies on strong resource constraints. If the assignment of an operation to the current control step would violate the design constaints, its assignment is deferred. In the context of this paper, we just consider checking (meta) port constraints for RAMs. This check is complicated by the fact that shared address meta ports can be assigned to two operations in the same control step if their address is identical.

*Example:* For the running example, TODOS would find that all memory references can be assigned to the same control step, because `reg` has enough meta ports. As a result, TODOS would assign all operations to the same control step. If we had put a tighter constraint on the number of ports, more control steps had been generated.

This consideration of library information during scheduling is not possible with any HLS system we are aware of. In fact, for available systems, the number of variable references per control step is even unbounded.

# 6 ALLOCATION

Next to scheduling, TODOS performs allocation of arithmetic components (ALUs). Automatic allocation of ALUs in TODOS is based upon the IP-formulation described in detail in Marwedel (1990).

*Example:* For the running example, we need to implement `+`, `-`, and `*` operations in the same control step. Given a library with types `aluType` (able to implement all three operations), `subType` (able to implement `-`) and `mulType` (able to implement `*`, `+` and `DIV`), TODOS could allocate, for example, one instance of each of the component types. Let us assume, they will be called `alu`, `sub`, and `mul`, respectively.

# 7 ASSIGNMENT

## 7.1 Problem description

The assignment phase in TODOS is special in that it simultaneously generates bindings for ALUs, immediate control fields and RAM meta ports. In this context, the term *immediate control field* denotes a set of controller output bits used to generate constants which are present in the behavioral description and address constants generated during variable to storage binding.

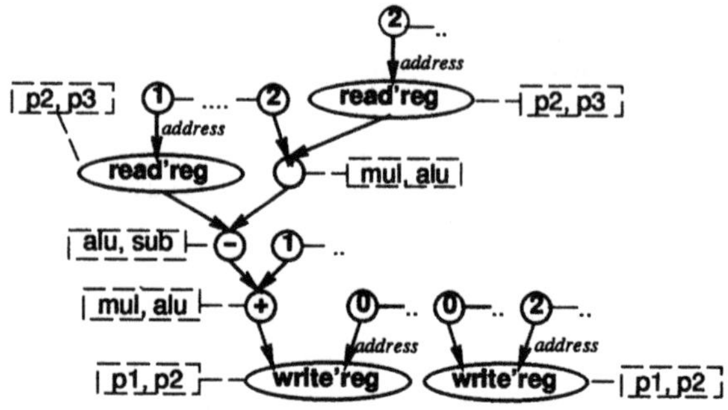


**Figure 1** Internal representation of the running example

*Example:* Consider our running example. In the internal representation of this example, all operations, including reads and writes, are made explicit (see fig. 1). This figure just describes the *data flow* within a single control step and in the general case, there would be one such figure for each control step. In this figure, operations are shown in bold. For reads and writes, the address is used as one of the arguments. For each of the operations, there is a set of *candidate hardware resources*. In fig. 1 candidate hardware resources are shown with thin lines. For reads and writes we have candidate meta ports, for +,- and * we have `alu`, `sub` and `mul` as candidate resources and for all the constants, we have candidate immediate control fields (all fields of the width of the constant, not shown in fig. 1). Hardware resources are represented by dashed lines, if they are not yet assigned to operations and by non-dashed lines otherwise.

The question is: which assignment of resources to operations minimizes interconnect? The assignment is not immediately obvious (at least not to the computer) and several assignments could be tried.

Simultaneous assignment for (meta) ports, ALUs and immediate fields is essential, because the generation of these assignments in sequence would be unable to optimize interconnect. If, for example, we would generate the bindings for ALUs first, we would have no information about the (meta) ports used to read the arguments and to store the results. Hence, we do not partition the assignment phase into subphases for ALU assignment, port assignment and interconnect assignment.

## 7.2 General assignment algorithm

The general assignment algorithm is based upon the observation that –except for the first few control steps– new control steps require only a small number of new interconnections. Therefore, we are using the following approach: We compute a lower bound on the number of missing interconnections. Starting with that lower bound, we execute a branch and bound procedure trying to assign resources to each of the operations, counting the number of required new interconnections. If the number of new connections exceeds the current lower bound, the assignment procedure is aborted and restarted with an increased lower bound. This process terminates as long as the scheduler has not ignored resource constraints.

Two special features are implemented:

- Commutative operations are exploited
- Common subexpressions can be computed either once or several times, depending on what fits resource constraints and required interconnections better. This feature is not implemented in any other HLS system.

## 7.3 Simplification rules

### *Removal of redundant control steps*

The TODOS assignment algorithm simultaneously generates bindings which are normally generated in sequential order. As a consequence, the complexity is very high. In fact, in the worst case, the running time of the assignment algorithm is an exponential function of the number of operations. In order to solve practical cases in reasonable time, control steps are considered step after step. Motivated by the fact that control steps with a large number of operations have the largest impact on the resulting data path, TODOS *sorts* control steps with respect to their number of operations. Assignment then proceeds from the "largest" control step towards the "smallest" control step.

In order to reduce the running time as well as storage requirements, one simplification is used: many of the control steps are equivalent in the sense that they will generate the same data path.

*Example:* All control steps containing just transfers between two storage units are data-path equivalent, regardless of source and destination addresses. For TODOS, the controller is responsible for generating these addresses.

It is sufficient to consider only one representative for each set of equivalent control steps. This idea is implemented in TODOS as follows:

*Definition:* Control steps $i$ and $j$ are *data-path equivalent* $\Longleftrightarrow$ their flow graphs –except for the value of constants– are identical.

This definition is conservative in the sense that it guarantees that TODOS generates the same data path for both control steps. The definition could, however, be extended to reduce the set of control steps that have to be considered further.

The actual procedure for control step ordering uses a list `csl` of control steps, sorted by the number of operations. Each newly generated control step is checked against all control steps with the same number of operations and ignored if data-path equivalent to an existing control step.

### *Exploitation of resource set cardinalities*

The running time of the assignment algorithm remains high, even if we consider only single control steps at a time. This is caused by the fact that in practical cases we found up to 80 operations per control step (this includes reads, writes, concatenations, constants and arithmetic operations). The assignment problem becomes seemingly complex especially if RAMs with several (meta) ports are used. Analysing practical cases, we found that the running time could be reduced by orders of magnitude, if the special structure of the assignment algorithm at hand is exploited using a few simplification rules. With these rules, the running time becomes acceptable for all the cases that we have studied so far.

*Definition:* Operation $o$ is said to be in relation $\approx$ with operation $o'$ $\Longleftrightarrow$ there exists a resource $r$ which is able to execute both $o$ and $o'$.

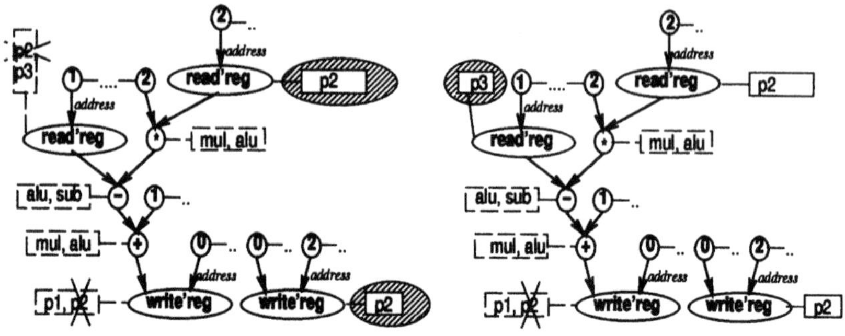

**Figure 2** Implied assignments

*Definition:* Relation $\approx^*$ is the transitive closure of relation $\approx$.

These relations are used in the following procedure which is iterated until no further simplification is possible:

```
FOR ALL operations op of the current control step DO
  BEGIN
    O := {o|o ≈* op};
    R := {r|r can execute at least one operation in O};
    IF |R| = |O| AND ∃r' ∈ R : r' can only perform operation o'
      THEN assign r' to o'
      ELSE IF |R| < |O| THEN --more operations than resources; shared ports required
        IF NumberOfSharedReads= |R| -- all shared ports required
          THEN reduce resource candidates of shared READS and WRITES to shared ports.
    IF |resource candidates for op| = 1 THEN assign resource to op.
  END;
```

*Example:* For the running example, the steps of the algorithm would be as follows:

- Let *op* denote the second **write**-operation. The set $O$ of operations will then contain all **reads** and **writes**, because the p2 is able to perform both operations, implying that the assignment of these operations is not independent. The corresponding set $R$ contains the three meta ports. Since $|R| < |O|$, some of the reads and writes must have a common address (otherwise the scheduler is faulty). Since exactly one read and one write have a common address (2 in this case) and since there is exactly one shared port, reading and writing of location 2 will be assigned to the shared port (see highlighted areas in fig. 2 (left)).
- The next operation considered will the `read'reg(1)`. Only a single resource (p3) is left for this operation and it is assigned (see highlighted area in fig. 2 (right)).

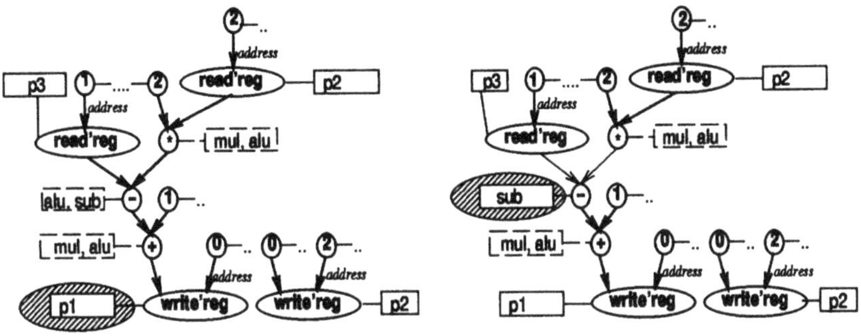

**Figure 3** More implied assignments

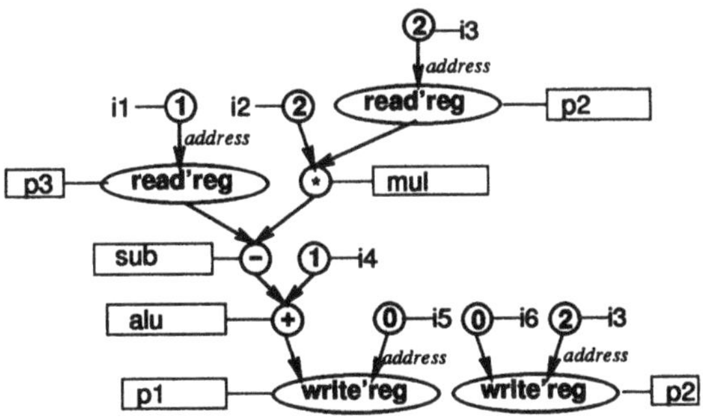


**Figure 4** Final assignments

- The same arguments hold for the first write operation and it is bound to `p1` (see highlighted area in fig. 3 (left)).
- Next, operation - could be considered. - is in relation $\approx^*$ with the other operations and cannot be assigned independently. $O$ will be set to `+, -, *`. $R$ will contain `mul, alu, sub`. We will have $|R| = |O|$. Hence, all resources within $R$ are required. Since `sub` is only able to perform -, it will be assigned (see highlighted area in fig. 3 (right)). Note that we have concluded, that we cannot use the subtraction mode of `alu`. This decision would not have been possible with a covering approach.
- Finally, resource assignments for operations + and * will be made such that interconnections between resources will be minimized, taking predefined connections into account if such connections exist. Concurrently, the same will be done for the immediate fields `i1` to `i6`. Fig. 4 contains possible assignments.

**Table 1** Comparison between register- and RAM-based structures generated by TODOS

| *Example* | *Input ports* | | *input ports * bitwidth* | |
|---|---|---|---|---|
| | *register-based* | *RAM-based* | *register-based* | *RAM-based* |
| PDP-8 | 58 | 51 | 796 | 603 |

**Table 2** Effect of eliminating data-path redundant control steps

| *Example* | *Total number of steps* | *Redundant steps (absolute)* | *Redundant steps (in % )* |
|---|---|---|---|
| Diffeq | 1-12 (several constraints) | 0 (for every set of constraints) | 0 % |
| Ref | 60 | 40 | 66.7 % |
| page68 | 86 | 63 | 73.2 % |
| PDP-8 | 101 | 47 | 46.5 % |
| mergesort | 161 | 134 | 83.2 % |

# 8 RESULTS

Table 1 shows that the PDP-8, if synthesized with TODOS, actually contains a reduced number of input ports if a RAM-based architectural style is chosen. The RAM based design was automatically generated with TODOS. Due to the limited support of register-based architectures in TODOS, the register-based architecture was generated after manually binding variables to registers.

Table 2 shows that 4 out of 5 examples contain data-path redundant control steps. From the table, it can be seen that large examples usually contain a higher percentage of data-path redundant control steps. Hence, this type of redundancy cannot be exploited for the usual small HLS benchmarks. It is, however, important for real applications. `Diffeq` is the standard differential equation solver, `Ref` and `page68` are available from the authors. `PDP-8` and `mergesort` are standard algorithms although we did not use the standard HLSW benchmark sources.

Fig. 5 shows how different constraints for the number of memory ports can be used to generate the fastest design under those constraints. A single RAM with multiple shared i/o (meta) ports has been used. The left part of fig. 5 shows the mergesort example. The right part shows results for the elliptical wave filter example (see e.g. Tsai (1992) for a flow graph). It is important to analyze the solutions a) and b). If the selected library component has 5 (meta) ports, TODOS immediately generates a solution with 10 control steps (solution a) ). Time-constrained scheduling with post-synthesis RAM generation could end up with solution b), which is as fast as solution a) but more expensive. The results cannot be directly compared with other synthesis systems because TODOS unfortunately does not support multi-cycle operations and the elliptical wave filter example has been partially prescheduled to a minimum of

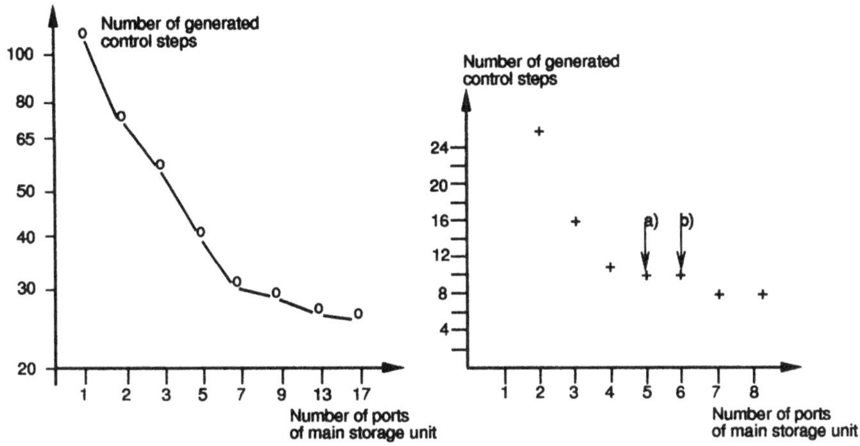


**Figure 5** Resource constrained scheduling for mergesort (left) and the elliptical wave filter (right)

8 control steps. Nevertheless, the essential point is clear: scheduling should take (meta) port constraints into account.

# 9 SUMMARY

This paper describes various aspects of handling RAM library components in the TODOS high-level synthesis system. The contributions of this paper are in the following areas:

- TODOS uses resource constrained scheduling taking characteristics available RAMs into account.
- The assignment phase in TODOS integrates operator-, port-, and immediate field assignments.
- The paper introduces the notion of data-path redundant control steps. Redundant control steps are eliminated in order to reduce the running time of the assignment algorithm.
- Furthermore, information about how to reduce the complexity of memory port assignment is given. The important case of *shared* I/O-ports is considered.

# REFERENCES


I. Ahmad and C.Y. Chen. Post-processor for data path synthesis using multiport memories. *Int. Conf. on Computer-Aided Design (ICCAD)*, pages 276–9, 1991.

A.K. Agrawala and T.G. Rauscher. *Foundations of Microprogramming*. Acadamic Press, New York, 1976.

M. Balakrishnan, A.K. Majumdar, D.K. Banerji, J.G. Linders, and J.C. Majithia. Allocation of multiport memories in data path synthesis. *IEEE Trans. on Computer-Aided Design*, pages 536–40, 1988.

N. Hendrich, J. Lohse, and R. Rauscher. Prototyping of microprogrammed VLSI-circuits with MIMOLA and SOLO-1400. *EUROMICRO*, 1992.

T. Kim and C.L. Liu. Utilization of memories in data path synthesis. *30th Design Automation Conference*, pages 298–302, 1993.

P. Marwedel. A new synthesis algorithm for the MIMOLA software system. *23rd Design Automation Conf.*, pages 271–7, 1986.

P. Marwedel. Matching system and component behavior in MIMOLA synthesis tools. *Proc. 1st EDAC*, pages 146–56, 1990.

M.C. McFarland, A.C. Parker, and R. Camposano. The high-level synthesis of digital systems. *Proc. of the IEEE, Vol. 78*, pages 301–18, 1990.

P.G. Paulin and J.P. Knight. Force-directed scheduling in automatic data path synthesis. *24th Design Automation Conference*, 1987.

Z. Peng and K. Kuchinski. Automated transformation of algorithms into register-transfer level implementations. *IEEE Trans. on Computer-Aided Design*, pages 150–66, 1994.

L. Stok. *Architectural Synthesis and Optimization of Digital Systems*. PhD thesis, Technische Universiteit Eindhoven, 1991.

F.-S. Tsai and Y-C. Hsu. STAR: An automatic data path allocator. *IEEE Trans. on Computer-Aided Design*, pages 1053–64, 1992.

## BIOGRAPHY

*Peter Marwedel* was born in Hamburg, Germany, in 1949. He received his PhD (*Dr. rer.nat.*) in physics from the University of Kiel, Germany, in 1974. Since that time, he has been working on CAD for VLSI. He started to work on high-level synthesis in 1976 and on retargetable microcode generation in 1980. In 1987, he received the *Dr. habil.* degree with a thesis on these two subjects. Currently, he is a professor at the Computer Science Department of the University of Dortmund, Germany. He finished serving as a Department Head in 1995.

*Birger Landwehr* was born in Herne, Germany, in 1967 Between 1986 and 1991 he studied computer science and theoretical medicine at the University of Dortmund and the Ruhruniversität Bochum, respectively. In 1991 he received his diploma degree in computer science from the University of Dortmund. He has been employed as an assistant in the research group "Methodology for computer-aided design of integrated circuits" (Informatik XII) since 1991. His current research activities are concerned with high-level synthesis and intelligent component library mapping.

24

# Module generators and their integration in an architectural systhesis system

*J.F.M. Theeuwen*
*Eindhoven Universisty of Technology*
*P.O. Box 513, 5600MB Eindhoven, The Netherlands.*
*Telephone: 31-40-473344. Fax: 31-40-464527.*
*email: J.F.M.Theeuwen@es.ele.tue.nl*

**Abstract**

*Currently a lot of research is going on in the field of architectural synthesis. Many problems are addressed in this field, like scheduling, allocation, module binding and register allocation. To be able to judge the value of an architectural synthesis system in real world and to make the system appropriate for real world applications an extended library of basic operations is required. Not only the number of basic operations has to be large enough, it should also be possible to make a trade off in terms of area, speed and required number of clock cycles for an operation. In this paper we show how flexible module generators can be connected to an architectural synthesis system. Moreover we will describe two module generators, a divider generator and a multiplier generator.

**Keywords**

Architectural synthesis, module generators.

## 1 INTRODUCTION

Currently a lot of research is going on in the field of architectural synthesis. Many problems are addressed, like scheduling, allocation, module binding and register allocation. There are a number of well known benchmark examples like FDCT (Fast Discrete Cosine Transform) [Mall90] and WDF (Fifth order elliptical filter )[dewi85]. For benchmarking purposes the library of basic operations that is used is rather simple and does not have much relation with the real world. Especially the notion about area and delay is too simple. To be able to judge the value of an architectural synthesis system in real world and to make the system appropriate for real world applications an extended library of basic operations is required. Not only the number of basic operations has to be large enough, it should also be possible to make a trade off in terms of area, speed and required number

*This research has been made possible by the support of the ESPRIT III BRA project 6855, called Link

of clock cycles for an operation. To make the above described requirements possible two things should be available:

(1) module generators that are able to generate basic functions in many different versions (speed, area, number of clock cycles, bit width).
(2) communication from the module generators to the architectural synthesis system, to inform the architectural synthesis system the ability of the module generators.

In the next chapter an overview of the developed architectural synthesis system will be given, with special emphasis on the description and selection of the library elements. After that the generation of a customized library for a certain design will be explained. At the two examples of module generators will be shown.

## 2 THE ARCHITECTURAL SYNTHESIS SYSTEM

The task of an Architectural Synthesis System is to translate a behavioral description of a chip into a digital network description. This network has to be composed of functional modules (like ALU's, multipliers, adders), storage modules (register files, ROM, RAM), and controllers. One of the requirements on these items is that there is an (automatic) way to generate a layout for the required module). The task of translating the behavioral description into a network is rather complicated, so architectural synthesis is partitioned into a number of sub steps:

- Selection: What kind of modules are required
- Allocation: How many resources are necessary
- Binding: Which operations have to be performed on which specific resource
- Scheduling: When should specific operations be activated

In general all these problems are NP complete and a lot of research is going on in these fields. To allow a large freedom for research in the field of architectural synthesis, an interface, the New Eindhoven Architectural synthesis Toolbox (called NEAT) has been developed, which is independent of the synthesis tools and the design trajectory chosen. A central item in this system is the way the design data are represented. In an architectural synthesis system three domains of data can be distinguished:

- Behavior:
  The behavior of a chip is often specified by an algorithm, written in a special language like VHDL, Silage, Ella or Hardware C. To resolve the different natures of the languages a central format is needed. This format should have a close relation with the nature of hardware, but should not impose a particular hardware structure. A Data Flow Graph (DFG) is chosen as the central format [ eijn92]. A DFG is a directed graph consisting of nodes representing operations, and edges representing transfer of values (tokens). The behavior of a data flow graph is defined by a so called token flow mechanism. Tokens are objects which can bear a particular value. Tokens are transported from origin nodes (producing a value) to destination nodes (consuming a value) by the edge which connects the two nodes. A node can start its execution when tokens are available

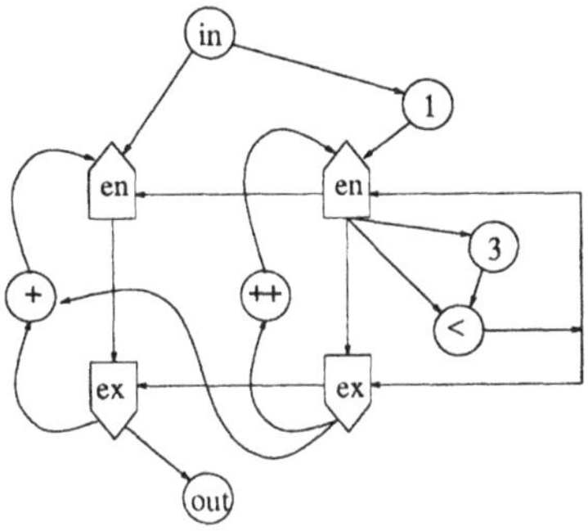


```
process example (b,c)
  in boolean b[8];
  out boolean c[8];
{
  a=1;
  while ( a < 3 )
  do {
    a++;
    b = b + a;
  }
  c = b;
}
```

**Figure 1** A hardware C program with the corresponding DFG

on all the incoming edges of the node. After execution the node produces tokens on all the outgoing edges of the node. To support special language constructs like loops and branches, nodes with a different execution model are introduced. In figure 1 an example of a Hardware C algorithm and its corresponding data flow graph are shown.

- Time:
  A control graph is used to specify the behavior of the controlling finite state machine. The nodes in the control graph correspond to states, and the edges denote possible state transitions. Control graphs are extended with some special constructs to explicitly represent conditionals, loops, multiple active states and hierarchy.
- Structure:
  A network graph is used to describe the resulting digital network. The nodes of the network graph correspond to the physical modules. Edges represent the interconnections between these modules.

## 2.1 Intra-domain relations

If a graph is used somewhere as an operation, it will appear as a single node. The node is called an instantiation of the graph. The type of the node is used to refer to the graph that is instantiated, and hence defines the behavior of the node. Standard operations or modules, like additions, multiplications, ALU's etc. can be created by instantiation of the corresponding graph. These operations or modules can be supplied to the system by libraries. In chapter "libraries in the NEAT system" we will discuss this item more extensively.

## 2.2 Inter-domain relations

In passing through some architectural synthesis processes, relations arise among objects of different domains. A scheduler for instance relates data flow nodes to particular states; a binder relates data flow nodes to functional modules. There are two kinds of such inter-domain relations:

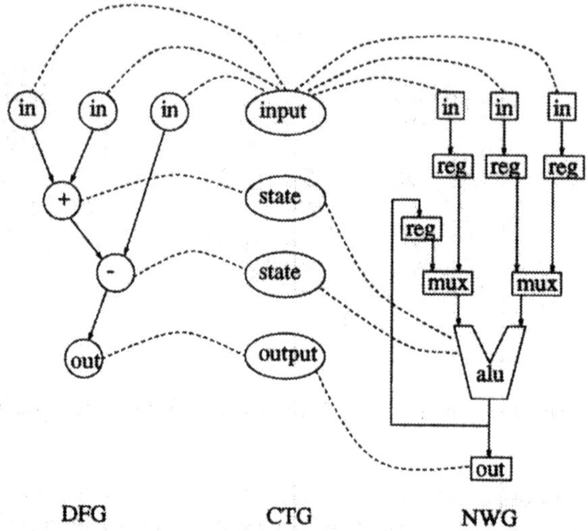


**Figure 2** Representation of a complete design

- Graph level:
  relates behavior, time and structure, i.e. which data flow graph can be implemented upon which network graph, and which control graph specifies the time behavior. The objects used to denote these relations are called graph links.
- Node level:
  relates operations, the states in which they are executed, and the modules upon which they are executed. The objects to denote these links are called node links. Node links are the fine grain relations among graphs; they denote relations between data flow nodes, control nodes and network nodes.

An example can be found in figure 2

## 3 LIBRARIES IN THE NEAT SYSTEM

As explained in a previous chapter the basic operations like addition multiplication etc. are stored in libraries. In these libraries all the properties of the available modules and operators have to be stored. There are three basic items that should be described in the libraries:

(1) The basic operators that are available in the system.
    Examples of operators are:+ , - , *, /, etc.

Together with the operation type the name and the type of the input and output terminals have to be annotated. These data are described in the data flow domain.

(2) The basic modules that are available in the system.
Examples are : adder, subtracter, ALU etc.
From these modules the input and output pins have to be described, ( type, name, bit width). Also properties like, speed and power consumption of the module can be described on this place. These data are described in the network domain.

(3) The relation between the operators and the basic modules.
The library must tell the system that a + operation can be performed on either an adder or on an ALU. This can be described by graph links between the data flow domain and the network domain. Also the relation between the terminals in the data flow domain and the pins in the network domain have to be described. This can be done by the node links.

## 3.1 Generation of customized libraries

As you can imagine the power of an architectural synthesis system is heavily influenced by the size and flexibility of the library. The more different kinds of modules for a given operator exist, the more freedom the synthesis tools have to select the appropriate module, and the more possibilities there will exist to share the functionality of that module with other operators. If for a given module different versions in the area-speed domain are available the architectural synthesis system can try to use the smallest version of a module still providing the required speed. It is also evident that for each module there have to exist versions for a lot of different bit widths. Taking into account all the above described properties that have to be present in the library it is clear that the library will be rather large. This has lead to the decision to generate a specialized library for each design. This specialized library will contain only those library elements that will be useful for the given design, i.e. only the appropriate modules with the appropriate bit widths will be described.

## 3.2 The library generator

The basic structure of the customized library is described in the so called template library. In this library the basic dependencies between operators and modules are described. For instance on this place it is stored that a + operator can be implemented by either an adder, an adder/subtracter or an ALU. By examining the data flow graph to be synthesized, the library generator obtains information about the required operations with the accompanying bit widths. It is evident that it is nearly impossible to design for all the modules many different versions for all the required bit widths and for the different points in the area time domain by hand. This would take a lot of work and it would be very tedious to keep the library up to date. Because of this, module generators for the required operators are written. These module generators not only produce the VHDL description that can be synthesized into layout but also generate data about the performance of the resulting modules in terms of area, delay and power consumption. These data are also used by the library generator to generate the customized library. The overall design and information flow is shown in figure 3.

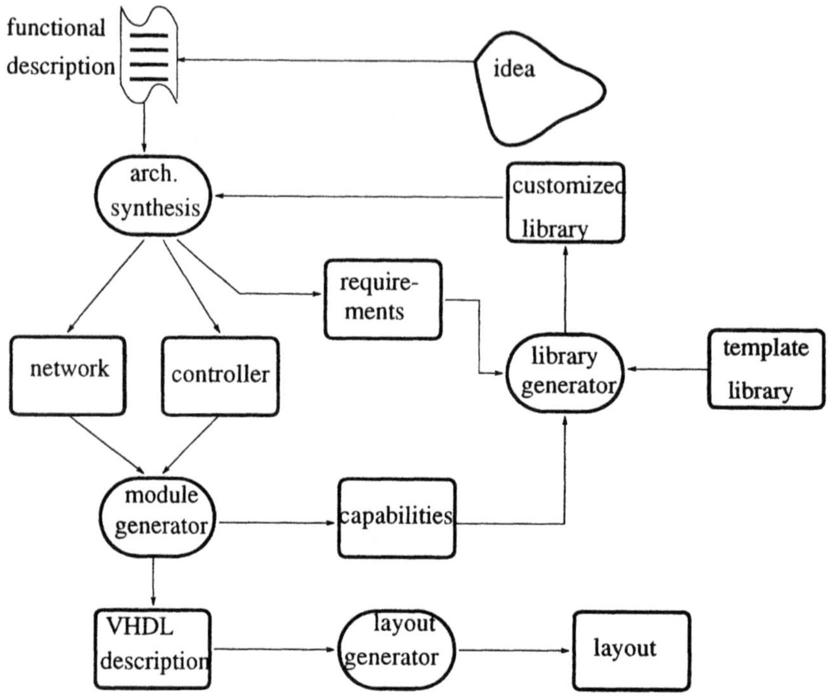


**Figure 3** The design flow

## 4 THE MODULE GENERATORS

In this chapter we will show two examples of module generators that are developed for use in our architectural synthesis system. To be able to use the module generators in as many design systems and technologies as possible the module generators produce a structural VHDL description that can be synthesized into layout by available systems like for instance the "asic synthesizer" from Compass Design automation.

### 4.1 The Block multiplier generator

The architecture used in the block multiplier generator is described in [arts91], [the93]. The basic idea consists of dividing both operands X and Y of Nx and Ny bits into kx and ky blocks of nx and ny bits (Nx = kx * nx, Ny = ky * ny). The blocks of nx and ny bits will be multiplied with each other in a carry save multiplier and the intermediate results are added in an accumulator. By changing Nx and Ny many different versions of a multiplier can be generated.

## 4.2 The block divider

The algorithm chosen resembles very much the tail division algorithm used by people performing a division by hand. First the divider and the dividend are aligned (roughly: most significant bits at the same position) Then during each clock cycle n bits of the result are going to be computed. Each cycle it is checked how many times the divider can be subtracted from the dividend. This is done by trying to subtract the multiples 1 to 2n -1 of the divider from the dividend. The largest multiple that results in a non negative result is subtracted and n bits of the result are known. After this the divider is shifted n places to the right and the process is repeated. Studying this algorithm it is clear that for larger values of n this algorithm takes a lot of hardware because all the 2n - 1 multiples of the divider are to be computed and stored. Moreover there need to be 2n - 1 subtracters to check which multiple of the divider can be subtracted from the dividend. Hardware could be saved here by doing the subtractions sequentially instead of parallel. Currently however this option is not exploited.

# 5 CONCLUSIONS

In this paper we showed that it is possible to make a tight but flexible connection between module generators and an architectural synthesis system. This connection gives the architectural synthesis system the possibility to make full advantage of the capabilities of the module generators. Currently the modules are generated in a standard cell layout style. In the future we will incorporate the use of data paths for some parts of the modules.

# 6 REFERENCES

[arts91]
Arts, H.M.A.M., Stok, L, Eijndhoven J.T.J. van, "Flexible block multiplier generation", Readers Digest of Technical Papers of the international conference on computer aided design, Santa Clara, November 1991, pp 106-110.
[eijn92]
J.T.J. van Eijndhoven and L. Stok, "A data flow graph exchange Standard", Proceedings of the European Conference on design automation, pp. 193-199, Brussels, March 1992
[dewi85]
Dewilde, P, Deprettere, E, Nouta, R,"parallel and pipelined VLSI implementations of signal processing algorithms, pp 258-264, Prentice Hall, Englewood Cliffs, 1985
[mall90]
Mallon. D.J., Denyer, P.B.,"A new approach to pipeline optimization", proceedings of the European conference on design automation, pp. 83-88, Glasgow, March 1990, IEEE Computer Society Press
[the93]
Theeuwen, J.F.M., Arts, H.M.A.M. Arts, Eindhoven van , J.T.J., Sleuters, H.J.H., Wijdeven, J.H.P.,"Module generation in an architectural synthesis environment" pp. 359-371, , IFIP Transactions: Synthesis for control dominated circuits, North Holland, 1993

25

# Towards Better Accounting of Physical Design Effects in High-Level Synthesis

*P. Jha* † *C. Ramachandran* ‡ *F. Kurdahi* § *N. Dutt* †
†*ICS Department, University of California, Irvine, CA 92717, USA*
§*ECE Department, University of California, Irvine, CA 92717, USA*
‡*Cadence Design Systems, San Jose, CA 95134, USA*

**Abstract**
High level synthesis has long relied on *point models* for RT level components, where the area and delay of a component are assumed to be constant for a given style. Nowadays, many synthesis algorithms attempt to incorporate physical design information into the design process so as to better guide the synthesis tasks. In this work, we explore the combined effect of style and aspect ratio variations on the area and delay of RT level components, and on RT level designs which use such components. Our results indicate that point models are inadequate for use in the synthesis process due to the large variations in the area and delay that occur when component styles and aspect ratios are varied. We believe that our results have some deep implications with respect to the flow of the design tasks during high level synthesis.

## 1 INTRODUCTION

High-level synthesis (HLS) has long relied on *point models* for RT-components that assume fixed area and delay values for a given component style. However, aspect ratio variations alone can result in substantially different area-delay characteristics for a component, and thus may result in large area-delay variations for a complete design composed of the same set of components by simply changing the aspect ratios and the ensuing feasible floorplans. In this work, we attempt to realistically explore the effects of varying aspect ratios on the area and delay of complete RT-designs generated by HLS using generic components.

In particular, we are interested in finding answers to the following questions.

1. What are the 1st order factors (parameters) that determine and/or influence the size, composition, and distribution of the "real" design space of a RT-component?

2. How would the nature of the design space of individual components (from 1) influence the overall design space of a "typical" complete design composed of several RT components?

3. The main question we wish to address is: are these "point models" of RT-components adequate for characterizing the resulting space of complete design implementations using

these RT-components? In other words, how much penalty in accuracy is incurred by using the "point model" instead of the "comprehensive model"?

In order to answer these questions, we performed an extensive set of experiments. We present sample results showing the area-delay variations at RT component level (Section 2) and at the complete design level (Section 3). Please refer to [7] for detailed results.

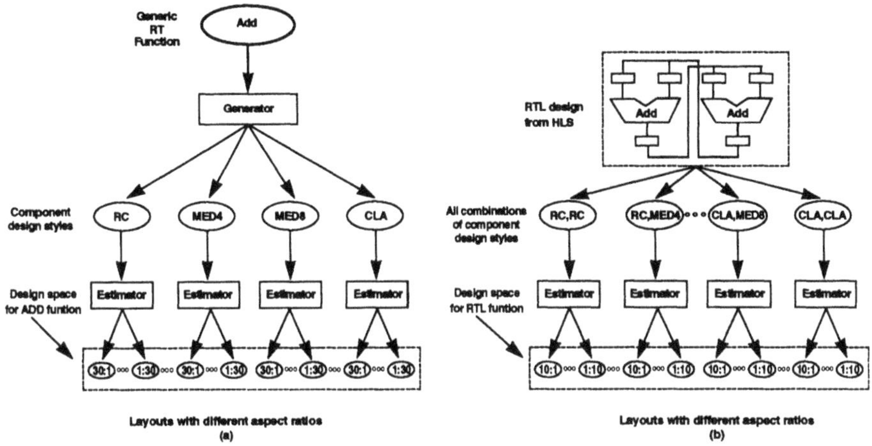


**Figure 1** Experimental setup for design space exploration of an RT component

## 2 EFFECT OF ASPECT RATIO VARIATION ON RT-COMPONENTS

In the first set of experiments, depicted in Figure 1(a), we varied the aspect ratios and styles for an individual RT-component and observed the resulting variations in the component's area and delay attributes.

1. We generated several ADDER(a commonly used generator) components by varying the styles of designs for each bit-width, and repeated this process for different bit-widths. In particular, we generated:

   - Ripple Carry adder (RC),
   - MED4 adder, (i.e., 4-bit carry-lookahead blocks rippled)
   - MED8 adder, (i.e., 8-bit carry-lookahead blocks rippled)
   - Full carry-lookahead adder (CLA), and
   - Carry save adder (CS).

2. For each implementation, we first estimated the area and delay of the corresponding design netlist assuming a unity aspect ratio (i.e. square layout). This essentially gave us a "point model" estimate. The estimation was done by first generating a standard

cell netlist of each implementation, and then used the highly accurate predictors LAST and TELE to obtain estimates of its layout area and delay.

3. We also used the "comprehensive model" to generate area/delay values for different aspect ratios of each RT-component. This capability is built into our model because it employs the constructive-analytical technique described in [1][2] when estimating both area and delay.
4. To analyze the effects of aspect-ratio variation on each RT component, we plotted the results of the experiment in several ways and attempted to capture the overall effects with figures indicating percentage variations in area and delay for each RT component, with respect to different design styles (e.g., ripple-carry or carry-lookahead) and bit-widths.

The area-delay variations across different aspect ratios for 16 bit adder are summarized by the graph in Figure 2(a). In this graph, the point model (corresponding to designs with unity aspect ratios) are represented by big circles.

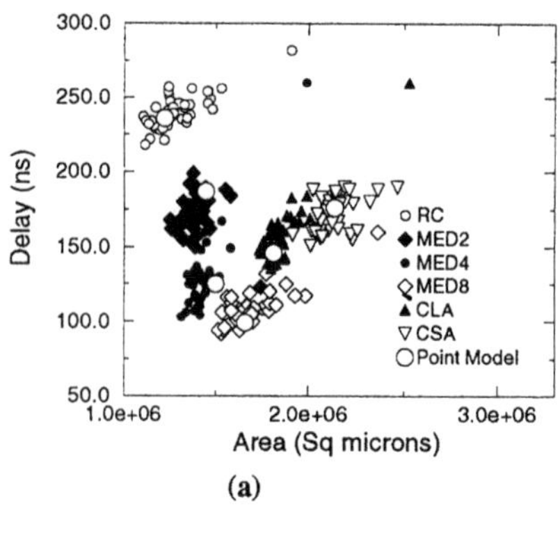


**(a)**

| Style | | Timer | Shift-add mult |
|---|---|---|---|
| *Ripple carry* | *area* | 13.33 | 14.02 |
| | *delay* | 51.68 | 27.68 |
| *4-bit CLA* | *area* | 23.51 | 13.43 |
| | *delay* | 59.36 | 35.79 |
| *8-bit CLA* | *area* | 20.88 | 16.71 |
| | *delay* | 53.88 | 32.95 |
| *Full CLA* | *area* | 17.28 | 11.70 |
| | *delay* | 51.97 | 33.12 |
| *Over all designs* | *area* | 27.86 | 26.58 |
| | *delay* | 92.37 | 64.47 |

**(b)**

**Figure 2** Area and delay variations with varying aspect ratios (a) 16-bit adder (b) some benchmark designs

From the area-delay graphs of Figure 2(a), we observe large variations in area and delay. In most cases, the variations exceed 30% in either area or delay by simply changing the component's aspect ratio, and hence one can easily see that the point models are not adequate for representing the RT-component attributes, and for use in High-Level Synthesis because:

1. RT components are building blocks that will be floorplanned together to form a design layout. Thus, the final aspect ratio (and hence the are-delay attributes) of each component cannot be known until physical design is started. However, point models implicitly assume a single aspect ratio configuration which yields a single area-delay

value for each component – in other words, they assume that area and delay do not vary under different aspect ratio configurations.

2. Since point models are not adequate for modeling the metrics of RT component, they cannot be used as a basis for module set selection prior to, during, or after scheduling and allocation. Instead, one must examine the full design space of these components, or alternatively, include the physical design information into the synthesis process. The latter approach means that component style *and aspect ratio* must be determined at the same time in order to obtain reliable measures of the attributes of the components.

# 3 EFFECT OF ASPECT RATIO VARIATIONS ON COMPLETE RT-DESIGNS OBTAINED FROM HIGH-LEVEL SYNTHESIS

In the second set of experiments, depicted in Figure 1(b), we took the output of high-level synthesis applied to some benchmark descriptions, and studied the effects of varying component styles and the design's overall aspect ratio on the area and delay of two examples: Shift-add multiplier [3] and an industrial timer example. The objective of these experiments was to study the overall RT area-delay design space as populated by different RT-component styles (e.g., CLA adder and RC adder), and by varying the overall aspect ratio of the final RT design, as shown in Figure 1(b). That is, our goal was to compare the effectiveness of the "point" model (using fixed aspect ratios) for covering the complete design space represented by the "comprehensive" model. The results are summarized using graphs that plot area vs. delay, aspect ratio vs. area, aspect ratio vs. delay, and aspect ratio vs AT, as well as tables that show percentage variations in area and delay attributes with respect to the "point" model for the complete RT designs.

The experiments were run in the following manner. For each RT design generated by a HLS tool ([4]) from a benchmark description :

1. We used the "point model" to generate the design space with a fixed aspect ratio using typical RT components, including layout and wiring which were estimated using LAST and TELE.
2. We then modified the aspect ratio of the final design and generated the RT design space for different combinations of RT component styles/implementations, and included the effects of wiring and PD. To make the comparisons realistic, we restricted the aspect ratio variations to between 1:10 through 10:1.

The results for the timer and shift-add multipler benchmarks are shown in Figure 3(a) and 3(b) respectively. Note that the "point model" (with unity aspect ratio) is represented by big circles in these figures. The percentage variation in area and delay with respect to the "point model" for the timer and the shift-add multiplier designs are shown in Figure 2(b). From these figures, we make the following observations:

- In all the cases, we observed a large size design space (as noted from the amount of variations in area and delay from Figure 2(b)). This design space tends to be densely-populated in some regions and sparse as area and delay increase. However, we note that the size of the densely populated region is large, indicating several not-so-obvious design alternatives for similar area/delay constraints. Furthermore, a seemingly inferior design point on the traditional Area-vs-Delay curve (i.e., a design with inferior area and inferior delay) may turn out to be the *only* feasible design due to the shape constraint in the final floorplan and layout.

- In comparing the point models to the overall design space, we observe that the space defined by the set of points in the "point model" is a small subset of the overall design space(Figure 3).
- Given any target delay or area for the system within a certain range, it is possible to find *some* design point which is quite close to that target. This stems from the fact that the design space is quite dense for a wide range of area and delay values. More analysis is needed to be able to quantize this density in general.
- We observe more swing in delay than area (2-3 times or more), which indicates that in contrast to area, delay is quite sensitive to aspect ratio variations (this agrees with the results in [5]).
- The Aspect-vs-Delay graph indicates that designs using the MED* & CLA adders have close delays, while design using the RC adders are clearly slower.
- The designs with MED adders are generally the best for medium-to-high performance, while RC-based designs dominate for area-efficiency, however,
- The A*T figure of merit gives almost equal scores to all implementations, except RC which is higher in most cases. What this indicates is that it may not be worth the small savings in area to go with RC-based designs because of the drastic degradation in performance which would result from such a choice.

Based on these observations, the experimental results show conclusively that aspect ratio variations cannot be ignored during HLS, since variations are quite large and that the "point model" is indeed restricting the design space. We need to consider the "comprehensive model" that factors in aspect ratio variations in order to capture a more realistic design space.

## 4 CONCLUSION

In this work, we attempted to explore the effects of varying aspect ratios on the overall RT design space. Although it is intuitively obvious that aspect ratio variations of RT-components affect the area-delay attributes of complete RT-level designs, we provided conclusive data derived from a fairly extensive set of experiments to verify this fact. We presented experimental data for aspect ratio variations for complete RT-level designs generated by HLS tools from some benchmarks. The variations in area and delay were observed to be significantly large to warrant early inclusion in HLS decisions.

Indeed, we believe that the results we obtained have some deep implications on the "traditional" flow of HLS design tasks where scheduling and allocation are subject to some area and/or delay constraints. Our results raise two interesting questions related to this approach:

1. Given these large variations in the final area/delay, how can one ensure that the area/delay constraints initially imposed on the overall design are indeed satisfied when the design is laid out? This is especially relevant when the synthesized design is part

of a chip which contains other blocks, thus the shape of the final design is not exactly known in advance.

2. Does it make sense to assume "point models" during HLS? Why constrain the HLS to assume a certain (area, delay) point for each component type? Why not have HLS optimize (or select) preferred (area,delay) for each component type since we can find *some* implementation which can be very close to those specs? This argument can be extended to more recent synthesis systems (e.g. [4] and [6]) which incorporate a mix-and-match strategy of component selection, except that here only the aspect ratio is the source of variation in results.

Clearly, these and other interesting questions remain to be answered in light of the results we presented. Future work needs to address the issue of how exactly to abstract our this information for use early in the design (for example High-Level Synthesis).

## REFERENCES

[1] F. Kurdahi and C. Ramachandran, "LAST: A layout area and shape function estimator for high level applications," in *European Conf. on Design Automation*, February 1991.

[2] C. Ramachandran and F. Kurdahi, "TELE: a timing evaluator using layout estimation for high level applications," in *EDAC-92*, 1992.

[3] F. Brewer and D. Gajski, "An expert system paradigm for design," in *The 23rd Design Automation Conference*, 1986.

[4] L. Ramachandran and D. Gajski, "An Algorithm for Component Selection in Performance Optimized Scheduling," *International Conference on Computer-Aided Design*, 1991.

[5] C. Ramachandran and F. Kurdahi, "A combined topological and functionality based delay estimation using a layout-driven approach for high level applications," in *The Euro-DAC*, September 1992.

[6] P. Gutberlet, J. Muller, H. Kramer, and W. Rosenstiel, "Automatic module allocation in high level synthesis," in *EuroDAC*, 1992.

[7] P. Jha, C. Ramachandran, N. Dutt, and F. Kurdahi, "The effects of variations in component styles and shapes on high-level synthesis," *Technical Report 92-115, University of California, Irvine, USA*, December 1992.

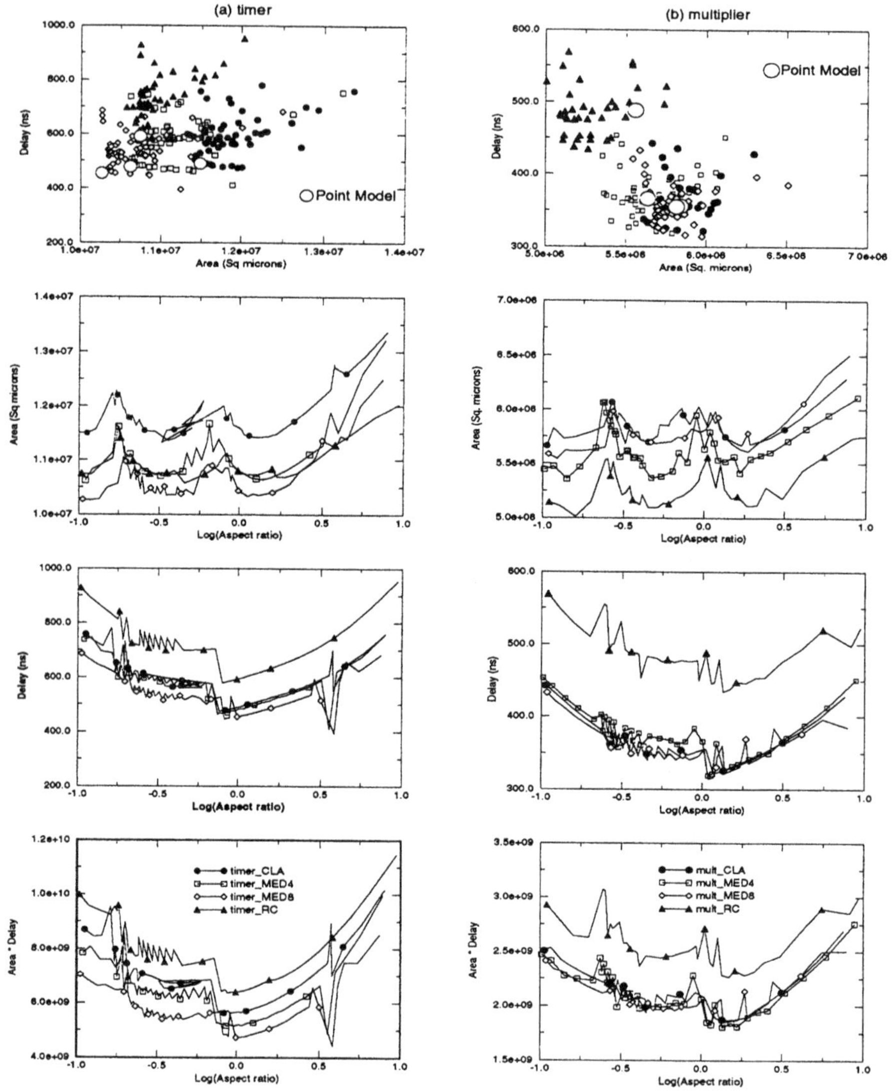


**Figure 3** Area and delay variations of with varying aspect ratios (a) timer (b) multiplier. Both designs assume 16-bit words.

PART NINE

# Operator Type Selection

26

# Delay/Area Trade-Off Exploration Using an Architectural Jiggling Algorithm

*C. J. Rouse and A. J. Carter*
*School of Electronic and Manufacturing Systems Engineering,*
*University of Westminster,*
*115 New Cavendish Street, LONDON, W1M 8JS, UK*
*Tel. +44 171 911 5083 Fax +44 171 580 4319*
*email : rouse@cmsa.westminster.ac.uk*

**Abstract**

This paper describes a novel technique for module-selection exploration. The approach selects from a library containing a wide range of implementations to provide Delay/Area Trade-Off information for the architecture in question. The *Jiggle* Algorithm, which is a modified natural algorithm, is based on Darwinian and accelerated evolution ideas and provides effective *computation time - results* trade-off. Instead of starting the *Jiggle* Algorithm with a single seed, a re-generative multiple-seed approach is taken which drastically reduces the chance of getting trapped in a local minimum. A more appropriate 'cost function' is used in the natural algorithm and a suitable metric is used to evaluate the quality of the DATO graphs produced.

**Keywords**

Module-selection, binding, trade-offs, natural algorithms

## 1. INTRODUCTION

With both the increase in the use of high level synthesis and the increase in size and complexity of VLSI and digital systems, designers need better ways to increase the effectiveness of the synthesis process. Most high level synthesis systems cannot perform extensive design space exploration: they normally restrict one or more parameters while trying to optimise the remaining. A technique is presented here that allows Delay/Area Trade-Off (DATO) assessment of a system by provide a plot of timing and area estimates for a range of architectural configurations.

## 2. DESIGN SPACE EXPLORATION

Design Space Exploration (DSE) includes assessing all aspects and all possible solutions to a particular problem, from the top level system to the device level. Architectural Exploration (Ae) is a sub-space concerned with looking at different architectures for a structural description, including the use of partitioning, scheduling, and allocation (Landman, 1993). A further sub-space can be identified as Module-Selection exploration (MSe), which in this paper is taken to mean experimenting with a fixed pre-determined architecture by changing the hardware module assignment.

Partitioning, resource allocation, and scheduling, can produce many different architectures, even before considering the design space at the functional block implementation level. Attempting to perform Ae and MSe at the same time complicates the process by introducing too many parameters to control. Currently the techniques described here concentrate on MSe so are used after a traditional high level synthesis algorithm has scheduled and allocated the functional units.

## 3. AN LDI IIR FILTER AS AN EXAMPLE SYSTEM

A lossless discrete integrator (LDI) ladder filter has been chosen to demonstrate the approaches proposed here. With the coefficients quantized to 12 bits (not including sign bit), the filter's transfer function closely matches the ideal characteristic (Turner, 1986) (Bruton, 1975). The LDI filter structure used here is shown in Figure 1. To ease the demonstration of the algorithms used here a one to one mapping of the SFG to functional blocks has been used.

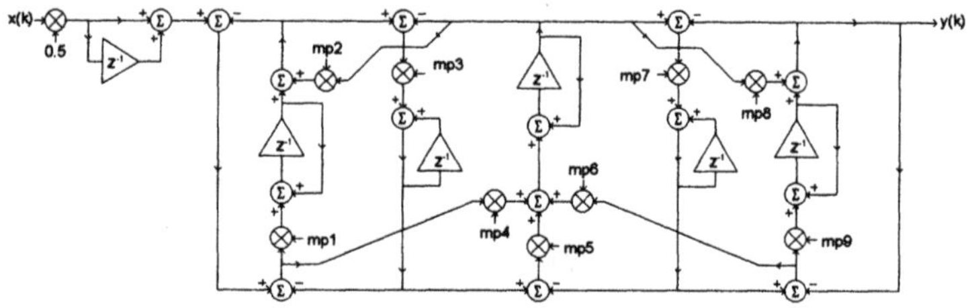


Figure 1 Fifth Order Digital Elliptic LDI Low Pass Filter from (Turner, 1986).

## 4. THE LIBRARY

For each function, an approach should be allowed to select from modules with a large variety of delays and areas, as well as other trade-off factors (Landman, 1993). This is demonstrated in (Timmer, 1993). All the multi-architecture functional blocks should contain all the 'tricks of the trade' (Rechtin, 1992) in different combinations, that the designer would use in a manual design. The novel approach described here relies on module generators to rapidly provide delay and area information (Bidet, 1993), as well as provide silicon layouts of the functional units when necessary.

The gate level implementation for constant multipliers can be optimised, the complexity being directly related to the number of nonzero digits in the coefficient. To show the type of delay/area information provided by the module generators, the coefficients for multiplier *mp2* and *mp9* in Figure 1 were fed into the multiplier generator. The DATO graphs are shown in Figures 2 and 3 respectively.

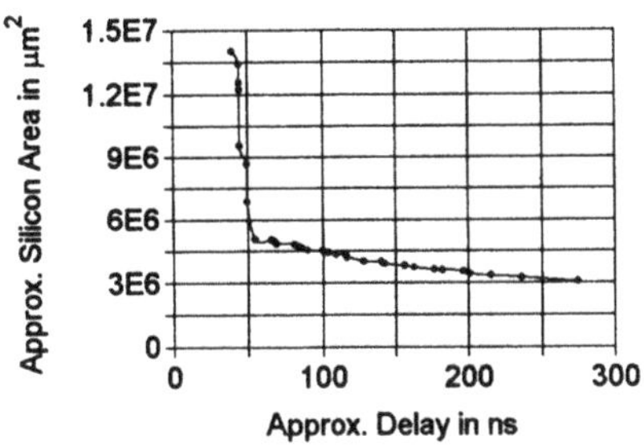


Figure 2 Thirty Five Possible Architectures for Multiplier *mp2*.

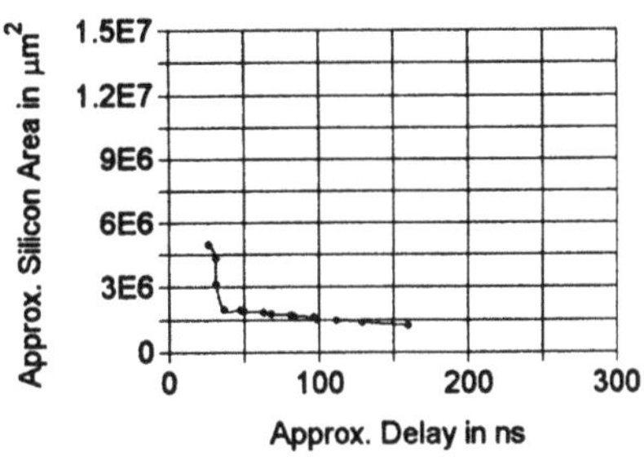


Figure 3 Fifteen Possible Architectures for Multiplier *mp9*.

The gate level implementation for each configuration is constructed from the most primitive standard gate logic elements for the 1.2μ ES2 CMOS process. The availability of specific hand crafted cells will provide much better results than those given here.(eg. adder cells with generate and propagate instead of carry outputs)

# 5. RANDOM SEARCH OF THE MODULE-SELECTION SPACE

For later comparison, this section discusses the use of a simple random search technique for finding an acceptable implementation. The technique uses a uniformly distributed pseudo random variable to select gate level implementations for each function in the SFG. The 'cost', which in this case is obviously speed and area, of the new architecture is evaluated, and to decide whether it should enter the DATO graph it is checked against every entry in turn. If both the speed and area are better, the one checked against is discarded and the new architecture substituted in its place. If either the speed or area is better, the new architecture is added to the DATO graph. If neither is better, the new architecture is discarded. The results of this search for the example are given later.

# 6. NATURAL ALGORITHMS FOR DSE

The *Jiggle* Algorithm used here is based on a Darwinian technique (Cain, 1990) which belongs to the class of natural algorithms, all of which model natural processes. Natural algorithms include among others, genetic algorithms (Wehn, 1991) and simulated annealing (Wehn, 1991) (Kale, 1993). Darwinian techniques are based on Darwin's theories of survival ONLY of the fittest, which implies (for MSe) never accepting an architecture that is slower and larger than one already discovered, even if it may lead to an even better architecture if it was kept and used for subsequent generations.

Natural algorithms are in general easier to implement and visualise than exact and heuristic algorithms, especially when other constraints have to be considered. The main characteristic of natural algorithms is that they are experimental in nature, ie. they experiment in the same way nature does in trying to overcome difficulties.

A combined cost function and single figure of merit are normally used to guide a random natural process. This type of cost function is abstract and detached from the reality of designing a system to a specification. For this reason the more appropriate 'cost' of separate delay and area has been chosen. The 'cost' function is the same as the criteria for entering the DATO graph for the random selection exploration.

## 7. RANDOM JIGGLE AND ACCELERATED EVOLUTION

The simplest of Random *Jiggle* Algorithms takes a seed architecture, randomly chooses a function from the SFG and exchanges its current gate level implementation with one that is either slightly faster or slightly slower (with corresponding change in area). Figure 4 shows the growth of a seed to form a DATO graph. The results of each jiggle are numbered and it should be noted that the result of jiggle 1 would eliminate the seed from the DATO graph. The result of jiggle 4 eliminates 1 and 2, while the sixth jiggle eliminates 3.

The algorithm used here is a modified version of the basic Random *Jiggle* version in that it incorporates the idea of accelerated evolution, it is referred to as the *Jiggle* Algorithm. If the change of a particular functional block implementation results in an improvement, then a similar change is made in the hope that an even better architecture is found. This further slight change in the same direction is attempted every time the architecture is better than the previous one. The *Jiggle* Algorithm locks onto an improvement and takes it to its limit. Accelerated evolution is demonstrated in Figure 5 where a new architecture is accelerated three times.

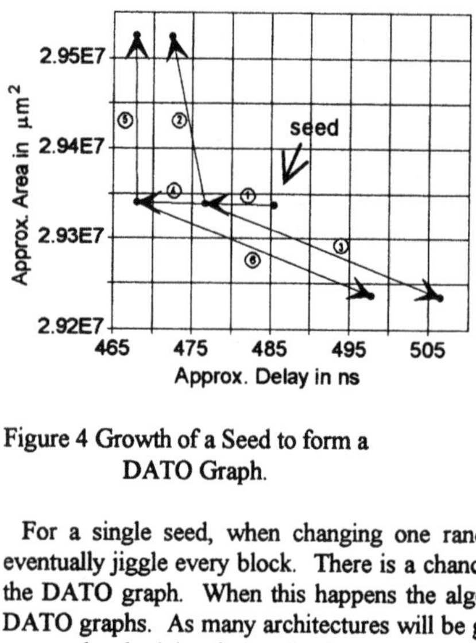


Figure 4 Growth of a Seed to form a DATO Graph.

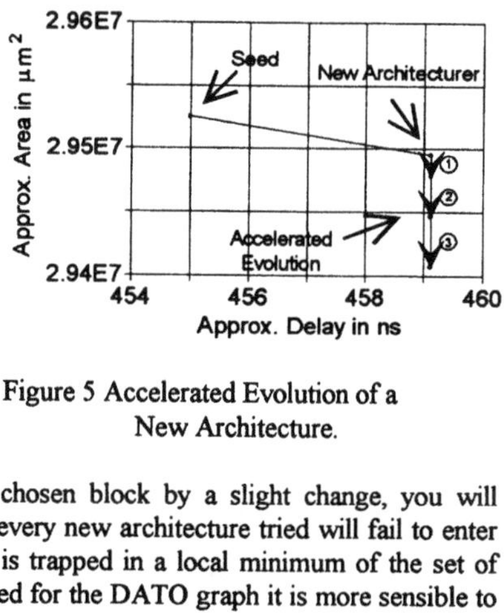


Figure 5 Accelerated Evolution of a New Architecture.

For a single seed, when changing one randomly chosen block by a slight change, you will eventually jiggle every block. There is a chance that every new architecture tried will fail to enter the DATO graph. When this happens the algorithm is trapped in a local minimum of the set of DATO graphs. As many architectures will be generated for the DATO graph it is more sensible to prevent local minima by totally changing the seed at some stage.

Instead of starting the *Jiggle* Algorithm with a single seed, a seed DATO graph is used. This contains the fastest possible and the smallest possible architecture, as well as a range of points in between. The points are produced by heuristic based module selection from the available implementations in the library. The *Jiggle* Algorithm goes through each architecture in the seed DATO graph in turn and jiggles it, accelerates it if possible, and then continues to the next DATO point when the current one fails to enter the graph. When all the points in the current graph have been jiggled another pass is performed if the required number of jiggles has not been reached. Each full DATO graph pass will generate extra points in the DATO graph which are then used as the seeds for the next pass. Using a constantly changing seed drastically reduces the chance of getting trapped in a local minimum of the set of DATO graphs.

# 8. RESULTS

Extensive trails were carried out on the demonstrator system. To show the *Jiggle* Algorithm in action and demonstrate its advantages, results for various numbers of iterations are shown and compared to the same number of iterations for the Random Search method. DATO graphs for 10 to 100 million iterations (in increasing powers of 10) for the Random Search method are shown in the cascade plot in Figure 6 and for the *Jiggle* Algorithm in Figure 7. (The graph at the back is the seed graph.)

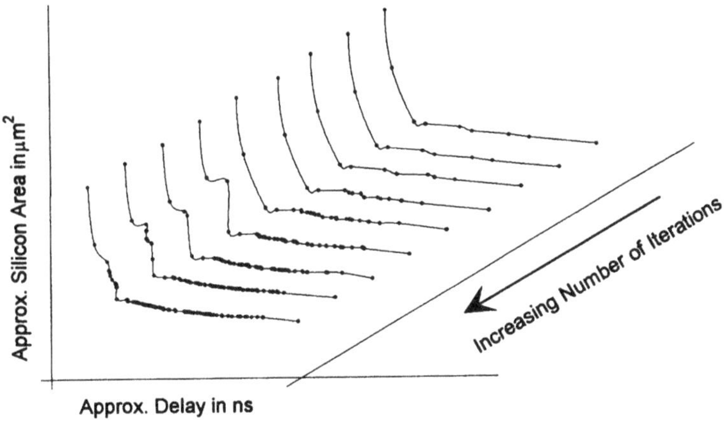


Figure 6 Cascaded DATO Graphs for LDI IIR Filter Using Random Search.

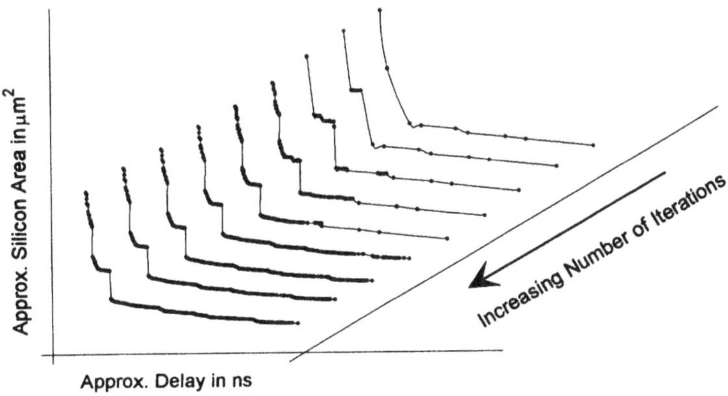


Figure 7 Cascaded DATO Graphs for LDI IIR Filter Using Jiggle Algorithm.

As can be seen the *Jiggle* Algorithm produces much superior DATO graphs in terms of accuracy at every number of iterations, but at the expense of extra computation time, as shown in Table 1.

Table 1 Computation Times for Jiggle Algorithm and Random Search (for a 50MHz 486)

| *Number of Iterations* | *Jiggle Algorithm* | *Random Search* |
|---|---|---|
| 10 | 0.05 sec | 0.11 sec |
| 100 | 0.16 sec | 0.22 sec |
| 1000 | 0.55 sec | 0.44 sec |
| 10 000 | 2.52 sec | 2.30 sec |
| 100 000 | 27.79 sec | 20.48 sec |
| 1 000 000 | 5 min 44 sec | 3 min 31 sec |
| 10 000 000 | 58 min 22 sec | 37 min 49 sec |
| 100 000 000 | 9 hr 45 min | 6 hr 52 min |

The metric of the number of DATO points in the graph is not appropriate if you are judging the quality of the DATO graph. The exact number of points has little relation to the quality of the graph as was demonstrated in Figure 4 where new architectures nullify previous entries. For this reason it has been chosen to use the area under the DATO graph as the metric for comparison. Figure 8 shows the areas under the DATO graphs when using the Random Search and *Jiggle* Algorithm. The entry at '1 iteration' is for the simple heuristic used as the seed for the *Jiggle* Algorithm and is also used to kick start the Random Search. This is done for comparison purposes.

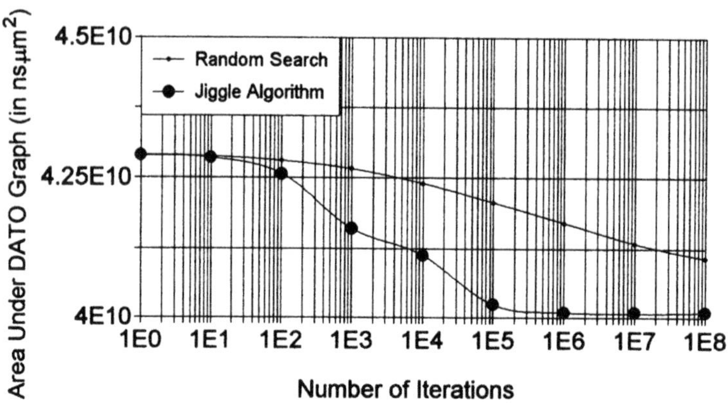


Figure 8 Area Under DATO Graphs.

## 9. CONCLUSIONS

An exhaustive search of all possibilities will take an excessively long time but a well chosen subset of these will be a useful guide for becoming familiar with the task in question. The basic shape and vital points of the curve are established quickly. Even 28 seconds of jiggling will give a valuable insight, showing that long run times of random based natural algorithms need not be tolerated, but for good results run times can be as long as the user desires or has to spare. This is useful to the designer who dose not want to wait long for a rough guide and designers who want a final solution.

The *Jiggle* Algorithm which includes a heuristic for accelerated evolution, is an easily implementable method that it is worth considering as an alternative to exact approaches. For finding an optimal solution immediately, exact and heuristic approaches (Sakouti, 1993) (Karkowski, 1993) are normally better, but at a cost, especially for very large or complex systems. These results demonstrate that natural algorithm based MSe is an acceptable method of design space exploration if sufficient DATO data is available for the individual blocks in the system. Module generators are of prime importance.

## 10. REFERENCES

Bidet E., C.Joanblanq and P.Senn, (1993) GENRIF: An integrated VLSI FIR Filter Compiler, European Design Automation Conference, 466-471.

Bruton, L.T., (1975) Low-Sensitivity Digital Ladder Filters, IEEE Transactions on Circuits and Systems, Vol. CAS-22, No. 3.

Cain G.D., A.Yardim, (1990) Darwinian Design of Digital Filters, Proc. IEEE Workshop on Genetic Algorithms, Simulated Annealing and Neural Nets Applied to Problems in Signal/Image Processing and Communications, 1-14.

Kale I., P.Belloni, G.D.Cain, E.Del Re, (1993) Guided Simulated Annealing with Tempering for FIR Filter Design Via Z-Plane Zero Manipulation, Proc. Int. Conf. on Digital Signal Processing, 390-395.

Karkowski I., (1993) Circuit Delay Optimization as a Multiple Choice Linear Knapsac Program, European Design Automation Conference, 419-423.

Landman P.E., J.M.Rabaey, (1993) Power Estimation for High Level Synthesis, European Design Automation Conference, 361-366.

Rechtin E., (1992) The Art of Systems Architecting, IEEE Spectrum, 66-69, October.

Sakouti, K., P.Abouzeid, M.Belrhiti, M.Crastes, and G.Saucier, (1993) Coherent Optimization Strategies for Multilevel Synthesis, European Design Automation Conference, 378-385.

Timmer A.H., M.J.M.Heijligers, L.Stok, and J.A.G.Jess, (1993) Module Selection and Scheduling using Unrestricted Libraries, European Design Automation Conference, 547-551.

Turner, L.E and B.K. Ramesh, (1986) Low Sensitivity Digital LDI Ladder Filters with Elliptic Magnitude Response, IEEE Transactions on Circuits and Systems, Vol. CAS-33, No. 7.

Wehn N., M.Held, and M.Glesner, (1991) A Novel Scheduling / Allocation Approach for Datapath Synthesis based on Genetic Paradigms, Proc. Int. IFIP Workshop on Logic and Architecture Synthesis, 47-56.

## 11. BIOGRAPHY

Chris J. Rouse received the B.Eng(Hon) degree in electronic engineering and the MSc degree in digital signal processing systems from the University of Westminster, London, in 1990 and 1992, respectively. He is currently working towards the PhD at the University of Westminster, London. His research interests include DA, DSP, and VLSI. He is a member of the IEEE, the ACM, and the IEE.

Alison J. Carter received the B.Sc degree in Mathematics from Imperial College, London, in 1977. She is currently a senior lecturer at the University of Westminster, London. Her research interests include all aspects of CAD.

27

# The Influence on Synthesis of Modern Computer Arithmetic

*Anne Mignotte, Jean Michel Muller and Olivier Peyran*
*URA CNRS 1398, Laboratoire de l'Informatique du Parallélisme*
*LIP, École Normale Supérieure de Lyon, 46 Allée d'Italie, 69364 Lyon Cedex 07 France, Telephone: (33) 72 72 85 49, Fax: (33) 72 72 80 80.*
*email:* `amignott, jmmuller, opeyran @lip.ens-lyon.fr`

**Abstract**

Taking into account the various possibilities offered by computer arithmetic during high-level synthesis may help to design much powerfull architectures. The arithmetic operators (mainly in floating point arithmetic) may be carefully tuned up to exactly fit the time and accuracy requirements. This, of course, may have an influence on the synthesis process. In this paper, we mainly address two topics: first of all the problem of numerical error control; and then the possible use of redundant number systems – which allow some nice features such as carry free addition or digit-serial, most significant digit first arithmetic operations. We then propose some potential applications and we discuss the new synthesis problems they involve.

**Keywords**

Arithmetic, redundant number systems, floating point number, high level synthesis

## 1 INTRODUCTION

Macro blocks libraries are essential parts in synthesis tools. They allow to hierarchically create a design, as blocks within a library can be synthesized separately. They also allow the use of an entity regardless of its architecture: the designer could change the architecture without modifying the whole design, or he could propose several architectures for the same entity: the main design would only deal with its specification (i.e its I/O ports). To handle such a top-down approach, every synthesis step has a link with those libraries. The higher the level of synthesis, the more functional the view of the library. Macro blocks will be used at different levels of granularity: system level synthesis will view an entity as a process, as RTL synthesis will see synchronised register transfers.

A library that no designer can avoid is the arithmetic one. For a simple address computation, one might need an adder. However, arithmetic libraries usually do not use all the capacity of computer arithmetic. Our aim is to allow CAD tools to synthesize behavioral descriptions using standard floating point and "exotic" arithmetics. As we will see, those arithmetics can not be handled only as macro blocks instanciation. However, even if one needs a global approach, the designer should not need any particular knowledge on those arithmetics: the synthesis tool will manage the communications to the outside in an usual number representation. Those new arithmetics imply new scalar types, new number representations and new operator architectures.

Up to now, the main used scalar types in VHDL were integer and real. However, even if real numbers were in a floating point representation, their definition was far from the IEEE standard: the mantissa and exponent have the size of the "single precision" standard, but there is no constraint on the accuracy of the operations using floating point numbers (see section 2 for the constraints imposed by the standard). That accuracy is the main difficulty in designing architecture using standard floating point numbers.

Concerning number encoding, synthesis tools only deal with conventional radix 2 representation. There are other interesting representations though, particulary the so-called redundant ones, which allow the definition of carry free adders.

In fact, some libraries propose operator architectures using internally redundant representations, but with I/O in a conventional representation. Yet, available operator architectures are generally most classical. If a designer wishes to implement a fast and inovative arithmetic, he should first create a new library of components, and then handle, in the designing of his circuit, all the communications that an original arithmetic implies. However, we will see that this is not satisfaying in terms of accuracy.

The first section discusses floating point arithmetic, the IEEE standard and a proposal to gloabally handle it. The second section treats exotic arithmetics using redundant systems, and the way to take into account those particular encodings in the synthesis steps. The last section finally deals with pipeline and its impact on scheduling, as redundant systems allow interesting digit-serial arithmetics.

# 2 FLOATING POINT ARITHMETIC

## 2.1 The IEEE 754 standard

The first system used for real numbers was called fixed point representation. In such a system, real number, $r$, is represented by an integer $n$, after the designer had implicitely choosen a *scale factor* K, which needs to be a *power of the radix* in which integers are represented. The integer $n$ satisfies $r = n \times K$ (for example 26.722 is represented by 26722 with scale factor $10^{-3}$). As the definition of K is implicit, this implies that every number uses the same scale. This is the reason why it is called *fixed* point representation. One might need to use real numbers in a large range though, and fixed point representation would require the use of huge registers.

Floating point number systems aim at representing very huge, as well as very small, numbers. The IEEE 754 standard implies a particular representation: a number $x$ is characterized by its sign $s$, its mantissa $x_0x_1 \ldots x_{M-1}$ and its exponent $e_0e_1 \ldots e_{k-1}$ such as $x = s \times 1.x_0x_1 \ldots x_{M-1} \times 2^{e_0e_1 \ldots e_{k-1}}$. The mantissa and the exponent can have two different sizes: either (respectively) 24 and 8 bits – this is called the single precision – or 53 and 11 bits – this is called double precision. One might notice that, if we add the sign bit, a real number in single precision would be encoded on 33 bits (65 in double precision), which seems not very clever. In fact, as the standard imposes the first bit to be 1, it is not

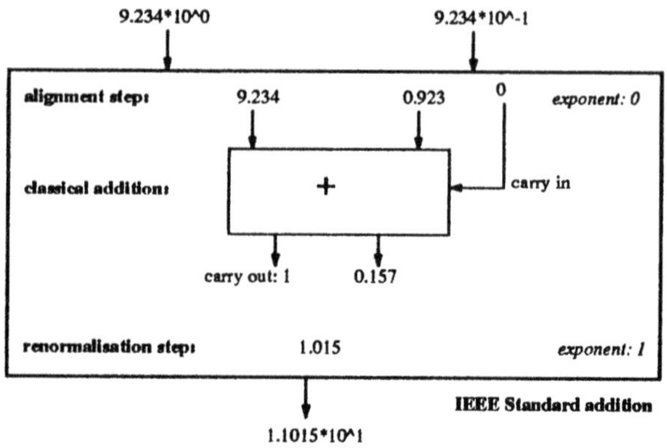


**Figure 1** Example of floating point addition. Numbers are in radix 10, with 4 bit mantissas

represented – and the mantissa is effectively encoded on 23 bits (52 in single precision). This implies a special representation for 0.

The arithmetic operations become slightly more difficult; let us examine the addition. As the two operands may not have the same exponent, the first step consists in an alignment of the operands: assume that the difference between the two exponents is $p$. The mantissa of the smallest operand is shifted by $p$ bits to the right. This means that the least $p$ significant bits are lost. Then, there is an addition of the two mantissas. The most significant bit of the result may be different from 1, or, contrarily, the operation may return a carry. Therefore, the last step is a renormalisation of the result: in the first case, the result is shifted to the left until the most significant bit becomes non-zero (and the exponent is consequently decreased). The least significant bits are replaced by 0. For the second case, the result is shifted by one position to the right, and the exponent is increased by one. The least significant bit is lost. Figure 1 shows an example.

## 2.2 Inaccuracy in floating point operations

During a floating point addition, if both operands have different exponents, it is obvious that there might be a loss of accuracy: the least significant bits might have been non-zero if they had not been lost in a precedent alignment step. In fact, that loss of accuracy could occur in a single addition. Let us consider non standard floating point numbers (to simplify the example): numbers are in radix 10, using a 2 digits mantissa. As an example, let us examine the addition of $A = 1.0 \times 10^0$ and $B = -9.9 \times 10^{-1}$ (1 - 0.99). The first step leads to the loss of the least significant bit of B. That phenomenon is called absorption. Then comes the addition of 1.0 and $-0.9$. The result is 0.1, and a normalisation is neccessary. The previous loss of the least significant bits now becomes really important, as the result is not a rounding of the exact result anymore. That phenomenon is called cancellation. The final result is $1.0 \times 10^{-1}$ (0.1 instead of 0.01)! Fortunately, that loss of accuracy is proportionnal to the size of the mantissa. However, even if, in most cases, the inaccuracy is bearable, its repetition may lead to nonsensical results. Therefore, the IEEE standard imposes basic operations to return a rounding of the *exact* result and proposes five different *rounding modes*: rounding towards zero, rounding up, rounding down, rounding towards infinity (i.e. rounding up if the number is positive and rounding down if the number is negative), and rounding to the nearest.

To satisfy this last constraint, one is forced to keep the last two bits (called guard and round bits) lost in the alignment step, and to compute another one, called sticky bit, which indicates either if the other lost bits were all zero, or if one were a non-zero. This computation is a logical *or* between the *effectively* lost bits (i.e the lost bits excepted the two "kept bits"). The standard also proposes special formats for representing particular cases such as plus or minus infinity — which can result from overflows or underflows —, plus or minus zero - which can happen when the exponent is below the minimum possible value -, Not-a-Number — e.g. $\sqrt{-1}$, $\infty - \infty$ or $0/0$. These special cases are also part of this "work of accuracy", as those special number can only appear after an arithmetic operation.

## 2.3 Implementation of floating point systems

Up to now, a designer who wishes to create a design using floating point arithmetic should himself define the three steps: alignment, mantissas addition and renormalisation. He will usually not follow the IEEE standard, as the constraints may be too drastic for particular applications. But as he knows, for example, if the $\infty$ or plus or minus zero cases can occur, he will create his design to be able to handle it. Usually, special formats are treated by software (for example, division by 0 is treated as a software interrupt in most of the arithmetic coprocessors). The designer also knows the range of the numbers he will have to use, and he can decide to use, in some parts of his design, larger representations, so that he would not lose any accuracy. In fact, using a larger accuracy can be considered as using a fixed point representation: $1.0 \times 10^0$ plus $-9.9 \times 10^{-1}$ would be handled as 1.00 plus $-0.99$; these are fixed point numbers, represented by 100 and $-99$, with a scale factor of $10^{-2}$. This approach cannot be generalized to a complete design, as the floating point numbers are used to represent reals in a very large range, a fixed point representation would imply huge register sizes.

A designer using a top-down approach should define a macro block library using the fully 754-IEEE standard. However, such an approach would not be efficient for two reasons. First, those blocks would need renormalisation and alignment parts. *Every* operation does not need those steps though. In fact, there would probably be some operators which would never need them, and the design would use very complex operators for conventional operations. A way to improve this would be the definition of operators with different levels of granularity and the instanciation of the useful blocks. For instance, the library could contain a sticky bit computation block, a rounding to zero block, ... Then the designer may know that the operator does not need the computation of the sticky bit, or that the result should always be normalized.

But this is not enough, and the second and main reason why macro block libraries is not a good solution is accuracy. Let us consider as an example:

$(4.200 \times 10^1 + 5.486 \times 10^{-2}) - (4.200 \times 10^1 - 1.213 \times 10^{-2})$ (rounding to the nearest is used).

The result should be $6.699 \times 10^{-2}$. Let us examine the different steps (see Figure 2).

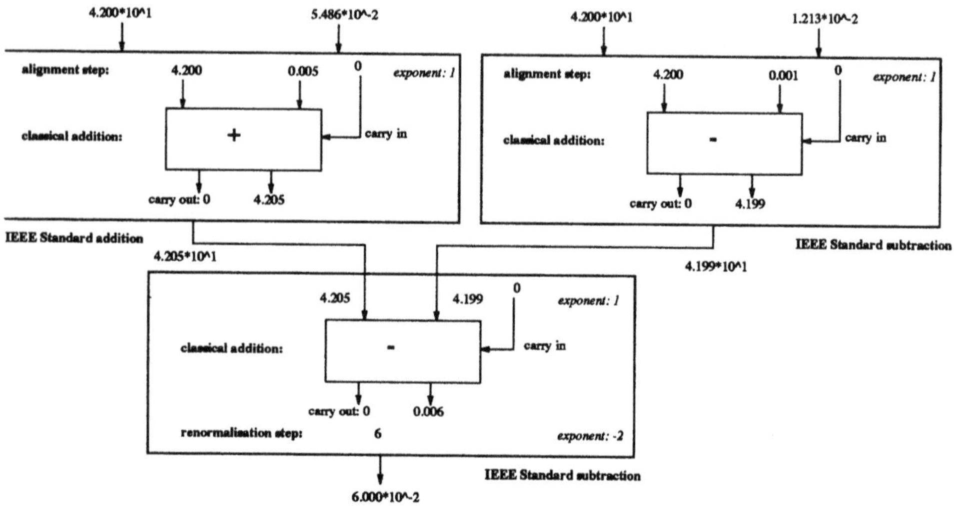


**Figure 2** Inaccuracy in floating point operations. Numbers are in radix 10, with 3 bits mantissa

The first addition, $4.200 \times 10^1 + 5.486 \times 10^{-2}$, needs an alignment: $5.486 \times 10^{-2}$ becomes $0.005 \times 10^1$ The result is $4.205 \times 10^1$, which is effectively a rounding of the exact result. In the same way, the subtraction returns $4.199 \times 10^1$, which also follows the IEEE standard. The final result will then, after renormalisation, be $6.000 \times 10^{-2}$, which is the *exact* result of the operation, but which is far from the exact result of the global operation ($6.699 \times 10^{-2}$, more than 10% inaccurate)! This simple example shows that accuracy has to be treated globally and not locally by rounding at each operation.

## 2.4 Handling floating point systems during synthesis

Our aim is to allow a designer who has defined floating point numbers to use those number just as if they were integers: he would not have to consider alignment and renormalisation steps; he would neither have to consider the need of accuracy each operation implies. The synthesis tool should handle all these steps. One solution would be, as it was explained before, the definition of a full 754-IEEE operator library. This could be adapted to any synthesis tool, though it would not remove the problem of accuracy.

Therefore, our approach is more general. The designer indicates an approximative range for his numbers and the synthesis defines parts within the ciruit which should use fixed point arithmetic, and most of all, the size of the integers used for the representation and the scale factor. Another idea would be to modify the arithmetic expression parenthesis, with respect of commutativity, associativity, ... properties, in order to group operands in the same range of exponents. The previous example would become:

$(4.200 \times 10^1 - 4.200 \times 10^1) + (5.486 \times 10^{-2} + 1.213 \times 10^{-2})$.

Moreover, as the apparition of the special cases, such as $\infty$ or plus or minus zero, usually implies special treatment, and as this usually means the end of the calculations (once $\infty$ has appeared, the following calculations would not have any meaning), these treatments could be avantagely handled by software. This brings synthesis to a codesign step. The designer indicates what special cases are to be treated, and the synthesis tool generates the appropriate software, and handles communication between software and hardware.

Our proposal is closed to the approach of a designer, and allows a more precise result (even more precise than what the IEEE standard imposes). Consecutive operations shall be seen as "black boxes" internally using fixed point arithmetic, but with inputs and outputs following the 754-IEEE standard. This imply a specific synthesis tool, but this is obviously the most satisfying solution.

Independently of the choice of using fixed or floating point arithmetic, there is the question of performing fast integer operations (addition of integers for fixed point numbers, of mantissa for floating point numbers...). Synthesis systems usually use conventionnal arithmetics, though there are much faster ones. Next sections will discuss two particular kind of arithmetics, redundant and on-line arithmetics.

# 3 REDUNDANT NUMBER SYSTEMS

## 3.1 Introduction and presentation

In 1961, A.Avizienis suggested to represent numbers, in base $B$, using digits between $-a$ and $+a$, where $a$ is an integer. As usual, the digit chain $x_n x_{n-1} \ldots x_0$ represents the number $x = \sum_{i=0}^{n} x_i B^i$. One can easily prove that all the numbers can be represented with such a number system, provided that $2a+1 \geq B$. If $2a+1 > B$, then some numbers have several possible representations: the number system is *redundant*. The main interest of the redundant number systems lies in the following result:

**Property 1** *If $2a \geq B+1$ and $a \leq B-1$, then the following algorithm makes it possible to compute the sum $s = s_n s_{n-1} \cdots s_0$ of $x = x_{n-1} \cdots x_0$ and $y = y_{n-1} \cdots y_0$.*

**Algorithm 1** *1. Compute for all $i = 0 \ldots n-1$ the values:*

$$\begin{aligned} t_{i+1} &= \begin{cases} -1 & \text{if} \quad x_i + y_i \leq -a \\ 0 & \text{if} \quad -a+1 \leq x_i + y_i \leq a-1 \\ 1 & \text{if} \quad x_i + y_i \geq a \end{cases} \\ w_i &= x_i + y_i - Bt_{i+1} \end{aligned}$$

*2. Compute for all $i = 0 \ldots n$ the values $s_i = w_i + t_i$, with $w_n = t_0 = 0$.*

This algorithm is not very different from our usual addition algorithm: the values $t_i$ are equivalent to the carries of the usual algorithm. The main difference is that $t_{i+1}$ *does not depend on* $t_i$, so the values $w_i$ and $t_i$, then $s_i$ can be computed in a fully parallel fashion, without any carry propagation. Conditions $2a \geq B+1$ and $a \leq B-1$ of the previous algorithm cannot be simultaneously satisfied in radix 2. However, it is possible to perform additions without carry propagation in radix 2, using digits -1, 0 and 1 (this is the *borrow save* number system), or digits 0, 1 and 2 (this is the well-known *carry save* number system). For instance, Figure 3 shows a parallel adder using the borrow-save system (each cell is a PPM cell: Plus Plus Minus cell). Thus the use of redundant arithmetic implies, *whatever* $n$ (the size of the operands), a critical path of *two* basic cells, instead of $n$ for a sequential adder. There is no longer any carry propagation.

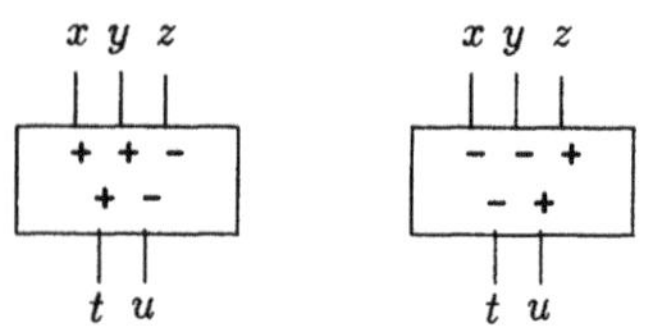

$u = x \oplus y \oplus z$

$t = \text{maj}\ (x, y, \bar{z})$

$x + y - z = 2t - u$

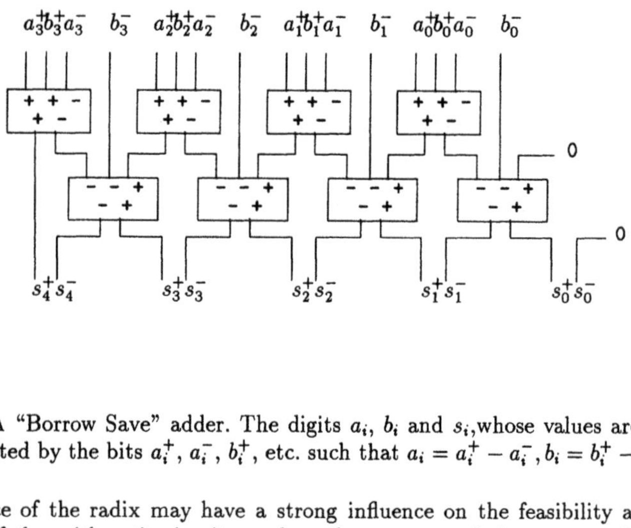

**Figure 3** A "Borrow Save" adder. The digits $a_i$, $b_i$ and $s_i$,whose values are -1, 0 or 1, are represented by the bits $a_i^+$, $a_i^-$, $b_i^+$, etc. such that $a_i = a_i^+ - a_i^-, b_i = b_i^+ - b_i^-, \ldots$

The choice of the radix may have a strong influence on the feasibility and the performances of the arithmetic circuits, and on the accuracy of the computations. The redundancy of Avizienis' number systems leads to the use of more memory and larger data paths. To represent all the integers that can be represented with $n$ bits in the usual binary number system, one needs $\frac{k+1}{k} \times n$ bits in a radix-$2^k$ redundant number system. Therefore, one could think that the best is to choose a radix as large as possible. Unfortunately, the time of computation and the difficulty of implementation greatly increase with the radix. This explains why most of the authors suggest the use of radices 2 and 4. One could design a circuit with parts of different radices, considering the constraints of speed and area.

Our goal is to build a synthesis tool allowing to instanciate operators using such a representation. But as there might be different types of encoding, our system will have to introduce conversion nodes. On one hand, conversion from non-redundant to redundant is quite simple. On the other hand, conversion from redundant to non-redundant is equivalent to a classical addition with carry propagation.

Redundant number representations are already used in complex arithmetic operators such as multipliers, square rooters or dividers. But their use in such operators does not need a general redundant approach, as their inputs and outputs are in classical representation. They are well-known-fast-operators using internally redundant representation, but they are used as "black-boxes".

## 3.2 Synthesis with redundant systems

Any high level system could locally use redundant arithmetic, by placing conversion nodes between every operations. However, a simple addition would become bigger and slower than the same in a non-redundant system, as the conversion node is equivalent to a non-redundant addition! Therefore, a synthesis tool has to consider redundancy globally.

Besides, the tool could neither place conversion nodes at the primary inputs and outputs of the design, as redundant representation is not always the best choice for any kind of operation.

These two aspects determine a caracteristic of the resulting architecture: the resulting design contains several operative parts surrounded by conversion nodes. Another caracteristic is the level of redundancy, choosen considering constraints of speed and area. The last one is the control part, as it might become more important because of the redundancy. Those three caracteristics imply the following synthesis flow:

- *Control part evaluation.*
- *Redundant/Non Redundant partitioning.*
- And For each part:
  - *Redundant system selection.*
  - *Conversion nodes insertion.*

As there is more parallelism, and as the tool will need a more complicated synchronisation, due to the fact that it will have to deal with slow conversion nodes and fast operators, the redundant arithmetics imply more control. Control might even become the slowest part, reducing to nothing our efforts in the operative part. We are currently performing some experiments to determine the effect of control part on the speed of circuits using redundant representation. Anyway, our technique is efficient for data dominated applications. The evaluation of the control part aims at determining if redundancy would not dramatically increase control.

The redundant/non-redundant partitioning determines which operations will be performed using a redundant number system, and above all, how the operands will be encoded. There is no systematic way to handle an operation; the tool always needs to take into account the precedent and following operations. Let us consider, as an example, three basic operations: multiplication, addition and comparison.

First the multiplication: define two numbers $a = \sum_{i=0}^{n-1} a_i 2^i$ and $b = \sum_{i=0}^{n-1} b_i 2^i$. $a$ and $b$ might be as well in a redundant or non redundant representation (in a non-redundant representation, $a_i$ and $b_i$ are equal to 0 or 1, while in a redondant representation, they are equal to 0, 1 or 2 (carry save) or -1, 0, 1 (borrow save)). The multiplication of $a$ by $b$ is peformed as follows:

1. *Generation of the partials products* $a_i b_j$
2. *Additions of the terms* $a_i b_j 2^{i+j}$. The result is in a redundant representation, as redundant operators are used, even if the inputs were non-redudant.
3. *Conversion of the result to a non-redundant system*

If the inputs were in a non-redundant form, the second part would lead to the manipulation of $n^2$ bits, whereas if they were in a redundant form, it would lead to the manipulation of $4n^2$ bits (and $2n^2$ bits if only one input was in redundant representation). As a consequence, multipliers with non-redundant *inputs* would be smaller and/or faster.

Let us assume now that we want to compute $a \times b + c$. If the result of $a \times b$ is in non-redundant representation, the addition will be performed in a classical way, with a carry propagation. But if the addition is performed with the redundant result of the second step (i.e in parallel), an addition with carry propagation will be saved. This means that, *in this case*, the best choice would be a multiplier with non-redundant inputs, and redundant output, followed by a redundant adder.

Even more complicated is the comparison between two values: since there are many ways to represent a number in a redundant system, the comparison becomes more complex. Thus, an "exact" comparison will require as much time in redundant or in non-redundant arithmetic. The choice will be done depending on the availability of the inputs representations.

The redundant system selection will result from a compromise between area and speed: higher radices lead to smaller operands (i.e smaller level of redundancy), but also to slowest operators, whereas small radices (e.g. 2 or 4) lead to larger operands (i.e higher level of redundancy), but faster operators.

Once the redundant system and the redundant parts have been located, the last step, concerning the introduction of conversion nodes, is then rather simple.

At the end, we obtain a description of the design made up of conversions, redundant and non-redundant operations. A textual description of the design should be generated, in VHDL for instance. This description will refer to specific redundant functions instead of VHDL operators (such as "+", "×", ...). To complete high level synthesis, three main steps have to be performed: operator type selection in a library, scheduling and ressource allocation.

In the context of redundant arithmetic, operator type selection has to be seen as a choice between pipeline or mono-cycle functional units. Each operand is either parallel or serial. Therefore, as we will see in section 4, scheduling is viewed as a retiming problem.

## 4 SERIAL ARITHMETIC

Surprisingly enough , one of the main applications of the redundant number systems that allow fully *parallel addition* is... *serial* arithmetic. Digit-serial operators receive their input values and give their results serially, one digit at a time. Usually, the first output digit is available much before the entrance of the last input digit: this makes possible to perform a digit-level pipeline. Unfortunately, such a pipeline is only possible if the digits circulate in the same direction in all the operators. This seems impossible: the usual algorithms give the result of an addition or a multiplication least significant digit first, and the result of a division most significant digit first. That problem is overcame using redundant number systems: with such systems, all the computations can be done most significant digit first. Serial operators working most significant digit first are called *on line* operators. They were introduced by Ercegovac and Trivedi in 1977. In on-line arithmetic, algorithms for addition, multiplication, division, square root, trigonometric functions (Asada 91, Duprat 93), and evaluation of some power series, (Bajard, 93), do exist.

A serial operator is said to have *delay* $\delta$ if it gives the $i$-th digit of the result after having received the first $i + \delta$ digits of its input operands. Among the advantages of these operators are On-line arithmetic implies a delay of the design and a specific clock period, as some registers are already placed within the operators (see the serial adder using borrow save system in Figure 4). Considering those two values (delay and period), we propose two operations: the first one tends to increase the period but decreases the delay, by taking 2 digits at a time. The idea is to group digits pairwise and to use particular operators (as a matter of fact, it can be viewed as a higher radix on-line computation). Those operators have a larger clock period, but they perform calculation in a better delay. The second operation tends to increase the delay, but decreases the period, by introducing new timing barriers, which can be considered as scheduling. We will now focuse on this second operation.

As an example, consider the design illustrated in Figure 4. The critical path (i.e the longest path between two registers) is decreased from 6 PPM cells to 3, and the delay is incremented by one (from 6 to 7). We propose a two-wave algorithm to find the best possible period for a given design by moving registers, but without changing the delay. The first wave tends to move registers toward the inputs as much as possible. Then the second wave move them toward the outputs with respect to a desired clock period. The system then binary search among an interval of available period'values (between the largest operator'period and the initial design'period) to find the smallest possible one, using the two-waves technique. The resulting design has the same delay, the same functionality and the best possible period. We have a special result when all the operator propagation times are equal: we first compute the optimal period, then we use the two-wave technique to move the registers. This algorithm might be usefull with on-line arithmetic, as the only operators used are half-adders and PPM cells. We finally propose another algorithm taking into account non-uniform propagation times within the operators.

So as to move the registers with respect of the initial functionality, our algorithm uses the retiming technique introduced by Leiserson and Saxe (Leiserson, 83). They also proposed an algorithm to find the optimal period (Leiserson, 91). However ours is faster, as our system only deals with direct acyclic graphs: it runs in time $O(|E|Lg(|V|)$ where $|E|$ is the number of connections between the different cells, and $|V|$ is the number of cells whereas Leiserson and Saxe's algorithm ran in time $O(|E|.|V|.Lg(|V|)$. We also used a technique proposed by Wehn, in 94, to minimize the number of registers with a given period and a given delay.

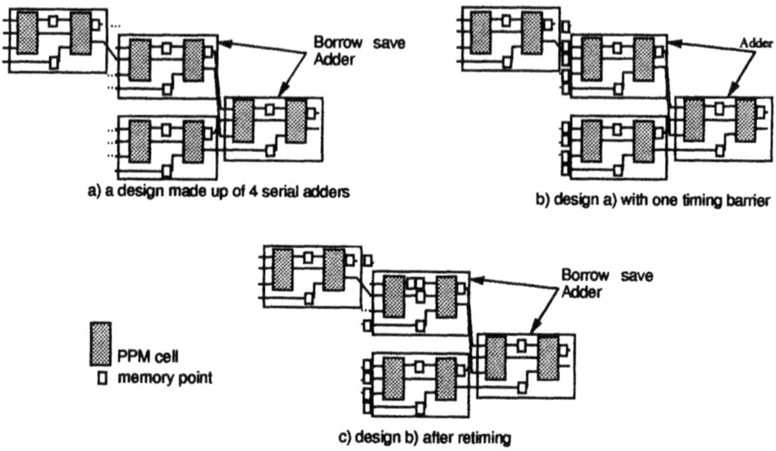


**Figure 4** Adding and moving registers to decrease the period

## 5 CONCLUSION

We have proposed a new way to deal with high-performance (in terms of speed or accuracy) computer arithmetic in synthesis tools. The main trick is to find the internal representation which is the best suited to the algorithm being implemented. Our current works are local determination of precision for floating point operations in order to globally satisfy the 754-IEEE constraints on accuracy; evaluation of the control part, in regard of utilisation of redundant systems; definition of a method to determine "redundant parts", and within those parts, the selection between serial or parallel operands to efficiently use redundant arithmetic.

## 6 REFERENCES

Asada T., Takagi N., and Yajima S. (1991) Redundant cordic methods with a constant scale factor, in *IEEE Transactions on Computers*, 40(9):989–95.

Avizienis, A (1990) Signed-digit number representations for fast parallel arithmetic, in *IRE Transactions on electronic computers*, 10:pp 389–400, Reprinted in E.E. Swartzlander, Computer Arithmetic, Vol. 2, IEEE Computer Society Press Tutorial, 1990.

Bajard, J.C. (1993) Evaluation de fonctions dans des systèmes redondants d'écriture des nombres, in Ph.D. dissertation, Ecole Normale Supérieure de Lyon and Lyon-1 University.

J.Duprat and J.M. Muller (1993) The cordic algorithm: new results for fast VLSI implementation, in *IEEE Transactions on Computers*, 42(2):168–78.

M.D. Ercegovac and K.S.Trivedi (1977) in *On line algorithms for division and multiplication, IEEE Trans. Comp.*, C-26(7):681–7.

C.E. Leiserson and J.B. Saxe (1983) Optimizing Synchronous Systems. in *Journal of VLSI and Computer Systems* , Vol. 1, No. 1, pp. 41-67.

C.E. Leiserson and J.B. Saxe (1991) Retiming Synchronous Circuitry. in *Algorithmica* , 6:5 - 35.

N. Wehn, J. Biesenack, T. Langmaier, M. Münch, M. Pilsl, S. Rumler, P. Duzy (1994) Scheduling of Behavioural VHDL by Retiming Techniques, in *Proceedings of the Euro-DAC '94*, pages 546-51.

## 7 BIOGRAPHY

Doctor Anne Mignotte is an assistant professor at IUT A of Université Claude Bernard of Lyon and a reseacher at LIP (Laboratoire de l'Informatique et du Parallélisme) within ENS (Ecole Normale Supérieure) - Lyon. She obtained a PhD degree from INPG (Institut National Polytechnique de Grenoble) on November 1992. Since 1989, Anne Mignotte works on architectural synthesis. From 1989 to 1992 she was working in CSI / INPG laboratory on scheduling and resource allocation for high level synthesis. Today, she is involved in LIP / ENS-Lyon laboratory where she is working on architectural synthesis with specific emphasis on floating point and exotic arithmetic.

Jean-Michel Muller was born in Grenoble, France, in 1961. He received the Engineer degree in applied mathematics and computer science in 1983, and the PhD in computer science in 1985, both from the Institut National Polytechnique de Grenoble, France. In 1986, he joined the CNRS (french national center for scientific research). He has been posted from 1986 to 1989 to Tim3-Imag laboratory, Grenoble, and then to LIP laboratory, Lyon. He teaches computer arithmetic in the Ecole Normale Supérieure de Lyon. His research interests include computer arithmetic and computer architecture. Dr Muller served as General Chairman of the 10th IEEE Symposium on Computer Arithmetic (Grenoble, france, June 1991).

Olivier Peyran received the engineer and M.S. degrees in computer science from the Ecole Nationale Superieure d'Informatique et de Mathematiques Appliqués de Grenoble, France, in 1993. Currently he is a doctoral student in computer science at the École Normale Supérieure de Lyon, France. His current research interests include multi-level synthesis and computer science arithmetics.

# Adders synthesis

*Alain Guyot*, Mohammed Belrhiti°, Gilles Bosco°*
** INPG/TIMA ° INPG/CSI*
*46, avenue Felix Viallet, 38031 Grenoble Cedex, FRANCE*
*Tel : (33) 76.57.47.34 Fax : (33) 76.50.34.21*
*E-mail: Alain.Guyot@imag.fr*

**Abstract**

The aim of this paper is not to improve the fastest adder, but instead to give a method to balance complexity and speed.

The paper is organized as follows: the Δ operator introduced by Brent and Kung (Brent-Kung 1982) is first recalled. Then it is used to design "standard" adders (Coren 1993), where the critical path is the longest path from any input of the adder to any output. Several well known adder structural organizations exhibiting different speed/area trade off are described by an unified formalism. By mixing different styles (Guyot 1993), the lowest cost for an imposed delay can be achieved. Then the Δ operators will be replaced with a ROBDDs (Reduced Ordered Binary Decision Diagramms) representation, and we shall see why this representation makes the adders easier to manipulate and to optimize for different technologies (Detjens-Gannot 1993)(Sakouti 1992).

**Keywords**

Adders, Δ-cells, ROBDDs, mapping.

## 1 INTRODUCTION

To avoid ambiguity, the (+,∗) symbols are used to indicate addition and multiplication, and the ($\wedge,\vee,\oplus$) for logical AND, OR and XOR. The digit positions are numbered from 0 to n-1, starting from the least significant position at the right of the figures. In an addition, at every position i, the next carry $c_{i+1}$ is either generated, i.e $c_{i+1} = 1$, killed, i.e. $c_{i+1} = 0$ or propagated, i.e $c_{i+1} = c_i$ according to the values of $a_i$ and $b_i$. Propagation is sometimes called transmission. So three signals can be defined, one for each case: $g_i = a_i \wedge b_i$, $p_i = a_i \oplus b_i$ and $k_i = \overline{a_i} \wedge \overline{b_i}$, but a minimum of only two signals is necessary. It turns out that $p_i$ is the most convenient as explained later, so we retain ($p_i$, $g_i$) though ($g_i$, $k_i$) is sometimes chosen (Guyot 1993).The combinatorial addition of two numbers $A = \sum_{i=0}^{n-1} a_i * 2^i$ and $B = \sum_{i=0}^{n-1} b_i * 2^i$ can be decomposed in three steps. First ( $g_i$ , $p_i$ ) are computed $\forall\ i \in \{0 .. n-1\}$. From them the carries $c_i$ are obtained. Finally the digits set $s_i = p_i \oplus c_i$ form the result is $S = \sum_{i=0}^{n-1} s_i * 2^i$.

We will assume that there is no carry input, that is $c_0 = 0$, and that we want the carry out $c_n$. If it were the other way around, we would take the positions from -1 to n-2, $g_{-1}$ being $c_0$, and then renumber them. If we want both the carry-in and the carry-out, we would just add a position, and remove one if neither is wanted.

## 2 GROUP PROPAGATE AND GROUP GENERATE

Let us note $\mathbf{P}_i^j$ the group propagate and $\mathbf{G}_i^j$ the group generate, with $n\text{-}1 \geq i \geq j \geq 0$.
$\mathbf{P}_i^j$ means that the carry propagates from position j up to position i, that is that $c_{i+1}$ is equal to $c_j$. $\mathbf{P}_i^j = \Pi_{n=i}^{j}\, p_n$. $\mathbf{G}_i^j$ means that a carry is generated somewhere between j and i and propagated from this location up to position i and yield $c_{i+1} = 1$. $\mathbf{G}_i^j = g_i \vee \Sigma_{n=i}^{j} (\mathbf{P}_i^{n-1} \wedge g_n)$. Note that the expression of $\mathbf{P}_i^j$ is simpler than the one of $\mathbf{G}_i^j$. Clearly we have $\mathbf{P}_i^i = p_i = a_i \oplus b_i$, $\mathbf{G}_i^i = g_i = a_i \wedge b_i$, $\mathbf{P}_i^i \wedge \mathbf{G}_i^i = 0$ and $c_{i+1} = \mathbf{G}_i^0$ (remember $c_0 = 0$). Let us note also $\mathbf{PG}_i^j$ the couple of bits $(\mathbf{P}_i^j, \mathbf{G}_i^j)$. For any k such that $n\text{-}1 \geq i \geq k \geq j \geq 0$, $\mathbf{PG}_i^j$ can be computed from $\mathbf{PG}_i^k$ and $\mathbf{PG}_{k-1}^j$ in the following way:
$\mathbf{PG}_i^j = (\mathbf{P}_i^j, \mathbf{G}_i^j) = (\mathbf{P}_i^k \wedge \mathbf{P}_{k-1}^j, \mathbf{G}_i^k \vee \mathbf{P}_i^k \wedge \mathbf{G}_{k-1}^j)$.
We note Δ the operator such that $\mathbf{PG}_i^j = \mathbf{PG}_i^k\ \Delta\ \mathbf{PG}_{k-1}^j$. In the figures the icon for Δ-cells is  .

It is easy to prove the following properties:

- Δ is associative: $\mathbf{PG}_i^j = (\mathbf{PG}_i^k\ \Delta\ \mathbf{PG}_{k-1}^l)\ \Delta\ \mathbf{PG}_{l-1}^j = \mathbf{PG}_i^k\ \Delta\ (\mathbf{PG}_{k-1}^l\ \Delta\ \mathbf{PG}_{l-1}^j)$
- Δ is non commutative $\mathbf{PG}_i^j \neq \mathbf{PG}_{k-1}^j\ \Delta\ \mathbf{PG}_i^k$
- Δ is idempotent $\mathbf{PG}_i^j = \mathbf{PG}_i^k\ \Delta\ \mathbf{PG}_l^j$ with $i \geq l \geq k\text{-}1 \geq j$
- Δ is non decreasing.

Any $\mathbf{PG}_i^j$ requires (i-j-1) Δ-cells to be computed from the inputs. Intermediate results from the Δ-cells may be reused, thus reducing the total number of Δ-cells, but increasing the fan-out of some of them.

## 3 SOME ADDER ORGANIZATIONS

Let us examine now some well-known organizations, their delay given by the number of Δ-cells along the critical path, the cost given by the total number of Δ-cells used, their regularity and their construction. Examples of 32-bit adders are provided.

### 3.1 Carry ripple adder

The ripple carry adder delay and its cost are in O(n-1). It is inefficient and easily constructed by abutment of Δ-cells.

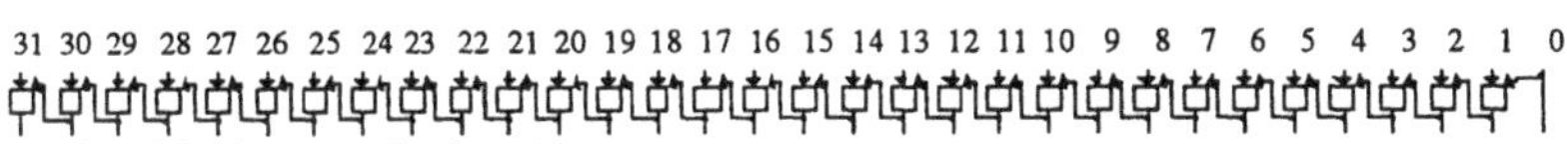


**Figure 1** A 32-bit carry ripple adder

### 3.2 Carry select adder (Sklansky 1960)

The two level carry-select-adder, also named conditional-sum-adder is based on the previous one truncated into blocks of varying sizes. Its cost is in O(2n) and delay $O\lceil\sqrt{2n}\,\rceil$, more precisely with k Δ-cells along the critical path, an adder can accommodate up to $1 + \sum_{i=1}^{k} i = 1 + \frac{k\,(k+1)}{2}$ bits.

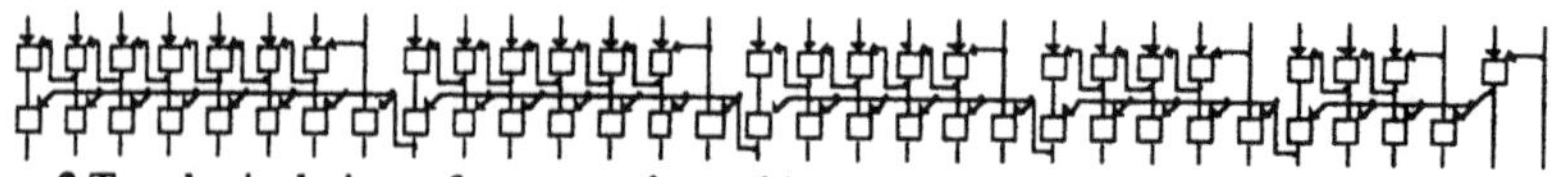

**Figure 2** Topological view of a carry select adder

### 3.3 Brent-Kung adder (Brent-Kung 1982)

The Brent-Kung adder is based on binary Δ-cell trees. The cost is O(2n), the delay $O(2\lceil\log_2(n)\rceil - 2)$.

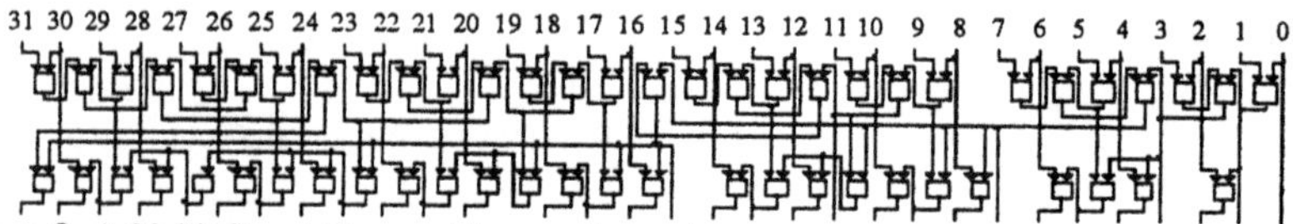


**Figure 3** A 32-bit Brent-Kung adder topology view

### 3.4 Von Neumann adder

The Von Neuman adder has been proved that it is the fastest structural approach (Guyot 1993). The cost of the Von Neumann is $O\lceil\frac{n\log_2(n)}{2}\rceil$, and the delay $O\lceil\log_2(n)\rceil$.

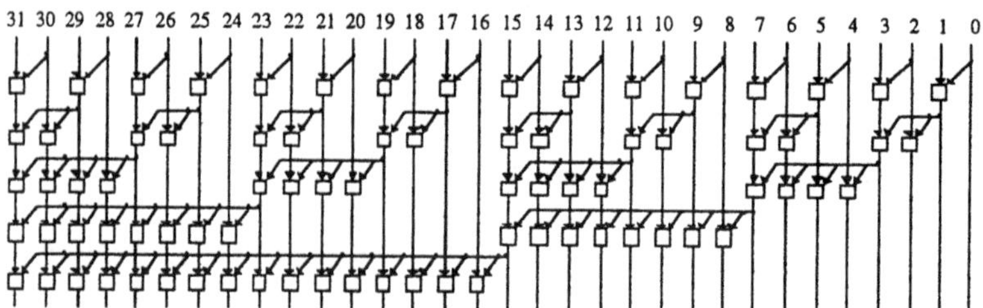


**Figure 4** A 32-bit Von Neumann adder

# 4 STRUCTURAL COMPARISONS

| Type of adder | # of Δ-cells | Delay (Δ-cell) | Max. fan-out | 32-bit example | | |
|---|---|---|---|---|---|---|
| Ripple | n-1 | n-1 | 1 | 31 | 31 | 1 |
| carry-select | $\lceil 2n-\sqrt{2n} \rceil$ | $\lceil \sqrt{2n} \rceil$ | $\lceil \sqrt{2n} \rceil$ | 56 | 8 | 8 |
| Brent-Kung | $\lceil 2n - \log_2(n)-2 \rceil$ | $\lceil 2 \log_2(n) -2 \rceil$ | $\lceil \log_2(n) \rceil$ | 57 | 8 | 5 |
| Von Neumann | $\lceil n/2 \log_2(n) \rceil$ | $\lceil \log_2(n) \rceil$ | n/2 | 80 | 5 | 16 |

# 5 FROM Δ-CELLS TO ROBDDS

## 5.1 Adder Generator

Let us try to find a good representation of the Δ-cells to make them easy to handle. The aim is to give a solution to build quickly the different adders organizations and make them mappable. We shall use the Reduced Ordered Binary Decision Diagrams (Bryant 1986), because they give the liberty to manipulate easily the Xor operations that are very present in the adders.
as we have mentioned in §1 :

$\mathbf{PG}_i^j = (\mathbf{P}_i^j, \mathbf{G}_i^j) = (\mathbf{P}_i^k \wedge \mathbf{P}_{k-1}^j, \mathbf{G}_i^k \vee \mathbf{P}_i^k \wedge \mathbf{G}_{k-1}^j)$.

so: $\mathbf{P}_i^j = (a_i \oplus b_i) \wedge \mathbf{P}_{i-1}^j$

& $\mathbf{G}_i^j = \mathbf{G}_i^i \vee \mathbf{P}_i^i \wedge \mathbf{G}_{i-1}^j = (a_i \wedge b_i) \vee ((a_i \oplus b_i) \wedge \mathbf{G}_{i-1}^j)$

$\mathbf{G}_i^i = a_i \wedge b_i$

$\mathbf{G}_i^j = a_i \wedge (b_i \vee (\overline{b_i} \wedge \mathbf{G}_{i-1}^j)) \vee \overline{a_i} \wedge (\overline{b_i} \wedge \mathbf{G}_{i-1}^j)$

We note $a_i$ Ai and $b_i$ Bi in the following figures. That will provide the following generators :

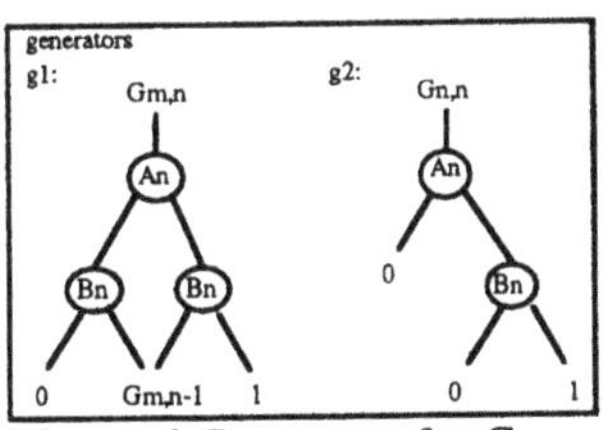


**Figure 4** Generators for G

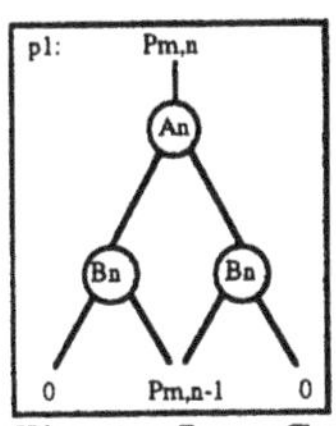


**Figure 5** Generator for P

Decomposition of $\mathbf{PG}_i^j$ for k such that $n-1 \geq i \geq k \geq j \geq 0$ .
We have got the formula: $\mathbf{PG}_i^j = (\mathbf{P}_i^j, \mathbf{G}_i^j) = (\mathbf{P}_i^k \wedge \mathbf{P}_{k-1}^j, \mathbf{G}_i^k \vee \mathbf{P}_i^k \wedge \mathbf{G}_{k-1}^j)$.

Let us consider we have the carry-ripple-adder ROBDDs corresponding to $(\mathbf{P}_i^j, \mathbf{G}_i^j)$; and let us see how we can easily compute $\mathbf{PG}_i^j$ from $\mathbf{PG}_i^k$ and $\mathbf{PG}_{k-1}^j$. The ROBDDs corresponding to $\mathbf{PG}_i^k$ and $\mathbf{PG}_{k-1}^j$ will be obtain by cutting the ROBDDs of $\mathbf{PG}_i^j$.

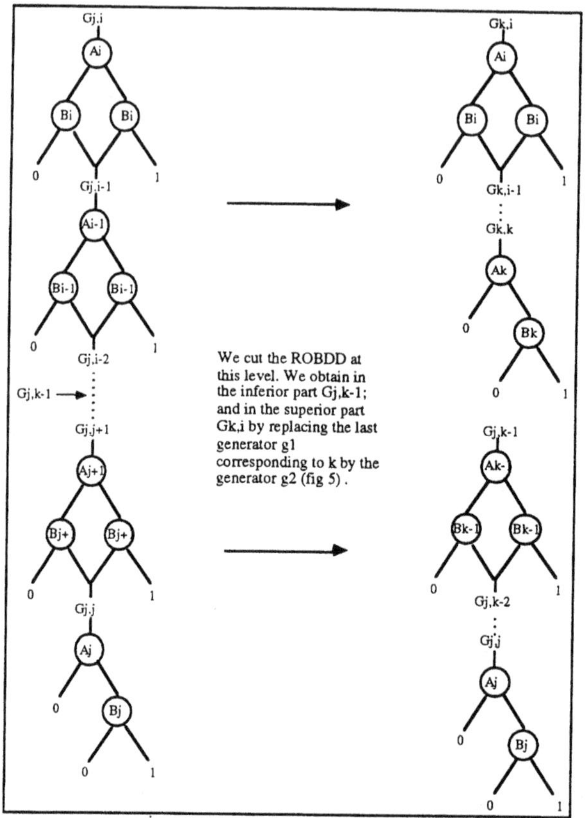


**Figure 7** decomposition of $\mathbf{PG}^j_i$.

For the ROBDD corresponding to P, the same operation will be realised.

Next, to compute the value of $\mathbf{PG}^j_i$, we have got to build the two ROBBDs corresponding to the two equations:

$$\mathbf{G}^j_i = \mathbf{G}^k_i \vee \mathbf{P}^k_i \wedge \mathbf{G}^j_{k-1}$$

$$\mathbf{P}^j_i = \mathbf{P}^k_i \wedge \mathbf{P}^j_{k-1}$$

by considering the values of $\mathbf{PG}^j_{k-1}$and $\mathbf{PG}^k_i$ as primary inputs, we can compute $\mathbf{PG}^j_i$ as follow:

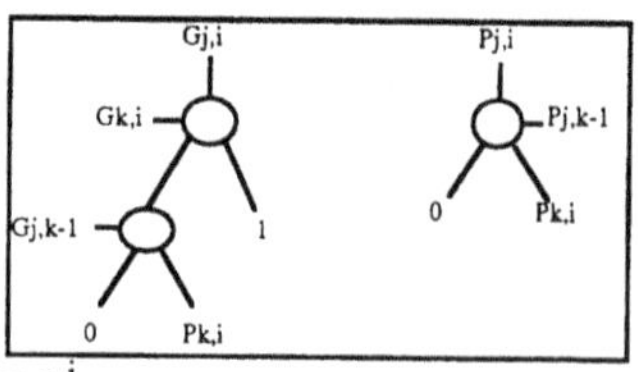


**Figure 8** computation of $\mathbf{PG}^j_i$

### 5.2 Mapping

By replacing the Δ-cells by ROBDDs. we can easily evaluate the more suitable adder for a technology because we can control better the number of entries and the fan-out of the nodes of the BDDs.

### 5.2 Advantages of this method

- representation very closed to the Δ-cells.
- linear size of representation for the adder.
- easiness to build and modify the adders.
- two possibilities for the mapping: directly on the ROBDDs (Brace-Rudel-Bryant 1990) (by using greedy algorithms) or by turning ROBDDs into a function (mapping algorithms on trees, tree matching (Lisanke-Bergly-Kedem 1988), dynamic programmation (Kentzer 1988) ).

## 6 EXPERIMENTAL RESULTS

The following results correspond to tests realised on Xilinx 4000. We gives the result after placement and routage performed by ppr (Xilinx placer and router).

**Table 1** Von Neumann adders:

| Size | Area (#Clbs) | Speed (ns) | level (#Clbs) |
|---|---|---|---|
| 4 | 2 | 33,8 | 3 |
| 8 | 9 | 41,0 | 4 |
| 32 | 25 | 52,4 | 5 |
| 32 | 58 | 60,9 | 6 |

**Table 2** Comparison between different structures for a 32 bits adder:

| | 2-levels carry-select adder | carry-ripple | Von Neumann |
|---|---|---|---|
| Area (#Clbs) | 54 | 47 | 58 |
| Speed (ns) | 73,8 | 208,2 | 60,9 |
| Level (#Clbs) | 9 | 32 | 6 |

**Table 3** Comparison between different structures for a 16 bits adder:

| | 2-levels carry-select adder | carry-ripple | Von Neumann |
|---|---|---|---|
| Area (#Clbs) | 24 | 22 | 25 |
| Speed (ns) | 65,9 | 113,9 | 52,4 |
| Level (#Clbs) | 7 | 16 | 5 |

## 7 CONCLUSION

First, we have a description of the adders by a unified formalism, and the modelisation allows the adders good regularity and flexibility. Then we can control the granularity of the adders by collapsing or clustering (Touati-Sajov-Brayton 1991) to implement it on a given technology.

## 8 REFERENCES

R.P. Brent and H.T. Kung, *A regular layout for parallel adders*, IEEE Trans on Computers, Vol.C31, pages 260-264, Mars 1982.

Israel Koren, *Computer Arithmetic Algorithms*, Prentice Hall, Englewood Cliffs, New Jersey, 1993.

Alain Guyot, *Computer Arithmetic*, E.N.S.I.M.A.G. 1993

E. Detjens, G. Gannot, *Tehcnology Mapping in MIS*, in Proc. Int. Conference on Computer Aided Design (ICCAD 1987), pp 116-119, Santa Clara, USA, Nov 1987.

J. Sklansky, *Conditional Sum Addition Logic*, IRE Trans. EC-9(2) (June 1960), pp226-231.

R.E Bryant, *Graph Based Algorithms for Boolean Functions Manipulation*, IEEE Trans on Computers, Vol C35, pages 667-692, August 1986.

K. Brace, R.Rudell, R.Bryant, *Efficient Implementation of BDD Package*, 27th DAC, June 1990.

R. Lisanke, F. Bergly, G. Kedem, *McMap : A Fast Technology Mapping Procedure for Multi-Level Logic Synthesis*, ICCD 1988, pp 252-256, Oct. 1988.

K. Kentzer, *Timing Optimisation in a Logic Synthesis System*, International workshop on logic and architecture synthesis for silicon compiler, Grenoble 1988.

H.J. Touati, H.Sajov, R.Brayton, *Delay Optimisation of Combitional Logic Circuit by Clustering and Partial Collapsing*, in Proc. Int. Conference on Computer Aided Design ICCAD 1991).

K. Sakouti, *Synthèse et Optimisation Temporelle de Réseaux Booléens*, PhD, Institut National Plolytechnoque de Grenoble, Dec 1992.

PART TEN

# High Level Synthesis

29

# High-Level Synthesis: A Critical Assessment*

*Norbert Wehn*
*Siemens AG*
*Semiconductor Division*
*D-81549 München*

*Darmstadt University of Technology*
*Institute of Microelectronic Systems*
*D-64283 Darmstadt*

**Abstract**

High-level synthesis is a very active research area in VLSI design automation upon which a lot of effort has been spent during the past. However, the high-level synthesis methodology has not yet received the same level of acceptance in industry as logic and RT synthesis. The purpose of this paper is not to give a tutorial[1], but rather to discuss some reasons for this lack of acceptance with respect to commercial issues, to analyze what requirements a high-level synthesis tool needs to fulfill to enable a similar boost in a designer's productivity as logic and RT synthesis tools have and finally to give an outlook on emerging challenges for high-level synthesis tools in the future.

## 1 INTRODUCTION

In the past ten years we have seen dramatic changes in the field of design automation (see [NEWTON 92] and the Special Universary Issue Electronic Design, Penton Publication [PENTON 92]). In less than a decade designers of complex circuits have come to rely on automatic or semiautomatic computer-aided design systems for physical design, logic synthesis and register-transfer synthesis. The transition from transistor-level entry to gate-level schematic entry enabled an order-of-magnitude improvement in designer productivity. Similarly, the transition from schematic entry to hardware description (HDL) based synthesis has again enabled another order-of-magnitude improvement in designer productivity.

In order to be competitive in today's marketplace (in terms of time-to-market and design-to-cost) CAD tools dominate system and chip design methodologies throughout

*This paper is a revised version of a paper from the IFIP Workshop on Logic and Architecture Synthesis 1993, Grenoble, France.

[1]For a tutorial introduction to high-level synthesis, the interested reader is referred to [MCFARLAND ET AL. 90] or one of the numerous books published on the topic. Some of them have in turn tutorial character [WALKER AND CAMPOSANO 91, MICHEL ET AL. 92, GAJSKI ET AL. 92], others are devoted to application specific synthesis [VANHOOF ET AL. 93, CATTHOOR AND SVENSSON 93] and dedicated synthesis approaches [THOMAS ET AL. 90, CAMPOSANO AND WOLF 91, KU AND DE MICHELI 92, GEBOTYS ET AL. 92].

the industry. Although the CAD market represents only a few percent ($1.3 billion at the beginning of the '90s) of the total electronic world market, CAD is essential for the electronics industry. The CAD industry is moving very fast. Start-up companies can grow very quickly, and they can disappear very quickly as well. Companies offering CAD synthesis support are credited with an an annual 45% growth rate. SYNOPSYS, one of the most successful synthesis companies, is even credited with a 50% growth rate [COURTOIS 93]. The reason is that the driving system technology is shifting from the computer industry to telecommunication and consumer electronics. Typical examples are digital television, digital video disc, ISDN, CD-player, digital compact cassette, digital mobile phone, mechatronics etc. These products have a short life cycle, but an extremely high degree of system complexity. As a consequence, the designer productivity has to be increased dramatically (by a factor of 15 by the year 2000). This productivity boost can only be achieved by massive use of advanced CAD support beyond the physical design level [DEMAN 92]. There is no doubt that the introduction of logic and RT-synthesis methodologies in industrial design environments was extremely successful and provided a large productivity increase.

# 2 HIGH-LEVEL SYNTHESIS AS THE NEXT PRODUCTIVITY BOOST?

## 2.1 High-Level Synthesis within the ASIC Design Flow

*High-Level synthesis* (there are many synonyms for high-level synthesis in the literature: architectural synthesis, behavioral synthesis etc.) has long been considered to be the successor to logic- and RT-synthesis [IEEE 90]. In the past ten years a lot of effort has been spent in the development of high-level synthesis techniques and there are still many ongoing research projects, both in academia and in industry.

The advantages of shifting to higher levels of abstraction are obvious [MCFARLAND ET AL. 90]: shorter design cycle, fewer design errors, exploration of the design space, the integrated circuit technology becomes available to more engineers ("meet-in-the-middle" approach), to name but a few of them. However, designers are still sceptical of the ability of high-level synthesis to deliver either higher productivity or the required circuit quality. Figure 1 reflects this situation.

The reasons for this situation are manifold. Synthesis on the layout level is primarily based on formal foundations in graph theory and graph algorithms [LENGAUER 90]. Logic synthesis originates from the well known formalism of Boolean algebras [BRAYTON ET AL. 84, ASHAR ET AL. 92]. Extending synthesis to higher levels of abstraction is quite difficult due to the lack of appropriate formalisms. But most important the design space grows tremendously, i.e. given a specification, the set of alternatives for an implementation is very huge (e.g. bit-serial or bit-parallel, pipeline or broadcast, shared or distributed memory, SIMD or MIMD architecture, hardwired or microprogrammed controller, synchronous or asynchronous, bus- or multiplexer-based etc.). Hence, a successful general approach able to manage all implementation alternatives will never exist. This leads to a revised definition of high-level synthesis as depicted in figure 2 [NEWTON 92].

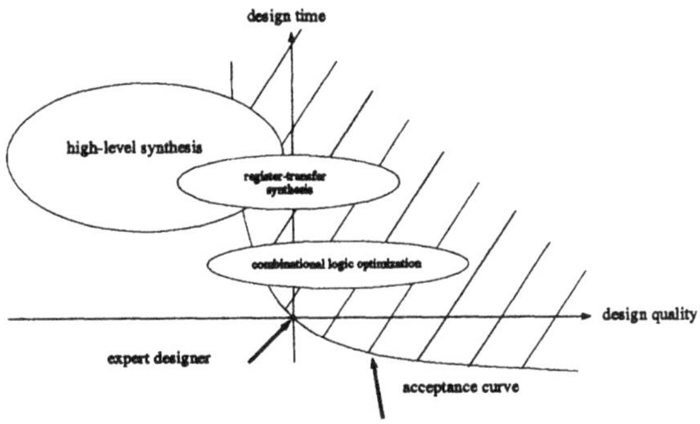


**Figure 1** Acceptance of synthesis techniques (Source: A. DeGeus/SYNOPSYS).

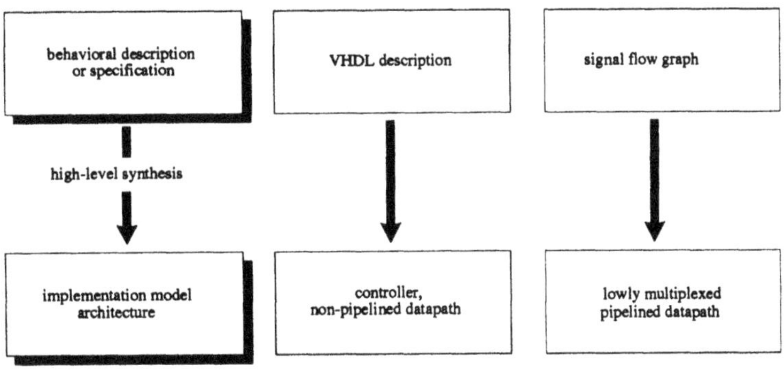


**Figure 2** Definition of High-Level Synthesis.

## 2.2 (Re-)Defining High-Level Synthesis

High-level synthesis is considered to be the task to map a specification or a description (we do not want to elaborate on the difference between *specification* and *description* in this paper) of a system or subsystem onto an *implementation model.* This implementation model is characterized by:

- the definition of a target architecture. Such an architecture typically consists of a set of data paths represented by register-transfer operators, storage units operating under a well-defined timing protocol and control units and their interconnections.
- the definition of a proper specification medium (textual or graphic) with corresponding implementation semantics.
- the identification of the tasks to be solved (allocation, scheduling and binding) and the derivation of their mathematical optimization problems.

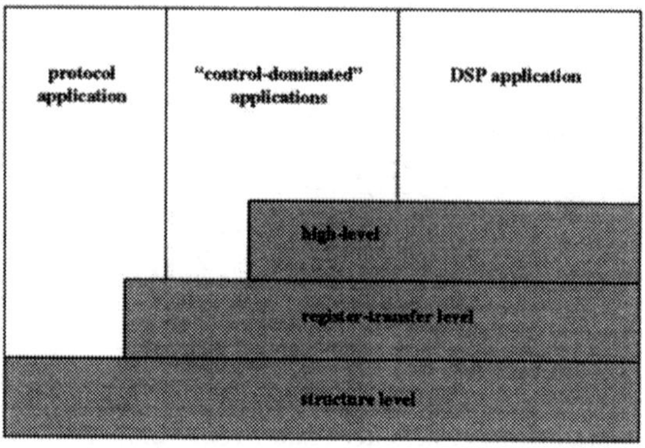


**Figure 3** Application domains and proper abstraction levels.

- the order of the different tasks and their interrelations (*synthesis script*).
- the selection of optimization techniques, e.g. global optimization vs. heuristics to solve mathematical optimization problems.

Consider, for instance, a single-process VHDL description which should be mapped onto an architecture consisting of a hardwired controller and a highly multiplexed non-pipelined data-path. Alternatively, consider a signal flow graph representing a filter which should be mapped onto an architecture consisting of a set of lowly multiplexed heavily pipelined datapaths with local controllers. It is obvious that the way of specification and the implementation model completely differ in both cases. *However, many CAD researchers falsely assume that moving to higher levels of abstraction leads them away from implementation/application specific issues.* Instead, the domain dependency of CAD increases when moving to higher levels of abstraction, i.e. the implementation model becomes strongly domain dependent. Experience in the past has shown that mostly tools that respect this rule were able to provide satisfying synthesis results, for instance the Cathedral tool suite [VANHOFF ET AL. 93].

Complex systems are implemented as heterogeneous architectures with e.g. asynchronous interfaces, embedded customized DSP-cores, accelerator datapath-units for computational intensive operations, large portions of RAM, cooperating controlunits and glue logic. Hence there is no single implementation model for such a system, instead a high-level synthesis-environment is necessary which provides various implementation models. Moreover not each part of a system should be specified on a high level of abstraction. High-level synthesis techniques are not suited for every application domain. Figure 3 gives an overview of different application domains and their proper levels of abstraction.

## 2.3 Requirements for CAD Research

Based on these experiences, we need to redefine the requirements for successful CAD tool research and development. Tool designers, system designers and system engineers have to *cothink* in order to understand each other's problems and should not consider high-level synthesis from the perspective of inventing yet another new algorithm, for instance for the scheduling problem. Even though scheduling, allocation and binding are indeed important issues in high-level synthesis, they are not the key problems in the high-level design process, especially if these problems are solved on RT-level granularity, where a single multiplexer input is traded off against registers. Global memory considerations and proper partitioning can influence the efficiency of an architecture much more than applying a scheduling algorithm which is optimized for the "fifth-order elliptic filter". In order to capture the "real world" high-level problems for a specific application, one has to be actively involved in the design process and needs to encapsulate application specific design knowlegde in order to be able to define the implementation model. However:

- academic researchers often do not have the possibility to access this design knowledge, since it is property of the system companies. Hence, many academic people spend all their efforts to solve very specific subproblems like scheduling or binding on sometimes unrealistic benchmarks which are tailored to these subproblems [WALKER AND CAMPOSANO 91, High-Level Synthesis Benchmark Suite 91] (the "fifth-order elliptic filter" syndrome). As a result, solutions are sometimes developed which do not contribute to real progress, or the problems addressed do not exist in the "real world".
- even CAD companies generally do not have this knowledge. Hence, it is questionable if the real innovation for the productivity increase will come from CAD vendors.

To overcome this problem, CAD research needs to be oriented more towards the "real life" design process. This requires a closer link between academia, system industry and CAD vendors (the so-called *Trinity* [DEMAN 92]). A shift from *CAd* towards *caD* has to be carried out, i.e. the *design aspect* has to be emphasized. Although this is not only restricted to high-level synthesis, it is especially true for design automation on higher abstraction levels. However, many system houses actually stop their CAD activities to rely on commercial systems. These CAD groups have to be re-oriented to keep their know-how in this field in order to efficiently interface CAD vendors, to match the specific needs of their company and to implement design methods by means of added value tools not supplied by CAD vendors [COURTOIS 93; DEMAN 92].

There is no doubt that high-level synthesis research carried out in the past has produced good results for specific problems, however, for a broader acceptance of high-level synthesis based design methodologies, the following aspects need to be taken into account in addition to the discussed issues:

- Time-to-market pressure and design risk in complex systems make designers to rely on a considerable amount of *reused* subsystems. Today complex ASICs consist of up to 50% of reused silicon, i.e. blocks adapted from previous successful designs, instead of being designed from scratch. In particular, these components must be represented in a suitable way such that high-level synthesis can access them. Hence, efficient design reuse methodologies will play a major role in future design systems [GIRCZYC AND CARLSON 93; DT ROUNDTABLE 94].
- Originally the goal of high-level synthesis was to automatically map a complete system specification down to silicon (ideal silicon compiler) [GAJSKI AND KUHN 83]. But in reality — as already pointed out — complex systems are of heterogeneous nature which combines different architectural styles and implementation paradigms on a single

ASIC and not each of its constituent subsystems is processed by high-level synthesis techniques.

This concludes that high-level synthesis can only be part of a complete design methodology and must be *well adapted* to the overall design environment. Those subcircuits of a complex system processed by high-level synthesis techniques have to satisfy stringent constraints in order to match the system environment. A correct and "user-friendly" specification of a system is a difficult task, since there is no unique specification model. There are various domain-specific entry mechanisms such as graphic statecharts which are based on the "design like you think" paradigm and are well suited for reactive control applications. On the other hand signal processing applications require efficient dataflow models. Hence the specification of complex systems requires the marriage of statechart-based and dataflow techniques.

High-level synthesis must fit into such an environment and must be able to consider timing or ordering constraints on the interfaces and throughput requirements. A consistent simulation of these subcircuits before and after synthesis in the system environment must be supported. This also implies that the circuit description before and after synthesis must be consistent with the remaining system representation. It is often stated that with the advent of VHDL [IEEE 88], the problem of a consistent description and simulation environment has been solved — at least for some specific application domains. However, VHDL was developed as a simulation language and its semantics are defined by means of an event-driven simulator which requires some notion of timing. There is a big difference between simulation and synthesis: simulation is an analysis task in which the timing (as for simulation purposes) has to be well defined, but synthesis is an identification task in which hardware primitives have to be extracted. In this context, we also speak of the "semantic gap" between simulation- and synthesis-related aspects of VHDL. As a consequence, each CAD tool vendor defines its own proprietary VHDL subset for synthesis resulting in non-portable VHDL descriptions. In addition, VHDL is not an adequate specification medium for each application as, for instance, filter specifications can be formulated efficiently in an applicative language such as Silage [GENIN ET AL. 90] instead of an imperative language such as VHDL.

Another important issue for successful high-level synthesis is the modeling of components and estimation of lower-level effects such as delay caused by interconnection. Technology and performance requirements progress so fast that the delay caused by interconnect dominates the overall circuit delay. The problem to decide what RT-level specification will result in a final chip satisfying all area and performance constraints has not yet been sufficiently solved — a prerequisite to do so is an efficient link between RT-level synthesis and layout synthesis tools. Only recently CAD tool vendors have started to tackle this problem. Moving up to a level of abstraction beyond RT-level makes this problem even more difficult. Powerful timing models and a closer link between floorplanning, RT-synthesis and high-level synthesis becomes mandatory. Many components are modelled in a very abstract way which does not reflect their real behavior. A typical case is the modeling of RAMs which in many high-level synthesis-approaches are assumed to behave like simple registers. However large memories are normally realized as DRAMs which can have different architectures (synchronous DRAMs, RAMBUS, EDO-DRAMs) and different access schemes such as burst mode or random access mode.

Interactivity is a further important issue. Due to the huge design space designer interactivity in the high-level design process becomes necessary. Designers usually have a good insight in a system behavior. This knowledge can help in the exploration of the design space. However to allow an efficient interactive high-level design process, fast and accurate estimators are necessary to feedback design decisions [JAIN ET AL. 92; WEHN ET AL. 93; OHM ET AL. 94; RABAEY AND POTKONJAK 94]. Some high-level synthesis-approaches provide powerful high-level transformations to improve the initial specification with regard to various cost criterions before starting the traditional process. Among them are HYPER [CHANDRAKASAN ET AL. 95], CATHEDRAL [VANHOOF ET AL. 93], SAW [WALKER AND THOMAS 89] and Flamel [TRICKEY 87]. These systems offer a wide variety of code transformations, partially well-known from the field of compiler construction [AHO ET AL. 86], such as loop transformations, code inlining, strength reduction, dead code elimination, constant propagation etc. Code is transformed as to optimize some objective function, for instance to improve performance by minimizing delay or, only recently in the HYPER system, to reduce power consumption.

This leads to a new approach in which the high-level design process can be considered as a source-level optimization process. One of the most recent approaches, VOTAN (VHDL Optimization, Transformation and ANalysis) directly operates on VHDL behavioral descriptions [WEHN ET AL. 94]. It provides a set of code transformations for optimizing control-dominated single-process descriptions subject to resource and performance constraints. Estimators permit evaluation of transformation steps without needing to go into the tedious task of RT-synthesis. In this context, scheduling is considered as yet another code transformation.

In the past application domains were mainly classified according to the predominance of dataflow or control flow. But there is a further category which is neither dataflow- nor control-dominated: *memory-dominated applications* in which memory organization and size influence the overall architecture. Typical examples are video processing algorithms such as video decompression. The overall control and troughput requirements on the various blocks of a decoding pipeline are heavily influenced by the memory bandwidth and size. In order to meet the system throughput requirement (e.g. fixed by the pixel clock), an efficient memory organization and mapping of data onto memories with a minimum number of page-overhead during reading or writing is necessary. The scheduling problem for such an application consists of managing the memory accesses from the different processing units (which are not simple ALUs, but complex blocks like an IDCT-unit or even DSP-cores) in the decoding pipeline. The processing units itself are either dataflow-dominated as for IDCT or picture resizing or control-dominated as in header extraction. This example shows the importance of real case studies. The high-level synthesis community spent a lot of effort in benchmarking which was often counterproductive by focusing research on the wrong issues. Case studies are a possible way to overcome this problem in order to address the real problems.

## 2.4 The Future — Challenges to CAD Tool Research

Many of the problems discussed above have not yet been sufficiently addressed by high-level synthesis research efforts. Besides, designers in industry are still educated to think and to specify circuits on the RT-level. The previous comments do not imply that the high-level synthesis story has failed. There are numerous successful approaches [CAMPOSANO

AND WOLF 91], especially in the field of DSP for low, medium and high throughput applications like the Cathedral synthesis environment [DEMAN ET AL. 90; CATTHOOR AND SVENSSON 93], the PIRAMID and PHIDEO systems [ROZA ET AL. 92], the HIS system [CAMPOSANO 91] for processor architectures, the CALLAS system [BIESENACK ET AL. 93], the OLYMPUS system [KU AND DE MICHELI 92], PUBSS [WOLF ET AL. 92] and the AMICAL system [CATTHOOR AND SVENSSON 93] for control dominated architectures and many other systems [WALKER AND CAMPOSANO 91]. Even CAD vendors have recently started to enter the high-level synthesis field. In this context, new challenges for high-level synthesis become obvious:

- Recently a trend towards specific application domain processors has become visible, especially in the field of telecommunications [PAULIN 92]. These processors are programmable by an instruction-set and optimized for *a set* of algorithms and not for *one* specific algorithm. This means that some application specific hardware is shifted into programmability and the problem of microcode synthesis for these processors has to be solved. High-level synthesis techniques can be used to efficiently solve the microcode synthesis problem for ASIPs [VAN PRAET ET AL. 94] and the derivation of ASPP architectures [ALOMARY ET AL. 93].
- Design reuse is another challenge for high-level synthesis, since designs specified on a high level of abstraction by an HDL and processed automatically by synthesis tools enhances their reuse potential. Hence, HDL and high-level synthesis methodologies change the vehicle of design reuse from a hardware to a software paradigm [GIRCZYC AND CARLSON 93; DT ROUNDTABLE 94].
- New implementation technologies like FPGAs pose new challenges to high-level synthesis due to their programmability on the hardware level. Hence, design space exploration on the hardware level and emulation in the system environment becomes practicable (rapid prototyping).
- The market for low-power electronic systems like portable computation and communication systems is growing explosively. Hence, power consumption must be considered on the architectural level and is a new facet in the design space exploration. Developing fast estimators for accurate power consumption estimation will be a primary problem, as these are the driving factors for successful search space pruning.
  Among the first systems to consider power issues was HYPER offering high-level estimators for power consumption along with behavioral transformations to reduce glitches and architectural transformations to preserve throughput while lowering supply voltage for low power.
- Low throughput signal processing tasks such as audio processing are more and more implemented on embedded DSP-cores. For design security the global controlling is often organized by an embedded RISC-core. Hence the boundary between hardware and software in complex systems is more and more blurring and the portion of embedded software increases continously. In order to meet the complete system requirements, hardware and software must be considered and developed hand in hand (hardware/software codesign).
- Moreover microelectronic circuits are used more and more in cooperation with other disciplines like in mechatronic- or microsystems. High-level synthesis has to cope with these new directions.

# 3 CONCLUSION

In this paper we gave a critical assessment on high-level synthesis research activities. A closer cooperation between system designers, CAD tool vendors and research groups could push the high-level synthesis methodology into industrial design environments. New challenges for high-level synthesis were outlined.

# 4 ACKNOWLEDGEMENTS

I would like to thank my colleagues from Siemens AG and Darmstadt University of Technology for their valuable contributions and helpful discussions. Special thanks go to Michael Münch, Darmstadt University of Technology, for his valuable suggestions.

# REFERENCES

A. V. Aho, R. Sethi, and J. D. Ullman. *Compilers — Principles, Techniques and Tools.* Addison Wesley, 1986.

A. Alomary, T. Nakata, Y. Honma, and J. Sato. PEAS-I: A hardware/Software Co-design System for ASIPs. In *Proceedings of the EURO-DAC'93*, pages 2–7, Hamburg, 1993.

P. Ashar, S. Devadas, and R. Newton. *Sequential Logic Synthesis.* Kluwer Academic Publishers, Norwell, MA, 1992.

J. Biesenack *et al.* The Siemens High-Level Synthesis System CALLAS. *IEEE Transactions on Very Large Scale Integration (VLSI) Systems*, 1(3):244–253, September 1993.

R. Brayton, G. Hachtel, C. McMullen, and A. Sangiovanni-Vincentelli. *Logic Minimization Algorithms for VLSI Synthesis.* Kluwer Academic Publishers, Norwell, MA, 1984.

R. Camposano. Path-Based Scheduling for Synthesis. *IEEE Transactions on Computer Aided Design*, 10(1):85–93, January 1991.

A. P. Chandrakasan, M. Potkonjak, R. Mehra, J. Rabaey, and R. W. Broderson. Optimizing Power Using Transformations. *IEEE Transactions on Computer Aided Design*, 14(1):12–31, January 1995.

B. Courtois. CAD and Testing of ICs and systems. Where are we going? INPG, 1993.

F. Catthoor and L. Svensson. *Application-Driven Architecture Synthesis.* Kluwer Academic Publishers, Norwell, MA, 1993.

R. Camposano and W. Wolf. *Trends in High-Level Synthesis.* Kluwer Academic Publishers, Norwell, MA, 1991.

H. De Man, F. Catthoor, G. Goossens, J. Vanhoof, J. Van Meerbergen, S. Note, and J. Huisken. Architecture-driven synthesis techniques for VLSI implementation of DSP algorithms. *Proceedings of the IEEE*, 78(2):319–335, February 1990.

H. De Man. Design Technology Research for the Nineties: More of the Same? In *Proceedings of the Euro-DAC*, pages 592–596, September 1992.

A D&T Roundtable. Design Reuse. *IEEE Design & Test of Computers*, 11(4):70–77, Winter 1994.

E. Girczyc and S. Carlson. Increasing Design Quality and Engineering Productivity through Design Reuse. In *Proceedings of the 30th DAC*, pages 48–54, June 1993.

D. Gajski, N. Dutt, A. Wu, and S. Lin. *High-Level Synthesis – Introduction to Chip and System Design.* Kluwer Academic Publishers, Norwell, MA, 1992.

D. Gajski and R. Kuhn. Guest Editors' Introduction: New VLSI Tools. *IEEE Computer*, 6(12):11–14, December 1983.

C. Gebotys *et. al. Optimal VLSI Architectural Synthesis.* Kluwer Academic Publishers, Norwell, MA, 1992.

D. Genin, P. Hilfinger, J. Rabaey, C. Scheers, and H. De Man. DSP Specification using the SILAGE language. In *Proceedings of the IEEE International Conference on Acoustics Speech and Signal Processing*, pages 1057–1060, 1990.

High-Level Synthesis Workshop Benchmark Suite. Available via **anonymous ftp** from **mcnc.mcnc.org** in **pub/benchmark**.

Proceedings of the IEEE, February 1990. Special Issue on CAD, Vol.78, No.2.

The Institute of Electrical and Electronical Engineers, Inc., 345 East 47th Street, New York, NY 10017-2394, USA. *IEEE Standard VHDL Language Reference Manual (IEEE Std 1076-1987)*, March 1988.

Rajiv Jain, Alice Parker, and Nohbyung Park. Predicting System-Level Area and Delay for Pipelined and Non-Pipelined Designs. *IEEE Transactions on Computer Aided Design*, 11(8):955–965, August 1992.

D. Ku and G. DeMicheli. *High Level Synthesis of ASICs Under Timing and Synchronization Constraints.* Kluwer Academic Publisher, Norwell, MA, 1992.

Th. Lengauer. *Combinatorial Algorithms for Integrated Circuit Layout.* John Wiley and Sons, New York, 1990.

P. Michel, U. Lauther, and P. Duzy. *The Synthesis Approach to Digital System Design.* Kluwer Academic Publishers, 1992.

M.C. McFarland, A.C. Parker, and R. Camposano. The High-Level Synthesis of Digital Systems. *Proceedings of the IEEE*, 78(2):301–318, February 1990.

R. Newton. Has CAD for VLSI reached a Dead End? *IFIP Transactions VLSI 91*, A-1:187–192, 1992.

S. Y. Ohm, F. J. Kurdahi, and N. Dutt. Comprehensive Lower Bound Estimation from Behavioral Descriptions. In *IEEE/ACM Digest of Technical Papers*, IEEE/ACM International Conference on CAD, pages 182–187, November 1994.

P. Paulin. DSP Design Tool Requirements for the Nineties: An Industrial Perspective. In *Proceedings of the 6th International Workshop on High-Level Synthesis*, Laguna Niguel, Nov 1992.

Special Universary Issue Electronic Design. Penton Publication, November 1992.

J. van Praet, G. Goossens, D. Lanneer, and H. De Man. Instruction Set Definition and Instruction Selection for ASIPs. In *Proceedings of the 7th International Symposium on High-Level Synthesis*, pages 11–16, May 1994.

J. M. Rabaey and M. Potkonjak. Estimating Implementation Bounds for Real Time DSP Application Specific Circuits. *IEEE Transactions on Computer Aided Design*, 13(6):669–683, June 1994.

E. Roza, J. Biesterbos, B. DeLoore, and J. VanMeerbergen. On the Application of Architectural Synthesis in the Design of High-Volume Production ICs for Consumer Applications. In *Proceedings of the Sixth International Worshop on High-Level Synthesis*, November 1992.

D. Thomas et. al. *Algorithmic and RTL Synthesis: The Systems Architect's Workbench.* Kluwer Academic Publishers, Norwell, MA, 1990.

H. Trickey. Flamel: A High-Level Hardware Compiler. *IEEE Transactions on Computer Aided Design*, 6(2):259–269, February 1987.

J. Vanhoof, K. Van Rompaey, I. Bolsens, G. Goosens, and H. DeMan. *High-Level Synthesis for Real-Time Digital Signal Processing*. Kluwer Academic Publishers, Norwell, MA., 1993.

R. A. Walker and R. Camposano, editors. *A Survey of High-Level Synthesis Systems*. Kluwer Academic Publishers, 1991.

R. A. Walker and D. E. Thomas. Behavioral Transformation for Algorithmic Level IC Design. *IEEE Transactions on Computer Aided Design*, 8(10):1115–1127, September 1989.

N. Wehn, M. Glesner, and C. Vielhauer. Estimating Lower Hardware Bounds in High-Level Synthesis. In *Proceedings of the International Conference on Very Large Scale Integration (VLSI) 93*, pages 6.4.1–6.4.10, September 1993.

N. Wehn, J. Biesenack, T. Langmaier, M. Münch, M. Pilsl, S. Rumler, and P. Duzy. Scheduling of Behavioral VHDL by Retiming Techniques. In *Proceedings of the Euro-DAC '94*, pages 546–551, September 1994.

W. Wolf, A. Takach, C. Y. Huang, and R. Manno. The Princeton University Behavioral Synthesis System. In *Proceedings of the 29th ACM/IEEE Design Automation Conference*, pages 182–187, June 1992.

# High Level Synthesis by Systematic Derivation of Vision Automata from Emulation Results

*I.C. Kraljić, G.M. Quénot* and B. Zavidovique*
*Laboratoire Système de Perception*
*DGA/Établissement Technique Central de l'Armement*
*16 bis avenue Prieur de la Côte d'Or*
*94114 ARCUEIL Cedex FRANCE*
*Tel: +33-1-42 31 97 07 Fax: +33-1-42 31 99 64*
*Email:* ik@etca.fr

**Abstract**
A methodology for deriving image processing ASICs from the results of their real-time emulation on the Data-Flow Functional Computer is presented. The aim of the method is to reduce the time and effort required for synthesizing and validating ASICs after emulation. Indeed, the same architecture which emulated the application is the entry point of the derivation process. Thus, the "synthesis" is restricted to the optimization at the Register-Transfer level of the validated emulator architecture, as a consequence the validation of derived ASICs is easier. The derivation process is demonstrated on the direction extraction macro of a defect detector application.

**Keywords**
High level synthesis, VLSI, image processing, emulation.

## 1 INTRODUCTION

High level (or architecture) synthesis takes as input an algorithm expressed in a high level language and generates an RTL netlist implementing the algorithm according to explicit constraints. The resulting netlist must be thoroughly simulated in order to validate its behavior. The increasing interest in rapid prototyping, formal verification or FPGA-based emulation indicates that simulation does not cover all the aspects of design validation. When real-time vision applications are concerned, the specification of the application is vague and imprecise and the enormous amount of input data requires weeks of traditional simulation (the computational requirements of real-time image processing range up to

*G.M. Quénot is currently at LIMSI-CNRS, BP133, 91403 Orsay Cedex FRANCE. Email: quenot@etca.fr

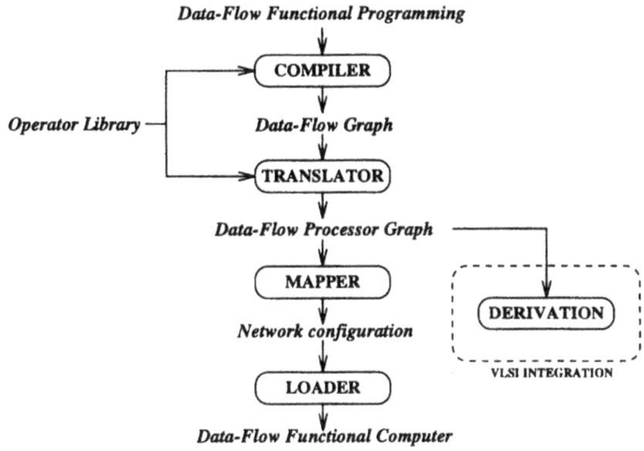


**Figure 1** Emulation flow.

billions of operations per second). Furthermore, the application should be validated in its environment (a problem seldom addressed by simulation).

This paper describes an approach called **derivation from emulation results** where the validation step precedes the synthesis step: the algorithm is firstly emulated in its environment (with input from video cameras) on a general image processing computer (the Data-Flow Functional Computer), then the actual computer configuration is analyzed, optimized and finally implemented as a VLSI chipset. The RTL netlist implementing the algorithm is thus directly derived from the hardware which emulated the algorithm: **emulation provides simultaneously the synthesis of an RTL netlist and its validation**. Hence the post-emulation "synthesis" step is reduced to the optimization of the emulator's active resources. Since data dependency is preserved, total equivalence of the derived RTL netlist's behavior with the results obtained during emulation is achieved (functional behavior and performance). As a consequence the amount of simulation required to validate the derived ASICs is restricted.

Firstly, the application is emulated in real time on the DFFC (Figure 1) [Quénot *et al.* 1994] [Sérot *et al.* 1993]. The DFFC is an 8×8×16 array of Data-Flow Processors (DFPs) and it processes digital video streams on the fly at a rate up to 25 MHz pixel. The algorithm is expressed in a functional programming language which is translated into a DFP graph using operators from a database (200 operators). Figure 2 summarizes the DFP's architecture. The programmable 3-stage pipelined datapath contains 3 input FIFO queues and 3 output queues, an $8 \times 8$-bit multiplier, a 16-bit 2901-type ALU, a $256 \times 9$ bits RAM and a 16-bit counter. The datapath is configured through a programmable controller (a $64 \times 32$ bits program is provided for its description). The DFP architecture has been designed to implement efficiently a wide range of low- to mid-level image processing operators (arithmetic/logic operations, 8/16-bit histogrammer, line/pixel delays and sums). Several significant applications have been successfully emulated, including colored object tracking, main direction follower, contrast enhancement and defect detection. In the following, we call a resource of a Data-Flow Processor any DFP datapath element, the DFP controller and its I/O ports.

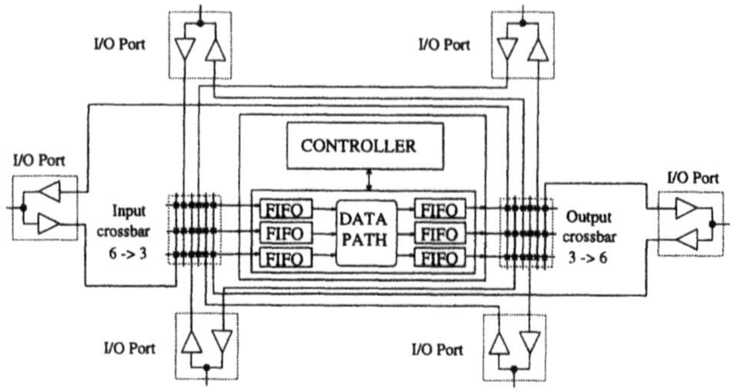


**Figure 2** Architecture of the Data-Flow Processor.

## 2 DERIVATION FROM EMULATION RESULTS

### *Principle*

The actual DFFC configuration implementing the algorithm is used as an input to the derivation process. The process is based on a processor-level analysis: identification of the effectively used resources and optimization of these resources. The resulting automaton is implemented as a VLSI chipset whose architecture is pipelined Finite State Machine with Datapath. The VLSI approach to derivation is highly facilitated by 1) the granularity of the Data-Flow Processor which is at the 8/16-bit operator level and 2) the relative simplicity of the DFP architecture which allows a deterministic knowledge of whether a resource is used or not. Derivation is similar to retargeting an FPGA implementation into an ASIC: the conversion of the design into a different technology (*e.g.* gate array) yields a smaller and faster implementation. Thanks to the coarse grain architecture of the Data-Flow Processor derivation is far more easily automated.

### *Justification*

One question that immediately arises is whether one could not reduce a graph of several dozens processors into a few chips. Table 1 shows the average (static) resource utilization (and actual area) of DFP resources computed on several non-trivial algorithms (3×3 convolution, Nagao-like filter, erosion, defect detector...). It is clear that the resource utilization of the presented algorithms is low. Furthermore, the average area of a DFP after resource reduction (i.e. removal of the unused resources) can be estimated. For each resource $i$, let us call $u_i$ its average static utilization rate and $a_i$ its actual area (in $mm^2$). The estimated average area of a DFP after resource reduction is

$$A = \sum_{i \in [resources]} u_i \times a_i.$$

Using Table 1 yields A = 7.35 $mm^2$ (1$\mu$m technology). Thus, considering only the lowest-level optimization step (discarding the unused resources) shall yield a theoretical reduction in area of 80%.

**Table 1** Average static utilization rate and area of DFP resources

| Resource | I/O Ports | I/O FIFOs | Data-Path | Data RAM | Controller |
|---|---|---|---|---|---|
| Utilization rate | 42.4% | 42.2% | 22.2% | 20.0% | 4.9% |
| Area (mm$^2$) | 9.34 | 3.74 | 4.85 | 1.60 | 8.55 |

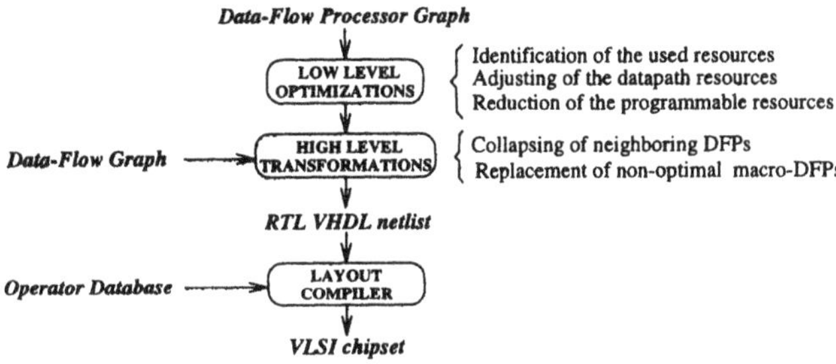


**Figure 3** Derivation flow.

# 3 THE DERIVATION PROCESS

This Section explains the optimizations performed during derivation. In accordance to the emulation flow steps, 3 different representations for an algorithm are available (from high to low level of abstraction): 1) data-flow graph, 2) Finite State Machine (FSM) network and 3) RTL netlist. Whereas most traditional high level synthesis systems use only the data-flow representation (*e.g.* [Verdier & Zavidovique 1993]), derivation will also use the other 2 representations: low level optimizations are performed on the RTL netlist and high level transformations are performed on the FSM representation and data-flow graph*. The block diagram of the derivation process is shown in Figure 3.

## *The low level optimizations*

Their aim is to reduce each DFP to its **minimal equivalent implementation** by discarding its unused resources and adjusting its remaining resources. These optimizations are performed on the RTL netlist with input from the Finite State Machine definition of each DFP. They are applied to each DFP and to connections between adjacent DFPs.

1. **Removal of the unused resources:** an analysis of the static and dynamic (FSM) definitions of the DFP operator determines precisely whether a datapath resource (ALU, MUX, FIFO...) is used or not. A resource is unused if it produces no output flow or if its output flow is equal to its input flow or if it produces a constant flow. Unused resources are removed or replaced by buses.
2. **Adjusting bitwidths and depths:** the bitwidth of each datapath resource and the depth of each I/O FIFO and of the DFP pipeline are restricted to their minimum value.

*Note that more and more high level synthesis systems perform post-synthesis optimizations on the synthesized RTL netlist [Eveking & Höreth 1993] [Papachristou *et al.* 1993].

3. **Reduction of the programmable resources:** reducing a programmable resource (controller, crossbar or I/O port) means "freezing" it to its specific function in each DFP operator. The controller which was implemented with a $64 \times 32$ program RAM is implemented in standard cells. The I/O ports of a DFP were connected to its FIFOs through a crossbar; in a derived DFP the (fixed) connections are implemented statically.

As an example, a DFP implementing a 16-bit adder whose 8 MSBs are not used is basically reduced to a "one-stage" pipeline containing an 8-bit adder plus 3 two-word I/O FIFOs. These immediate optimization steps bring huge area reduction.

### *The high level transformations*

The low level optimizations reduce each DFP to its minimal equivalent implementation. This implementation is a data-flow operator, hence it still includes costly resources dedicated to flow management (I/O FIFOs and ports). These resources are eliminated through collapsing of neighboring DFPs into macro-DFPs according to the function they implement or to user specification. For instance, the **atan** arc tangent operator of the extraction macro (Figure 4a) required a $1024\times8$ Look-Up Table (LUT) and was emulated using 9 DFPs (4 LUTs, 5 selectors) due to the DFP limitations (a DFP contains a $256\times9$ RAM). This operator was derived into a macro-DFP containing a $1024\times8$ ROM.

The derived ASICs are validated by simulation as follows: 1) each derived DFP (or collapsed macro-DFP) is validated **if it has not already been derived and validated** and 2) the whole derived ASIC is validated. Typically a few dozens input vectors are enough to validate the function of each derived circuit.

## 4 RESULTS

Output of the derivation process is a VHDL netlist at the Register-Transfer Level which is fed into a commercial layout compiler (COMPASS). Derived circuits consist of both single and macro Data-Flow Processors implemented in standard cells and datapath compiler cells. Considering a $1\mu$m CMOS technology the average size of a derived DFP featuring only low level optimizations is 4.4 mm$^2$ (Table 2). This implies the 3 low level optimizations, thus it explains the difference with the theoretical area computed in Section 2 which included only the first low level optimization. The I/O FIFOs occupy a significant part (about 43%) of the derived processors' area, this justifies the need for high level transformations that deal with this problem. The high level transformations decrease the area by about 60%. Hence 50 derived DFPs can be integrated on a single chip. The use of a $0.7\mu$m technology shall raise this density up to 100 DFPs/chip. If the derived circuit must be the smallest possible for a large application, one can envision the use of a multichip module technology. Currently only the I/O FIFOs are implemented using COMPASS' datapath compiler, while the rest is implemented in standard cells. Subsequent use of the datapath compiler shall further reduce the datapath's area [Leveugle & Safinia 1993].

The direction extraction macro of the defect detector has been successfully derived using the derivation tool and COMPASS' back-end tools. The direction extraction chip (Figure 4) contains 26 initial DFPs in 50.7 mm$^2$ (21K gates, core size 35.7 mm$^2$, $1\mu$m CMOS technology). The chip has been derived with both low and high level optimizations, it features a gain in area of 60% compared to the extraction chip derived with only low level

**Table 2** Initial/derived Data-Flow Processors in a 1$\mu$m technology (areas in mm$^2$)

| Operator | Initial DFP | abs | add | and | max | 256-w FIFO | Pixel delay | Line delay | Histogram |
|---|---|---|---|---|---|---|---|---|---|
| Total area | 35.64 | 2.6 | 4.8 | 3.6 | 3.9 | 4.9 | 4.1 | 3.9 | 7.1 |
| Control/DP | 1.76 | 0.24 | 0.21 | 0.17 | 0.12 | 0.07 | 0.43 | 0.47 | 0.28 |
| FIFO/Op | 0.11 | 0.48 | 0.52 | 0.50 | 0.47 | 0.25 | 0.45 | 0.47 | 0.26 |

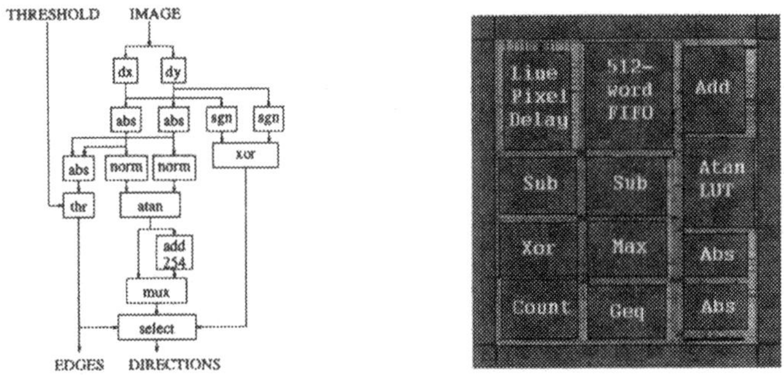


(a) Data-flow graph. (b) Derived chip.

**Figure 4** The direction extraction macro.

optimizations. The chip has been validated by simulation using only a few dozens input vectors at a clock frequency of 12 MHz. The chip alone can process images with at most 512 pixels/line due to the on-board 512-word FIFO; but the connection of an external FIFO relaxes this constraint. The computing power of the chip is about 110 MIPS.

## 5 CONCLUSION

An approach to high level synthesis called derivation from emulation results has been presented. It exploits the results of the emulation of an algorithm in real time and in its environment. The emulation step provides simultaneously the synthesis of an RTL netlist (the emulator's active resources) and its validation; then the derivation step optimizes this RTL netlist and a commercial layout compiler is used to derive a chipset implementing the algorithm. The derivation concept has been implemented for real-time vision applications: a general vision computer has been built (the 1024-processor Data-Flow Functional Computer), the computer is programmed in a functional programming language.

Our current work consists in improving the derivation process: apart from the already cited generalization of datapath compiler cells in derived circuits, an area of research is the optimal partitioning of Data-Flow Processor graphs into derived chips. The natural partitioning of DFP graphs (according to the data-flow graph hierarchy) does not necessarily yield an optimal result in terms of area/performance of the derived chipset.

## 6 REFERENCES

Allart, E. and Zavidovique, B. (1987) Functional Image Processing through Implementation of Regular Data-Flow Graphs. In *21st Annual Asilomar Conference on Signals, Systems on Computers, Pacific Grove, CA.*

Eveking, H. and Höreth, S. (1993) Optimization and Resynthesis of Complex Data-Paths. In *Proc. of the 30th ACM/IEEE Design Automation Conference*, pages 637–41.

Leveugle, R. and Safinia, C. (1993) Generation of optimized datapaths: bit-slice versus standard cells. In G. Saucier and J. Trilhe, editors, *Synthesis for Control Dominated Circuits (IFIP Transactions A-22)*, Elsevier, pages 153–66.

Papachristou, C. and Harmanani, H. and Nourani, M. (1993) An Approach for Redesigning in Data Path Synthesis. In *Proc. of the 30th ACM/IEEE Design Automation Conference*, pages 419–23.

Quénot, G.M. and Kraljić, I.C. and Sérot, J. and Zavidovique, B. (1994) A Reconfigurable Compute Engine for Real-Time Vision Automata Prototyping. In *IEEE Workshop on FPGAs for Custom Computing Machines*, Napa, CA, USA, pages 91–100.

Sérot, J. and Quénot, G.M. and Zavidovique, B. (1993) Functional Programming on a Data-flow Architecture: Applications in Real-Time Image Processing. *International Journal of Machine Vision and Applications*, pages 44–56.

Verdier, F. and Zavidovique, B. (1993) A Complete Environment for Global Architecture Synthesis. In *Proc. of Computer Architectures for Machine Perception* , pages 77–81.

## 7 BIOGRAPHY

**I.C. Kraljić** was born in Paris, France, February 1970. He is a Ph.D. student in the Laboratoire Système de Perception at the Établissement Technique Central de l'Armement. He received an M.S. degree in Electrical Engineering from University Paris XI, Orsay, France in 1992. His current research interests include high level synthesis, rapid ASIC prototyping and image processing.

**G.M. Quénot** was born in Casablanca, Morocco, May 1960. He received the degree of Engineer from École Polytechnique in 1983, and a Ph.D. in Computer Science from University Paris XI, Orsay, France, in 1988. From 1988 to 1991 he was in charge of the hardware activities for real-time image processing at ETCA. He is now a researcher at LIMSI, a French National Agency for Scientific Research (CNRS) Laboratory. He is mainly interested in parallel computer architectures, VLSI circuit design and image processing.

**B. Zavidovique** is professor in the EE Department at University Paris XI, and a scientific advisor at ETCA for the problems of real-time processing and robotics. He was invited at UC Berkeley-EECS, in 1986. His research interests include perception systems, real-time architectures and programming methods within the frame of circuit integration. Since his Doctorate of Sciences in Robot Vision from the University of Franche Comté (France) in 1981, he has published more than two hundred scientific papers in image processing, sensor fusion, computer architecture, intelligent control and learning.

31

# Synthesis : From Digital Signal Processing Specifications to Layout

*PHILIPPE J.- L.[1], SENTIEYS O.[1], DIGUET J.-P.[1], MARTIN E.[2]*
*[1]Laboratoire d'Analyse des Systèmes de Traitement de l'Information,*
*ENSSAT 6 rue de Kerampont 22300 LANNION FRANCE*
*Tel: (33)/96/46/50/30; FAX: (33)/96/37/01/99;*
*Email Name@merlin.enssat.fr.*
*[2]IUP LORIENT 10 rue Jean ZAY 56100 LORIENT FRANCE*
*Tel: (33)/97/87/28/61; FAX: (33)/97/87/28/15;*

**Abstract**

To be efficient, a design methodology has to take into account the following user requirements: fast prototyping in order to satisfy the time to market constraint, the production of efficient architecture, the comparison of the complexity of different DSP algorithmic solutions. These requirements may be associated to user constraints such as real time processing, cost or power consumption, the re-use of previous developments.

Such a design methodology has been developed and integrated into the tool GAUT. From the VHDL behavioural specification of a DSP algorithm, a time constraints and a technologic library (FPGA, standard cells), the dedicated pipeline architecture is synthesized. This VHDL description is then computed by a logic synthesis tool. This design flow, which produce quickly efficient dedicated architectures, is validated on some digital signal processing applications like acoustic echo cancellation and ADPCM.

**Keywords**

High-Level Synthesis

## 1 INTRODUCTION

The implementation cycle of an application has to consider three main points: the architecture, the algorithm, and the characteristics of the application itselve.

For one algorithm, you have to select which architecture you need. If the architecture already exists, you have to decide which part of the algorithm you have to implement on which part of the architecture. If the architecture doesn't exist, you have to design it in respect with the algorithm you have selected for the application. This one is very often: time, consumption, or cost constrained. So you have to verify that these constraints are verified all along the design flow. The constraints can be used together with classical critera, as the efficiency, the quality, in order to select an algorithm, or a class of algorithm, for the application.

Our approach looks to define a methodology which allows to take into account these three main points. It has three main characteristics. On the first hand, it is dedicated to signal processing application with real time and cost constraints. On the other hand, the method is based on the interaction between architectural and logical synthesis tools. Finally, in order to fit with the application field and reduce the complexity of the approach, the architectural model has been restricted to DSP. The paper will be organized around four main points: an overview of the tool, the link with the logical synthesis tool, results, and work currently in progress.

## 2 AN OVERVIEW OF GAUT

The VHDL compiler is integrated as a front end of the framework GAUT (Martin 1993). With regard to the implementation of procedures, the designer has two options: a hardware component or a software description. A hardware solution has the advantage of a good area-time trade-off. However, a logical solution leads to the selection of standard operators that can be re-used for other operations. This finally achieves a smaller global area of the architecture. In this case the procedures will be inlined during the synthesis. The best method depends on the use rate of the macro-function.

After dependence analysis, the front end translates the entry point into a signal flow graph representation. During this step, several transformations are realized such as procedure inlining, fixed-bound loops unrolling, variable propagation,.... This provides a unique assignment representation. Other transformations are realized on the signal flow representation, such as the transformation of conditional constructs in order to obtain a deterministic flow graph. From this internal representation, the operators are selected, the parallelism is reduced, the operations are assigned to operators, and some optimizations are carried out. Next, the description of the dedicated architecture and its controler will be produced. The architecture satisfies the real time constraint that was specified at the beginning of the synthesis.

## 3 HIGH LEVEL SYNTHESIS AND ITS CONNECTION WITH A LOGICAL SYNTHESIS TOOL

The design flow is represented in Figure 1. The component library contains various operators (multidelay, multifunctional, pipeline). The generic aspect of the components is computed by using the logical synthesis tool. Generic parameters give information such as area, delay, bit size, functions available (e.g. "+","-","shift", ... for an ALU), pipe stages, etc...

Architectural synthesis enables the re-use of operators. So, after the creation of a data flow graph from the VHDL specification, after transformations, the allocation-selection stage included in GAUT selects from the library the type and the number of operators in order to respect a real time constraint with a minimal cost area. It enables the use of an efficient and exhaustive research in relatively complex libraries. With a branch and bound method, the optimal solution is reached with a very short computation time. This task is realized before the time-scheduling and the assigment that minimize the cost of registers and interconnections.

After the selection of the optimal operators and the scheduling of the flow graph, GAUT produces a VHDL architecture. This description includes the behavioural description of the memory unit and the structural description of the the processing unit. The control unit is described as a flattened finite state machine with the help of behavioural VHDL. Its up now, to the logical synthesis to optimize and map out a gate level hardware structure.

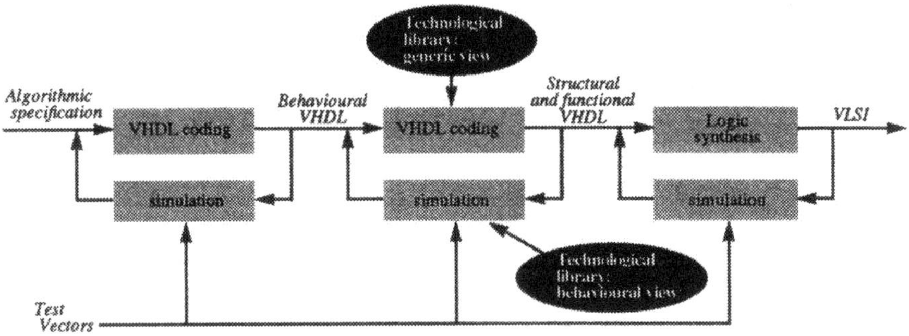


**Figure 1** Design flow

The test vectors used in the simulations of the design flow depend on the application. The processing unit is organized around cells (see figure 2) : a cell may integrate many registers or share a register with other cells through the use of multiplexer/demultiplexer and tristate modules.

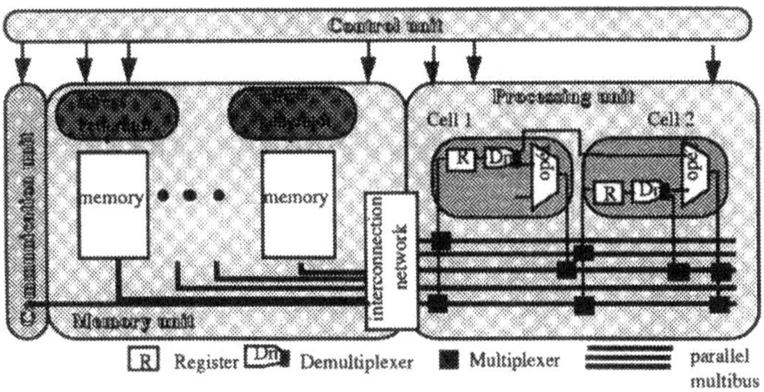


**Figure 2** Generic architecture of a dedicated processor.

The control model of the architectures synthesized by GAUT integrates two features : the use of the pipeline for registers and operators, and the use of a multi-phase clock to control the events in order to optimize the efficiency of each cell . It is implemented with a fast clock representing the minimal phase of the architecture (GCD of the time of the set of operators and of the memory access time). This minimal phase can be controled by the user. The regularity of the program is used in order to reduce the time of the logical synthesis and the complexity of the routing.

Digital Signal Processing applications all have the distinctive feature of producing FSM descriptions with a lot of states but with some regularities in the control of the datapath. Reducing the number of states can be done by eliminating the No OPerations of the description or by searching the repetitive sequence of the FSM graph in order to transform them into cycles. We introduce a counter in our generic model to implement these optimizations. Let us consider the following example in which the processing time of the operator 1 is four times greater than the processing time of the operator 2.

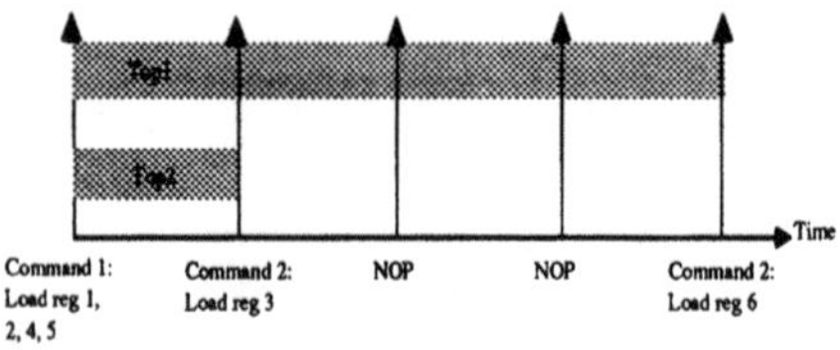


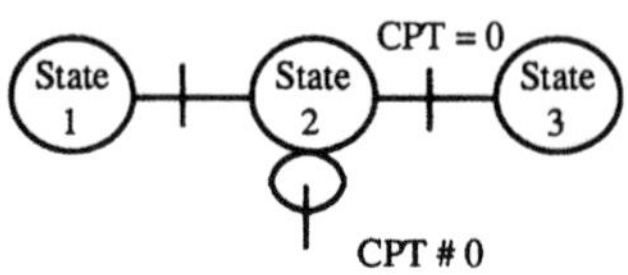


**Figure 3-a** Result of the synthesis: Sequencing of the operations on the data path (5 states before optimization).

**Figure 3-b** Result of the synthesis: 3 states by using a counter after optimization

In figure 2 we see that the control of one cell need several signals : multiplexer, demultiplexer and tristate control, synchronous load of registers, control of pipeline and multifunction operators. Therefore, the routing area between the control unit and the processing unit may increase significantly with the complexity of the data path. Almost of DSP applications need in fact a very few set of commands. This instruction will be routed on the circuit in order to be decoded by the cell itselve to decrease the layout synthesis task and to optimize the area cost of the chip.

The control unit is organized around the generic model of the figure 4. The Moore finite state machine, the decoders and the counter are described in VHDL in order to be synthesized by logic synthesis tool (Compass, Altera, ...). The user can easily control the floorplanning of the all circuit.

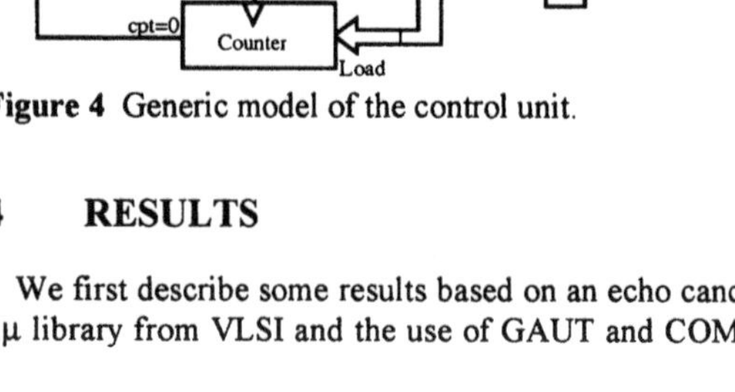


**Figure 4** Generic model of the control unit.

## 4 RESULTS

We first describe some results based on an echo cancellation application (Gilloire 1987), a 1μ library from VLSI and the use of GAUT and COMPASS. Figure 5 gives the layout of a

LMS (1024 weights, fmax=16kHz) obtained from the cooperation of GAUT and COMPASS. The generic aspects of some library components were computed by using the Compass datapath synthesis tool (e.g. the multiplier), or the Compass VHDL synthesis tool (e.g. the adder/substracter). This architecture has been automatically produced from a behavioural description and component library (fig. 6).

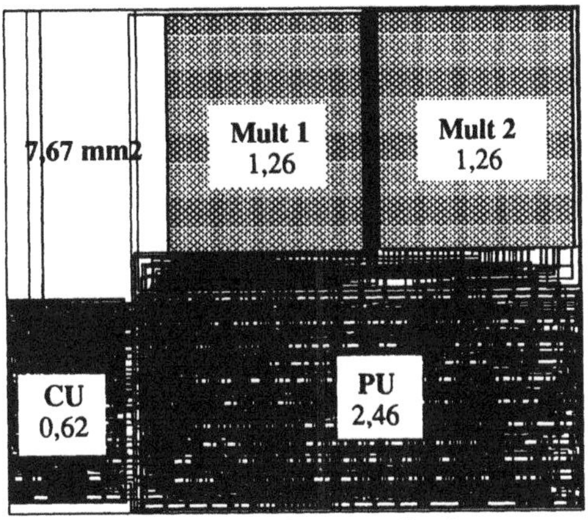


**Figure 5** LMS 1024 weights, sampling frequency : 16kHz, with improved routing

| BEHAVIORAL DESCRIPTION | LIBRARY DESCRIPTION |
|---|---|

```
ENTITY lms IS
GENERIC (latency_time : time := 62500);
PORT (xt,yt:IN integer; yct:OUT integer);
END lms;

ARCHITECTURE behavioural OF lms IS
  CONSTANT N: integer := 1000;
   TYPE vec is array (0 TO N) OF integer;
  SIGNAL x, H : vec;
BEGIN
  PROCESS
    VARIABLE adapt,mu, y: INTEGER;
    VARIABLE i: CONTROL;
   BEGIN
    y := xt * H(0);
    FOR i IN 1 TO N-1 LOOP   -- filtering
      y:= y + x(i) * H(N-i-1);
    END LOOP
    yct <= y;
    adapt := (yt - xt) * mu;
    FOR i IN 0 TO N-1 LOOP -- adaptation
       H(i) <= H(i) + adapt * x(N-i-1);
     END LOOP;
     FOR i IN 1 to N-2 LOOP -- signal variation
       x(N-i) <= x(N-i-1);
     END LOOP;
     x(1) <= xt;
    WAIT FOR latency_time
   END PROCESS;
END behavioural;
```

```
PACKAGE VLSI_GAUT IS
   CONSTANT nb_bit : integer:= 16;
COMPONENT register16
   GENERIC (area : integer:= 139;
            real_func: one_function:= (reg);
            time_func: proc_time := (6);
            nb_input : integer:= 2;
                 nb_pipe: integer:= 1););
END COMPONENT;
COMPONENT Adder_Substracter16
   GENERIC ( area : integer:= 252;
            real_func: two_function:= (add,sub);
            time_func: proc_time := (15,15);
            nb_input : integer:= 2;
            nb_pipe: integer:= 1);
END COMPONENT;
COMPONENT multiplier16
   GENERIC ( area : integer:= 2032;
            real_func : one_function := (mul);
            time_func : proc_time := 60;
            nb_input  : integer:= 2;
            nb_pipe : integer:= 1);
END COMPONENT;
...
END VLSI_GAUT  ;
```

**Figure 6** *Implementation of a LMS component: behavioural specification and library.*

To control the data path, among one thousand states, only twelve different commands are necessary. This is a result of the program's regularity. In order to reduce the routing between the control and processing units, a coding/decoding mechanism has been automatically introduced by GAUT. The number of states is reduced through the use of a counter. After these optimizations, the time for logic synthesis and routing has been reduced almost by half.

Others applications or technologies can be used. The figure 7 gives some results of a ADPCM algorithm used in speech coding (CCITT 1988). The results are given for two technologies and two specifications (one using multiplication operators, the other booth functions ).

| | **ALTERA 8 bits Library** | **COMPASS Library 1µm** |
|---|---|---|
| **ADPCM 1** | 1 multiplier, 1 add_sub,<br>1 comparator, 26 registers,<br>5 mux, 24 demux, 1 step pipeline<br>**area : 825 cells** | 1 multiplier, 1 adder, 1 substractor<br>1 shifter, 1 comparator, 43 registers<br>14 mux, 43 demux, 1 step pipeline<br>**area : 14,3 mm²** |
| **ADPCM 2:**<br>with booth functions | 1 add_sub, 2 comparators,<br>49 registers, 1 shifter, 15 mux,<br>49 demux, 1 step pipeline<br>**area : 1441 cells** | 1 add_sub, 2 comparator, 1 shifter,<br>1 alu (+ ; - ; and), 39 registers,<br>14 mux , 39 demux, 1 step pipeline<br>**area : 11,2 mm²** |

**Figure 7** ADPCM algorithm synthesis; two algorithmic specifications and two libraries

## 5 CONCLUSION AND FUTUR PROSPECTS

The methodology which has been presented is oriented towards fast prototyping. It allows the re-use of the results of previous developments and the specification of user constraints such as real time processing. It has been tested on real applications.

The work on ADPCM (Diguet 94) underlines the importance of the style of algorithmic specification. Different studies are currently in progress on the adequacy between the algorithm and the architecture to modelize the links between algorithmic style and hardware complexity. From an analysis of an architectural synthesis result, the purpose is to introduce transformations at different levels. The transformations have to be done to improve the adequacy algorithm-architecture in respecting a real time constraint. These transformations of the specifications can occur at several levels:

Algorithmic level: A DSP application can be realized by using various algorithmic equivalent processes, for instance, the choice between a lattice or a block version of an adaptive filter.

Structural level: The techniques which we are using are for the most part known. They are based on the properties of associativity, distributivity and commutativity. In addition we are using the fusion of equivalent nodes, the elimination of redundancy and loop transformations. We are working on transformations to improve the regularity of a flow graph.

Functional level: A given operation can be mapped on different kinds of operators or sets of operators. For example, a multiplication can be executed by a hardware multiplier or by the association of a shifter, an adder and a software function. Another example: it can be profitable for a convolution to replace a multiplier and an adder by a multiplier-adder ("mad"). The interest of such modifications depends on the regularity and the rate of use of the transformed nodes among the operations of the algorithm.

Another purpose is to extend the approach in terms of constraints and classes of architecture. This is currently being done by taking into account the consumption aspect, and a model of architecture based on programmable DSP and ASICs.

## 5 REFERENCES

CCITT (1988) Rec.G727 Melbourne.

Gilloire A., Jullien J.-P. (1987) L'acoustique des salles dans les télécommunications. *L'écho des RECHERCHES* ,N°127.

Martin E., Sentieys O., Dubois H., Philippe J.-L. (1993) GAUT: An Architectural synthesis Tool for Dedicated Signal Processors *EURO-DAC*.

Philippe J.-L, Sentieys O., Diguet J.-P., Martin E. (1994) Adequacy architecture algorithm, an experiment in signal processing by using FGGA *VHDL forum*, Tremezzo, Italy.

Sentieys O., Martin E., Philippe J.L. (1994) VLSI architectural synthesis for an acoustic echo cancellation application *VLSI signal processing VI* page 84- 92.

Diguet J.P., Sentieys O., Philippe J.L., Martin E.; "How Specify an Algorithm in VLSI architectural Synthesis, a Vocal Coding Application"; *VLSI Signal Processing VII*; La Jolla-San Diego; October 94; pp 346-355.

## 6 BIOGRAPHY

*Jean- Luc Philippe* received the Ph.D. in signal processing from the University of Rennes in 1984. Since 1992 he is working as an assistant professor at ENSSAT, University of Rennes, conducting a research group in CAD for VLSI design. His research interests include signal and image processing systems, VLSI and high level synthesis.

*Olivier Sentieys* graduated from ENSSAT as a senior executive engineer in Electronics and Signal Processing Engineering in 1990, and received the Ph.D. degree in Signal Processing and Telecommunications in 1993 from the University of Rennes. Since 1990 he has been working as an Assistant Professor at ENSSAT, and is involved in research of high level synthesis of VLSI systems dedicated to signal processing , and in the development of the synthesis tool GAUT.

*Eric Martin* received the professor agregation at ENS de Cachan in 1984, the Ph.D at university of Orsay in 1986 and the Habilitation at University of Rennes in 1993. Since 1994 he is Professor at the UBS (Universite de Bretagne Sud), and director of the research laboratory LESTER. His research activities are in the field of high level CAD tool development for real time signal and image processing implementation.

*Jean-Philippe Diguet* graduated from ESEO as a senior executive engineer in Electronics and Signal Processing Engineering in 1992, and received the DEA degree in Signal Processing and Telecommunications in 1993 from the University of Rennes.Presently he is pursuing a Ph.D.degree in the LASTI at ENSSAT, and is involved in research of high level synthesis of VLSI systems dedicated to signal processing and adequacy algorithm - architecture.

# BDD application to mutual exclusion testing in high-level synthesis

*Vasily A.Syngaevsky*
*Research Institute of VLSI CAD Systems,*
*8a, Mazhorov per., Moscow E-023, 105023 RUSSIA,*
*phone: 7 (095) 369 19 75 fax: 7 (095) 369 51 07*
*e-mail: syng@sapran.msk.su syng@redline.ru*

**Abstract**
This paper outlines extensions to an approach of M.Rim and R.Jain (Rim,1992) for mutual exclusion testing in High-Level Synthesis. These extensions enable to deal with more complicated control statements such as CASE with overlapping branches, LOOPs with several exits and even GOTO. Shared Reduced Ordered Binary Decision Diagrams are proposed to represent mutual exclusion relation and to construct simple algorithms that inherit all the power and reliability of existing BDD-packages
**Keywords**
Binary Decision Diagrams, mutual-exclusion testing, scheduling, high-level synthesis

## 1 INTRODUCTION

High-Level Synthesis (HLS) (McFarland,1988; McFarland,1990) is a commonly used term to represent a large list of problems that have to be solved to obtain a Register Transfer Level description of a system (or subsystem) from a given algorithmic specification. At present time HLS seems to be of interest for industry in two main application domains: Digital Signal Processing and Control-Dominated Applications (DeMan,1990; Payer,1992). In this paper we consider a problem that arise during scheduling of a Control-Dominated specification. It is often referred to as Conditional Resource Sharing (Kramer,1990; Prabhu,1993).

Algorithmic specification of a control-dominated design contain a lot of nested loops with several exits, IF and CASE statements (Stoll,1992; Gutberlet,1993). This leads to a situation when not all operations of the algorithm will be executed during the particular execution path. Hence, a mutual exclusion relation is introduced on a set of all operations: Two operations are mutually exclusive if during any particular execution path at most one of them is executed. This relation can be used during scheduling and binding to decrease an amount of hardware required to implement specification because mutually exclusive operations can share resources without violating correctness of implementation.

So the problem of testing for mutual exclusion of operations is of great importance for synthesis of control-dominated applications. Several approaches to perform mutual exclusion testing were reported in literature. N.Park and A.Parker used a kind of operations' coloring technique in their SEHWA synthesis system (Park,1988). K.Wakabayashi and T.Yoshimira proposed and K.Wakabayashi and H.Tanaka further developed an approach based on assigning a Condition Vector to each operation of the algorithm (Wakabayashi,1989; Wakabayashi,1992). M.Rim and R.Jain analyzed these mentioned above approaches, outlined their limitations, found even contra-examples to show that they are not always correct; and proposed their own algorithm for mutual exclusion testing based on a new data structure to represent Control-Data Flow Graph (CDFG) (Rim,1992).

In this paper we extend an approach of M.Rim and R.Jain to deal with more complicated control statements such as CASE with overlapping branches, loops with several exits and even GOTO. Paper has the following structure. In Section 2 we briefly review an approach of M.Rim and R.Jain as a basis for our extensions. In Section 3 we propose our extensions and show how they can be used to deal with complicated control statements. In Section 4 we introduce a ROBDD as a tool to represent mutual exclusion relation and to construct simple and reliable algorithms. Section 5 concludes the paper.

## 2 CDFG MODEL AND TAGGING ALGORITHM

In their paper (Rim,1992) M.Rim and R.Jain proposed the following model for CDFG. There are three types of edges in the model:

Strong edge (S-edge - thin lines) represent precedence relation between two nodes of CDFG that is imposed by a direct data transfer between these nodes, i.e. one node "consumes" the value "produced" by another one.

Pure edge (P-edge - bold lines) represent precedence relation between two nodes that is not imposed by data transfer but is required for correct execution, e.g. it must connect CB- and CE-nodes of each conditional branch (explained later).

Weak edges (W-edges - dashed lines) represent relations needed for correct representation of conditional branches. They carry an attribute TRUE or FALSE denoting true and false conditional paths respectively. Precedence relations imposed by W-edges can be violated for correct execution.

```
... /* n0 */
if( x1 ) {
        if( x2 )  ...; /* n1 */
        else      ...; /* n2 */
        if( x3 )  ...; /* n3 */
        else      ...; /* n4 */
}else{
        if( x4 )  ...; /* n5 */
        else      ...; /* n6 */
}
```

Figure 1. Piece of code with nested IFs.

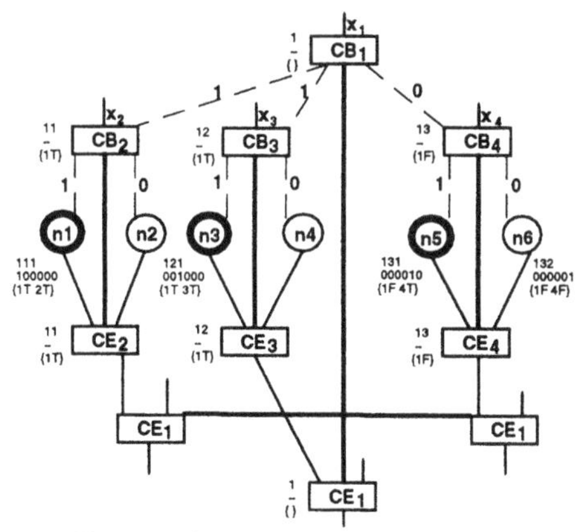


Figure 2. Control-Data Flow Graph

These types of edges are used in a different way during different tasks of HLS. P-edges are not used during mutual exclusion test; during scheduling W-edges are not used. During binding only S-edges are of the interest.

Each conditional is represented by one CB (conditional branch begin) node and several CE (conditional branch end) nodes indicating the termination of a path in the conditional branch.

A piece of code with nested IF statements and its CDFG are shown on Figures 1 and 2 respectively as in (Rim,1992).

This model enables to construct a simple algorithm for mutual exclusion test based on assigning a tag to each operation in DFG. Each tag is an unsorted set of elements. Each element consists of two fields: condition field and selection field. Condition field points to the Boolean value that drives conditional branch; selection filed - represents the value that activates the node. For example, (1T 2F) means that node is activated when 1st condition is TRUE and 2d is FALSE.

Mutual exclusion test is based on the following proposition: If tags of two nodes have elements with the same condition fields and different selection fields then these two nodes are mutually exclusive (Rim,1992).

Tags for CDFG on Figure 2, Colors and Condition Vectors are shown in Table 1.

**Table 1: Different algorithms for mutual exclusion testing: summary**

| Method | CB 1 | CB 2 | CB 3 | CB 4 | n1 | n2 | n3 | n4 | n5 | n6 | CE 1 | CE 2 | CE 3 | CE 4 |
|---|---|---|---|---|---|---|---|---|---|---|---|---|---|---|
| Colors | 1 | 11 | 12 | 13 | 111 | 112 | 121 | 122 | 131 | 132 | 1 | 11 | 12 | 13 |
| C. Vectors | - | - | - | - | 100000 | 010000 | 001000 | 000100 | 000010 | 000001 | - | - | - | - |
| Tags | | 1T | 1T | 1F | 1T 2T | 1T 2F | 1T 3T | 1T 3F | 1F4T | 1F 4F | | 1T | 1T | 1F |
| Conjunc. | 1 | $x_1$ | $x_1$ | $\overline{x}_1$ | $x_1 x_2$ | $x_1 \overline{x_2}$ | $x_1 x_3$ | $x_1 \overline{x_3}$ | $\overline{x_1} x_4$ | $\overline{x_1 x_4}$ | 1 | $x_1$ | $x_1$ | $\overline{x}_1$ |
| Function | $F_1$ | $F_2$ | $F_2$ | $F_3$ | $F_4$ | $F_5$ | $F_6$ | $F_7$ | $F_8$ | $F_9$ | $F_1$ | $F_2$ | $F_2$ | $F_3$ |

## 3 USING BOOLEAN FUNCTIONS FOR MUTUAL EXCLUSION TEST

As it is shown in Table 1 tagging proposed in (Rim,1992) is nothing more but assigning a Boolean function from a restricted class of functions (a class of simple conjunctions) to each node of CDFG.

This restriction of a class of Boolean functions to be used is very strong. This class is enough only to represent mutual exclusion relation imposed by nested IF statements, but it limits the use of more complicated control statements. It is not sufficient to represent, for example, CASE statements with overlapping branches (CASE statement which branches do not overlap is equivalent to a number of IF statements). To explain this fact let us consider a piece of code shown on Figure 3.

It is obvious that 2-branch and 3-branch overlap, i.e. control flow can result in line 6 either from condition test (line 2) or from the previous line 5. This situation can not be modelled by tagging of M.Rim and R.Jain as we need one more Boolean operation in addition to AND and NOT that are implicitly used in tagging: we need an OR operation.

Let us consider one more example. Algorithmic specification of control-dominated applications can contain a lot of loops with several exits and can even contain GOTO. As an exit from the loop's body can be modelled by condition test and GOTO we shall consider only the GOTO. Figure 4 shows one more piece of code that can not be modelled by tagging. Again control flow can result in line 4 either as a result of a jump from line 2 or as the next step from line 3. And again we need an OR Boolean operation to model the situation.

```
1:...; /* n0 */                              - F0
2:switch( x ) {                              - F1
3:        case 1   : ...; /* n1 */           - F2
4:                   break;
5:        case 2   : ...; /* n2 */           - F3
6:        case 3   : ...; /* n3 */           - F4
7:                   break;
8:};
9:...; /* n4 */                              - F5
```

Figure 3. CASE with overlapped branches

```
1:        ...; /* n0 */              - F0
2:        if( x ) goto label;        - F1
3:        ...; /* n1 */              - F2
4:label:  ...; /* n2 */              - F3
```

Figure 4. Condition test with GOTO

So our extensions are the following:

1. Let us allow CB-node to be driven by a set of Boolean variables instead of one.
2. Let us extend the set of possible attributes of W-edges in such a way that different branches of a CASE statement can be indicated: let W-edges carry a set of Boolean variables and their phases: $x_i^\alpha \Leftrightarrow x_i^1 \equiv x_i \,;\, x_i^0 \equiv \overline{x_i}$ .This is to code branches of CASE statement.
3. Let us use P-edges to represent overlapping control flow and, hence, consider them during mutual exclusion testing.
4. Let us assign such a Boolean function of a general form to each operation of a CDFG that it is TRUE on a given combination of arguments if for this combination the corresponding node is on the execution path, and is FALSE in the opposite case. These Boolean functions depend on a set of all Boolean variables that drive condition selections in the whole specification.

Functions are assigned to nodes according to the following rules:

1. All nodes with no predecessors are assigned the Boolean Function ONE.
2. All operational nodes are assigned functions according to following recursive formula:

$$F_{O-node} = \Big(\bigvee_{\forall P-pred} F_{P-pred} \vee \bigvee_{\forall S-pred} F_{S-pred} \vee \bigvee_{\forall W-pred} \big(F_{W-pred} \wedge \bigwedge_{\forall edge\ attributes} x_j^\alpha\big)\Big) \wedge \bigwedge_{\forall S-pred} F_{S-pred} \qquad (1)$$

3. All control nodes are assigned functions according to the following recursive formula:

$$F_{C-node} = \bigvee_{\forall P-pred} F_{P-pred} \vee \bigvee_{\forall S-pred} F_{S-pred} \vee \bigvee_{\forall W-pred} \big(F_{W-pred} \wedge \bigwedge_{\forall edge\ attributes} x_j^\alpha\big) \qquad (2)$$

Using these extensions an example shown on Figure 3 will result in a CDFG shown on Figure 5a and system of Boolean functions shown on Figure 5b that correspond to this CDFG.

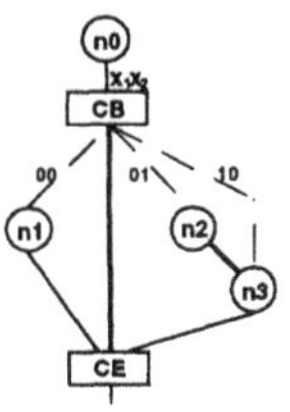


Figure 5a. CDFG for CASE statement

$$F_0 = 1$$
$$F_1 = F_0 = 1$$
$$F_2 = F_1\, \overline{x_1 x_2} = \overline{x_1 x_2}$$
$$F_3 = F_1\, \overline{x_1} x_2 = \overline{x_1} x_2$$
$$F_4 = F_1\, x_1 \overline{x_2} \vee F_3 = x_1 \overline{x_2} \vee \overline{x_1} x_2$$
$$F_5 = F_1 = 1$$

Figure 5b. System of Boolean functions

Here we had to code branches of a CASE statement with Boolean variables. The total number of these variables is less then $\lceil \log_2 n \rceil$, where ***n*** is a number of branches.

Example on Figure 4 has a CDFG shown on Figure 6a and the system of Boolean functions shown on Figure 6b that correspond to this CDFG:

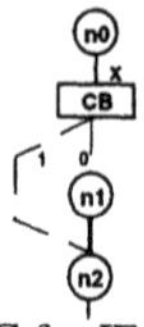


$$F_0 = 1$$
$$F_1 = F_0 = 1$$
$$F_2 = F_1 \bar{x}$$
$$F_3 = F_1 x \vee F_2 = 1$$

Figure 6a. CDFG for IF+GOTO statement Figure 6b. System of Boolean functions

As this example is very simple it seems meaningless, but in more complicated cases it can be useful, e.g. when there are several GOTOs and several labels that go in the different order as it is shown on Figures 7 and 8.

Now the mutual exclusion test can be reduced to the following question: Does there exist a combination of arguments (values of condition-driving variables) such that both assigned functions are TRUE?

It can be written as:

$$(x_0, \ldots, x_n) \rightarrow F_M(x_0, \ldots, x_n) \wedge F_K(x_0, \ldots, x_n)$$

If functions are already assigned then to test two nodes for mutual exclusion we simply have to get the function $F_{tmp} = \overline{F_M \wedge F_K}$ and to test whether $F_{tmp}$ is a tautology; if it is - then two nodes are mutually exclusive, if not - there exist such a particular execution path that both nodes are executed.

```
1:        ...; /* n0 */
2:        if( x1 ) goto label_1;
3:                ...; /* n1 */
4:        if( x2 ) goto label_2;
5:label_1:...; /* n2 */
6:label_2:...; /* n3 */
```

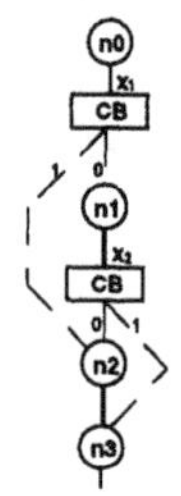


Figure 7. Condition tests with two GOTOs Figure 8. CDFG for two IF+GOTO statements

## 4 SHARED ROBDD AS A TOOL FOR MUTUAL EXCLUSION TEST

The previous section left the questions:

1. How to represent the assigned functions in a compact form?
2. How to manipulate them?

For both questions the answer is - using a software package for Shared Reduced Ordered Binary Decision Diagram (SROBDD) representation and manipulation.

From the last row of Table 1 you can see that functions are frequently repeated: $F_1$ - two instances, $F_2$ - four instances, $F_3$ - two instances. Moreover (see the 5th row of Table 1), some functions are the sub functions of another: $F_2$ is a sub function of $F_4$, $F_5$, $F_6$, $F_7$; $F_3$ - of $F_8$, $F_9$. This is also confirmed by a structure of a SROBDD constructed for this system of Boolean functions (Figure 9). Important is the fact that this SROBDD was constructed using the natural order of Boolean variables imposed by the algorithmic specification, hence, there is no problem with selecting variables' order that have a great impact on the size of BDD. As functions are frequently repeated, size of BDD will remain reasonable.

| Vertex CB1<br>[1] | | Vertex CE1<br>[1] |
|---|---|---|
| Vertex CB2<br>(x1:123412) | Vertex CB3<br>(x1:123412) | Vertex CB4<br>!(x1:123412) |
| Vertex n1<br>(x1:123524<br>(x2:123428)<br>[0]) | Vertex n3<br>(x1:123556<br>(x3:123444)<br>[0]) | Vertex n5<br>!(x1:123508<br>[1]<br>!(x4:123460)) |
| Vertex n2<br>!(x1:123540<br>(x2:123428)<br>[1]) | Vertex n4<br>!(x1:123492<br>(x3:123444)<br>[1]) | Vertex n6<br>!(x1:123476<br>[1]<br>(x4:123460)) |
| Vertex CE2<br>(x1:123412) | Vertex CE3<br>(x1:123412) | Vertex CE4<br>!(x1:123412) |

Figure 9. BDD for the example on Figure 1

This representation is compact enough and the more nested will be the algorithmic specification - the more shared will be the ROBDD. BDD technology also enables to perform simple mutual exclusion test as the only thing we need to do is to apply AND operation of a BDD package to two nodes, apply NOT operation to the result, and at last - check the latest result to be a tautology. All these operations as well as those that are needed to construct a shared ROBDD representing a system of assigned functions are implemented in existing BDD-packages. In our experiments we used Carnegie-Mellon's BDD-package (version 3.34) and also tested a Berkeley's BDD-package.

Algorithms 1 and 2 outline the procedure of ROBDD construction and mutual exclusion testing. Algorithm to assign Boolean functions (nodes of ROBDD) to nodes of CDFG traverse the CDFG similar to ASAP-scheduler's algorithm and could even be integrated into ASAP-scheduler. Algorithm for mutual exclusion test simply applies operations of BDD-package to ROBDD nodes assigned to operations.

Algorithm 1. Assignment of Boolean functions

```
Construct_ROBDD( CDFG )
{
  READY=TMP=∅
  Add all no-pred. vertices to READY
  while(READY≠∅) {
    for all nodes in READY do {
      create BDD-node according (1) or (2)
      associate new BDD-node with CDFG-node
      for all successors of CDFG-node
        if(all pred. already assigned funcs)
          add successor to TMP
    }
    READY=TMP
    TMP=∅
  }
}
```

Algorithm 2. Mutual exclusion test

```
IF_mut_exclusive( Op1, Op2 )
{
  bdd1=BDD-node associated with Op1
  bdd2=BDD-node associated with Op2
  bddT=bdd_and(bdd1,bdd2)
  return bdd_is_tautology(bddT,inversed)
}
```

## 5 CONCLUSIONS

In this paper we proposed extensions of a known approach for representing conditional branches in HLS and for mutual exclusion test. Our extensions enable to deal with more complicated control statements in algorithmic specification. We also proposed the ROBDD as a tool for representing and manipulating mutual exclusion relation. Algorithms to construct the corresponding ROBDD and to perform the mutual exclusion test were outlined. Proposed extensions improve the applicability of mutual exclusion test to control-dominated designs and relies on the power and reliability of existing BDD-packages.

## REFERENCES

McFarland M., Parker A.C., Camposano R. (1988) Tutorial on high-level synthesis, *Proceedings of 25th Design Automation Conference.*

McFarland M., Parker A.C., Camposano R. (1990) The high-level synthesis of digital systems, *Proceedings of the IEEE, Vol.78, No.2, February 1990.*

DeMan H. (1990) CAD for real time information processing systems: challenges and opportunities, *Proceedings of the Synthesis and Simulation Meeting and International Interchange, October 23-25, 1990.- Kyoto.*

Payer M., Duzy P., Michel P. (1992) Specification and synthesis of electronic systems at Siemens, *Proceedings of the Synthesis and Simulation Meeting and International Interchange, April 6-8, 1992.- Kobe, 303-312.*

Kramer H., Rosenstiel W. (1990) System synthesis using behavioral descriptions, *Proceedings of European Design Automation Conference (EDAC).*

Prabhu U., Pangrle B. (1993) Conditional and unconditional hardware sharing in pipeline synthesis, *Proceedings of European Design Automation Conference (EuroDAC), 238-243.*

Stoll A., Duzy P. (1992) High-Level Synthesis from VHDL with Exact Timing Constraints, *Proceedings of Design Automation Conference.*

Gutberlet P., Rosenstiel W. (1993) Interface specification and synthesis for VHDL processes, *Proceedings of European Design Automation Conference (EuroDAC), 152-157.*

Park N., Parker A.C. (1988) Sehwa: A software package for synthesis of pipelines from behavioral specifications, *IEEE Transactions on CAD, vol.7, No.3, March 1988, 356-370.*

Wakabayashi K., Yoshimura T. (1989) A resource sharing and control synthesis method for conditional branches, *Proceedings of International Conference on CAD, 62-65.*

Wakabayashi K., Tanaka H. (1992) Global scheduling independent of control dependencies based on condition vectors, *Proceedings of 29th Design Automation Conference, 112-115.*

Rim M., Jain R. (1992) Representing conditional branches for high-level synthesis applications, *Proceedings of 29th Design Automation Conference, 106-111.*

Vasily A. Syngaevsky received his honors diploma (M.Sc.) in electrical engineering from Moscow Technical University of Radio-engineering, Electronics and Automatics in 1991. Since 1991 he is with Research Institute of VLSI CAD Systems of Academy of Sciences of Russia. He worked in circuits' simulation and parametric optimization field. Now his field of research is logic and high-level synthesis, BDD technology and low-power design methodologies.

PART ELEVEN

# Applicative Studies

33

# VHDL-based behavioral synthesis : can it pay off for telecom ASICs ?

*P. Vanoostende, G. Van Wauwe*
*Alcatel-Bell, Advanced CAD for VLSI*
*F. Wellesplein 1, B-2018 Antwerpen, Belgium*

**Abstract**

VHDL-based behavioral synthesis is appearing on the market but it still has to prove that it can have a significant impact. In the past, most applications for behavioral synthesis came from the DSP area and from the academic world. In contrast, this paper describes the results of an investigation we carried out on recent designs of Alcatel-Bell, leading to a more detailed study of 5 industrial telecom non-DSP circuits, that were suitable for behavioral synthesis.

Our conclusion is that there is distinctly a market potential. However, we experienced that efficient use for telecom non-DSP circuits requires functionality that goes beyond simply generating an RTL-synthesizable description. This functionality is also discussed.

**Keywords**

VHDL, behavioral synthesis, telecom

## 1 INTRODUCTION

Behavioral synthesis has been an active field of research for about one decade (McFarland, 1988), leading in particular to a number of academic demonstrators (see e.g. Owall, 1993 and Vanhoof, 1994).

Currently, commercial behavioral synthesis tools are appearing (Schultz, 1994), but they still have to prove that they can have a significant impact, i.e. that they reduce the design time for a significant portion of an average design.

In the past, confrontation with industrial practice has often led to unexpected applications for CAD technologies. We believe the same holds for behavioral synthesis. The purpose of this paper is therefore to describe the lessons we learned during initial experiments with behavioral synthesis for our telecom applications, which are mostly non-DSP. These applications don't process signals, but consist of actions like bitmanipulations, counting, detecting specific word-values, storing/retrieving datawords etc.

More specifically we will discuss the following questions :

- For which telecom applications can behavioral synthesis be used, with the current state-of-the-art? What additional functionality is needed to handle "real-life" ?

- What functionality does the user need to control and understand the results of the synthesis?

Our conclusions follow from an investigation of 5 designs that were in the past implemented at the RTL-level and that are candidates for implementation with behavioral compilation. Typical complexity is 10-20k gates of logic (excluding RAM).

# 2 THE USE OF BEHAVIORAL SYNTHESIS FOR TELECOM

## 2.1 Type of applications

In telecom, the digital circuitry can be roughly subdivided into 3 classes :

- logic that implements the interface with the on-board controller or that stores the state of the ASIC (e.g. the number of errors of a certain type).
- logic that is basically doing the same job each cycle, e.g. a scrambler/descrambler.
- logic that starts processing once every X clock cycles. This processing takes a certain number of clock cycles. Note that telecom data is typically structured into cells or frames, e.g. the ATM cell of 53 bytes, and the SDH STM1-frame of 2430 bytes. Per cell or frame, a number of tasks must typically be performed.

We believe that it is this last class that is suited for implementation with behavioral synthesis. It is considered in the sequel. Inside this class, datapath-dominated DSP-systems have been the main focus of research (see e.g. Vanhoof, 1994). However, as mentioned above, our applications are mostly non-DSP and we investigated the applicability for behavioral synthesis outside the DSP area.

## 2.2 Traditional implementation

Due to the high data-throughput requirements, the following hardware architecture is typically used to implement the third class of circuits of Section 2.1 :

- an application-specific computation unit. It consists of tuned datapaths, registers and busses or multiplexers to implement the connectivity. The variety of such computation units is typically much larger than for the computation units in the highly-multiplexed architectures that are currently being considered in the research area of retargetable codegeneration (Marwedel, 1995).
- a controller for the computation unit. Traditionally it is a simple sequencer, that describes the actions during each counter-step. These actions are executed conditionally.

These architectures typically use very long instruction words (VLIW), and have only a limited number of control steps (e.g. 10-100).

## 2.3 Problems with the traditional implementation

To implement a design as described in Section 2.2, the designer typically writes two kinds of VHDL :

For *the computation unit*, the VHDL code typically describes the structure, consisting of datapaths, registers and busses or multiplexes, augmented with a behavioral description per datapath (e.g. an adder).

To develop the code for the computation unit, an analysis is needed of the operations that must be performed (e.g. finding the first zero in a bitvector). The resulting code contains a limited number of VHDL-lines, is structured, and easy to become reacquainted with during redesign.

For *the sequencer*, the VHDL code typically has the following structure :

```
case Counter is
        when 10 =>
                if mode = do_command1 then
                        connect_regA_ramAddressPin <= '1'
                ...
        when 11
  ..
```

To develop the code for the sequencer, the elementary actions (e.g. write-to-memory) must be scheduled by trial and error, and, for each control-step, the micro-actions (e.g. set a register load signal to true) must be written explicitly.

The resulting code contains a lot of VHDL-lines that describe the design at a low level of detail. They are hard to debug and error-prone, and don't allow reusability. Redesign of such code is particularly difficult : the code is hard to understand, and rework.

As a consequence, we believe that *the main potential for behavioral synthesis lies in avoiding the manual description of the controller.* In addition, synthesis can help during the design of the computation unit, by requiring from the user only the definition of the key features of the computation unit architecture (e.g. the application-specific datapaths) and by optimizing the resource usage.

# 3 FUNCTIONALITY NEEDED FOR A TYPICAL EXAMPLE

## 3.1 Example : mid_manager

This section describes a typical telecom non-DSP application that is suitable for implementation with behavioral synthesis.

The mid_manager block (Therasse, 1993) keeps track of the message identifiers (MIDs) that are allocated by a user of the considered telecom system. Each user can allocate up to 1024 MIDs. The status of each MID is represented by one bit in a bit-map arranged as a matrix of 32 rows by 32 columns. Each row is stored in one word of an external memory array. When a MID is requested, one must scan the bit-map to find a free MID.

To reduce the number of memory accesses, the bit-map is augmented by a summary word. Each bit in this word gives the status of the corresponding row, i.e. if all MIDs in this row are allocated or if at least one MID is still free. When a MID is requested, the summary word is read and one searches for a bit equal to zero. The column number of this bit gives the row number of a row containing a free MID. This row is read and one searches a bit equal to zero. This algorithm can find a free MID using two reads and two searches. After the allocation, the row is updated, and, if needed, the summary is updated.

Each user has two bit-maps. A bitmap can be active or inactive. When a bit-map is active, it is used to allocate and deallocate the MIDs. The two bit-maps can be both active. In this case, a MID can be allocated if and only if it is marked as being free in the two bit-maps. This can be implemented easily by taking the bit-wise OR of the two summary words and of the two selected bit-map rows.

This example highlights what we believe to be a typical characteristic of non-DSP telecom applications for behavioral synthesis : *the most critical resource is the RAM.* The RAM-accesses are typically the limiting factor when scheduling such an application.

## 3.2 Basic functionality

To provide a basis for the discussion, the basic functionality of VHDL-based behavioral synthesis is summarized :

```
in : - Behavioral VHDL description
     - min/max delay between particular operations ("compilation script")
     - library of operators (with mappings to hardware blocks)
out : - RTL VHDL description
```

In this investigation, we considered VHDL-based behavioral synthesis. We like the use of VHDL as design-entry language, because VHDL has most constructs required for a behavioral description and the use of VHDL fits nicely into the existing design methodology : it allows to use the installed base of VHDL-simulators, and the know-how built up in the past by the designers. As a consequence, the threshold for introducing behavioral synthesis is reduced.

To be suitable for design-entry, VHDL should offer support for structured programming. This seems to be the case. When writing a C-like description, we only encountered the limitation that in a procedure call, the signal actual must be a static signal : for example, tab(i) can not be passed on as argument, one must use a static signal like tab(0).

In addition, the synthesis tool can impose limitations on the VHDL (cf. logic synthesis tools), e.g. by forbiding the use of multi-dimensional variables as arguments of procedures, which will lead to a less structured design description.

## 3.3 Additional functionality needed for efficient use

Basically, it is sufficient for the behavioral synthesis tool to generate a description that can be read into a logic synthesis tool.

However, for real-life use the following additional characteristics are needed :

- The generated solution must be easy to understand : clear RTL-VHDL and complete reporting.
- The user must be able to have impact on the solution : specifying constraints and providing key elements of the computation unit.

The first requirement has an impact on the type of VHDL that must be generated by the behavioral synthesis tool. We experienced that it is important that this VHDL provides a clear and concise description of the design : separate entities for the computation unit and for the controller, and the architecture for the computation unit should show its structure in terms of hardware resources (datapaths, registers, busses, ...).

In addition, our impression is that in some cases debugging at the level of the RTL-code will still be necessary, e.g. to identify the cause of an error that is encountered after fabrication of the ASIC. This provides an additional argument for clear RTL-VHDL. With respect to this point, one can draw a parallel with the VHDL generation by graphical FSM-design tools (Vanoostende, 1994).

Apart from the RTL-VHDL, the tool should provide complete reporting. For a VHDL-description containing loops, two cyclecount reports should be provided : a 'control-step' report, where each control-step corresponds to one step of the controller, and an 'overall cyclecount' report that indicates the number of cycles needed to execute the complete VHDL-algorithm (considering e.g. that a particular loopbody is executed 3 times).

In the second requirement (cf. above), we indicated the need for the user to partially describe the architecture. This is related to the fact that in principle the behavioral VHDL contains a mix of low-level RTL-statements (to describe the detailed behavior of the operations used, e.g. finding the position of the first zero in a bitvector) and high-level constructs (e.g. perform, under certain conditions. a complex operation on certain data).

For the synthesis tool it is difficult to identify automatically to which type a certain piece of VHDL code belongs. Therefore, the user can help by indicating that a certain VHDL procedure describes a basic operation and that it must be implemented as a separate datapath that becomes part of the computation unit.

In the synthesis tool that we evaluated this was essential to obtain a result within a reasonable CPU time.

## 4 COMPLICATIONS

The previous section described what we consider to be the prerequisite functionality for efficient use of behavioral synthesis for a significant number of non-DSP telecom applications.

However, during our investigation we also encountered more complex issues that intervene :

- loop folding (see e.g. McFarland, 1988)
- the use of a resource external to the VHDL-entity that is being considered by the behavioral synthesis tool.
- complex resources.

The latter two issues are discussed below.

## 4.1 Resource external to the considered entity

During our investigation, it appeared that often a single RAM is shared by a number of weakly interacting VHDL-processes. There are a number of reasons why in some cases one wants a resource to be external to the VHDL-entity that is being considered by the behavioral synthesis tool :

- The size of the entity that is analysed at once by the behavioral compiler is e.g. limited by the fact that the RTL-level description of the complete entity (or at least the complete computation unit or controller) must still be synthesized at once by the logic synthesis tool. Commercial logic synthesis tools tend to be limited to handling 10k to 30k gates at once.
- The storage of a number of independent processes is grouped into a single RAM. This is mainly done when an off-chip RAM is used : the motivation is to limit the number of ASIC I/O pins. In that case, the natural hierarchy is to have a single RAM and a number of entities each containing a VHDL-process.

To handle e.g. the case where 2 processes share 1 RAM, we propose the following scheme :

- Initially schedule both processes as if 2 RAMs are available. This produces 2 schedules that have conflicting RAM-accesses.
- Based on the schedule for process 2, set constraints for the scheduling of process 1 of the type "block resource RAM between cycle i and cycle j". If the tool does not allow such constraints, a similar behavior can be obtained by constraining the minimum cycles between the start of the process and a certain RAM-access.
- Based on the modified schedule obtained for process 1, set constraints for the scheduling of process 2, and reschedule process 2. This mechanism continues until a non-conflicting schedule is found.

Behavorial simulation of a system with multiple entities that share a single RAM also introduces complications in the synthesis tool we evaluated, because at the behavioral level the RAM is represented by a array that is local to a process. Therefore, addition of VHDL code that is not synthesized is necessary for simulation, e.g. to allow a process to read data that another process wrote in the RAM.

A related problem is the handling of multiple processes in which sequencing constraints occur between operations : e.g. process1 and process2 can both be modifying a certain field of information (e.g. the bandwidth used by a certain user). In that case, process1 must have completed its update, before process2 reads this data. This problem can also be handled as described above.

We are aware that the proposed solution is a work-around : the clean handling of an external resource or interacting processing remains a challenge for commercial behavioral synthesis.

## 4.2 Complex resources

Simple resources consist of combinational logic, or have a limited number of pipeline stages. However, from our investigation it appeared that complex resources are used in our applications.

For example, a division resource is used to implement both an 8-bit division and a 12-bit division. The 8-bit division takes 8 cycles, the 12-bit division takes 12 cycles, and during one execution of the scheduled process both an 8-bit and a 12-bit division are executed on the resource. Therefore, the behavioral synthesis environment should allow mapping multiple operators on the same resource, where the cyclecounts and bitwidths of the operators can differ.

# 5 CONCLUSIONS

This paper has described our experiences with VHDL-based behavioral synthesis for non-DSP telecom. The conclusion is that in this area there is distinctly a market potential.

However, we indicated that it is not sufficient for the behavioral synthesis tool to simply generate an RTL-synthesizable solution : this solution should be easy to understand (clear RTL-VHDL and complete reporting) and the user must be able to have impact on the solution (specifying constraints and providing key elements of the computation unit).

In addition, we have described a number of real-life complications that such a tool must be able to handle if it wants to be applicable to a broad class of applications and if it wants to obtain results close to the efficiency of manual designs.

*The authors would like to acknowledge the interesting discussions with our collegues at Alcatel-Bell, and with K. Croes and J. Vanhoof of IMEC (Leuven, Belgium).*

# REFERENCES

Marwedel P., Goossens G. (1995) "Code generation for embedded processors", Kluwer Ac. Publishers.

McFarland M.C., Parker A and Camposano R. (1988) "Tutorial on High-Level Synthesis". Proc. Design Automation Conf..

Owall V., Andreani P., Brange L., Nilsson P., Wass A. and Torkelson M. (1993) "Custom DSP implementation of a GSM speech coder"., Proc EDAC/EuroAsic user forum, pp. 162-165, 1993.

Schultz S. (1994) "Behavioral synthesis : concept to silicon". ASIC & EDA, pp. 12-26, Aug. 1994.

Therasse Y, Petit G.H. and Delvaux, M. (1993) "VLSI architecture of a SMDS/ATM router". Annals of telecommunications, 48, nr. 3-4, pp. 166-180. 1993.

Vanhoof B., Bolsens I., De Troch S., Philips L., Vanhoof J. and De Man H. (1994) "Design of a voice coder ASIC with the CATHEDRAL II silicon compiler", Proc EDAC/EuroAsic user forum, pp. 150-153.

Vanoostende P., Moerman E. and Van Wauwe, G. (1994) "Graphical design-entry above VHDL : myth or reality?" Proc EDAC/EuroAsic user forum, pp. 189-193, 1994.

34

# An Asynchronous Microprocessor in Gallium Arsenide

*José A. Tierno*
*Alain J. Martin*
*Drazen Borkovic*
*Tak K. Lee*
*California Institute of Technology*
*Pasadena, CA 91125*
*USA*

**Abstract**

This paper describes how the design of an asynchronous microprocessor previously implemented in CMOS was ported to Gallium Arsenide. Techniques for designing asynchronous circuits in Gallium Arsenide are presented. New circuits, used to implement specific functions necessary to the design of a full microprocessor, are shown. A novel completion tree, and a novel general circuit structure for operators specified by production rules are introduced. These circuit were used and tested in a variety of designs, including two asynchronous microprocessors and two asynchronous static RAM's. One of the microprocessor runs at over 100 MIPS with a power consumption of 2 W.

**Keywords**

Microprocessor, Asynchronous Design, Gallium Arsenide.

## 1 INTRODUCTION

We have developed a design method for asynchronous VLSI that is, to a large extent, technology independent [Martin 86]. Thus, it should be straightforward to port a design from one technology to another. Also, since the circuits designed are quasi-delay-insensitive, they are more robust with respect to variations in physical parameters. Hence, the method makes it easier to design in a demanding technology such as GaAs, where parameters of the fabrication process — particularly threshold voltages — are difficult to control. Finally, since asynchronous circuits do not use a clock, we avoid the complexities of high-speed clocking schemes. Adapting our method to GaAs design is an excellent demonstration of the advantages of the method.

From the onset, our intention was to port to GaAs the asynchronous microprocessor we designed in CMOS in 1989 [Martin *et al* 89]. At the same time, we would demonstrate the portability of our approach across vastly different technologies, and the efficiency and robustness of the design method.

With an electron mobility about six times that of silicon at room temperature, and with a lower parasitic capacitance due to semi-insulating substrate, GaAs is potentially faster than silicon. Up until recently, however, GaAs was not available to the VLSI community at large because of inherent fabrication difficulties. These difficulties have been overcome to a large extent. Several foundries are now offering GaAs fabrication lines under conditions similar to CMOS, with chip size limited to about 100,000 transistors. In particular, Vitesse Semiconductors is offering fabrication through MOSIS to the academic community in the United States.

At the moment, the transistor of choice for GaAs digital VLSI is the MESFET. There is no oxide insulating the gate of a MESFET transistor from source and drain; therefore, the logic families available in GaAs are much less attractive than CMOS or even *n*MOS. Direct Coupled Fet Logic (DCFL) has been adapted to GaAs, but has very reduced noise margins, and restricted fanin and fanout. With no complementary transistor available, the logic is ratioed. As a result, a considerable fraction of the speed advantage is lost due to the complexity of the available logic families relative to CMOS.

To bypass these limitations, we have designed several new circuits, and adapted to our purpose a number of old circuits. We present these circuits in this paper, together with their use in the microprocessor.

## 2 DESCRIPTION OF THE PROCESSOR

The microprocessor is a 16-bit, pipelined, RISC-style processor. It is a modified version of the CMOS design described in [Martin *et al* 89]. Instructions are issued in order, but may complete out of order. The processor has 16 general-purpose registers, with four buses, two for read and two for write. Registers have individual locks to solve read-after-write and write-after-write conflicts.

In addition to the ALU, there is one adder in the program counter (PC) unit, for relative branching and incrementing the program counter register. For simplicity, the memory-address adder is omitted, which reduces the number of required buses from five to four, making the datapath smaller in comparison with the CMOS processor. Other modifications include a revised pipelining of the ALU unit, and a revised control so as to equalize delays in all pipeline stages.

The processor is initially specified as a set of concurrent processes. The text of these processes is later transformed into a signal transition language, or handshaking expansion, and then compiled into "production rules" (the gate netlist) for the circuit.

## 3 GAAS TECHNOLOGY MAPPING

The MEtal Semiconductor Field Effect Transistor, or MESFET, is the transistor of choice for GaAs VLSI applications. It is the easiest to manufacture, provides the highest density (about 100,000 transistors on a chip), and has a very good power-delay product that, for clock frequencies of more than 200MHz, is better than the power-delay product of CMOS. When compared with ECL, GaAs is slightly faster, uses far less power, and has higher circuit density for a similar cost. An important application of GaAs is to provide replacements for ECL parts, such as fast RAM's and other LSI circuits.

MESFET's are junction FET's. The gate is built with metal, and forms a Schottky junction with the transistor channel. There is no insulation between gate and channel; the Schottky junction creates a diode from gate to source and from gate to drain. This diode imposes severe limitations on the type of gates that can be used for digital circuits. As in $n$MOS, n-type transistors come in two flavors: depletion-mode and enhancement-mode.

Even though we can map the processor design into DCFL gates [Long Butner 90], the specific requirements of asynchronous circuits makes that choice impractical. DCFL has low noise margins, and practical gates are restricted to nor-gates with a small number of inputs. Asynchronous operation requires monotone signal transitions and is sensitive to noise and charge sharing. Also, the compilation process sometimes generates complex gates with a large number of inputs. Decomposing these gates into smaller gates in a delay-insensitive way is a delicate task.

To solve the problem of large operators, we have investigated several alternative logic configurations. We opted for different solutions for the datapath and for the control circuits.

### 3.1 Datapath

The datapath includes three different units: the ALU, the PC unit for manipulations of the program counter, and the memory unit for execution of load and store operations. Datapath delay is determined primarily by carry-chain, control signal, and bus delays [Martin 92]. With no feed-back loops, noise problems are less severe, since noise does not get amplified in the loop.

The datapath is optimized for size and power. In the datapath most signals are local, with the exception of control lines and buses, and delays and power depend directly on

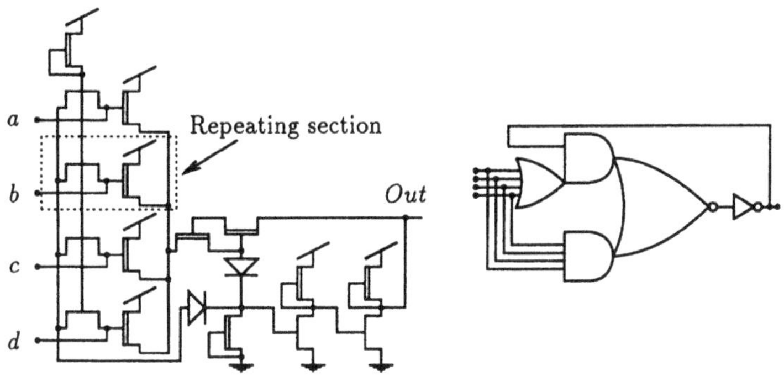


(a) (b)

Figure 1: Multi-input C-element. (a) Transistor schematic, and (b) logical diagram

the physical dimensions of the datapath. In this design, 70% of all power is spent in the datapath, 15% in the register file, and 15% in the control logic (Pad driver power is excluded from this computation).

To satisfy size and power constraints, all datapath gates are DCFL, except nand gates and buffers, and completion-detection circuits. DCFL gates are small and power efficient. Also, the datapath contains only simple gates available in DCFL.

To generate completion signals from the datapath, we use C-elements with a large number of inputs. They can be built from smaller C-elements connected in a tree [Martin 90], or, as in this case, as a single logic gate (see Figure 1(a) and (b)) [Tierno 92].

## 3.2 Control Logic

Control logic takes care of the sequencing of actions in the processor. After compilation, each control signal is assigned a set of "production rules", of the form:

$$\begin{array}{ccc} G & \rightarrow & z\uparrow \\ H & \rightarrow & z\downarrow \end{array}$$

where $G$ and $H$ are boolean expressions in terms of the other signals. $G$ and $H$ do not have to be complementary. In fact, most operators in the control are state holding, that is, $G \vee H$ does not hold. There are different direct implementations of these production rules. One, Source Follower Fet Logic, is given in [Tierno 92]. That paper presents a systematic way of generating any operator described by production rules. This method was applied

(a)

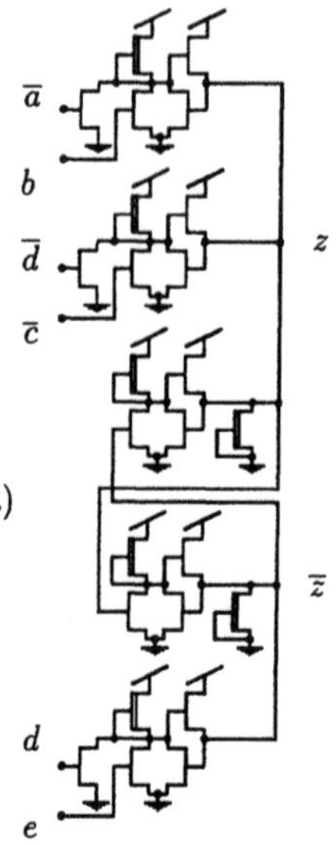


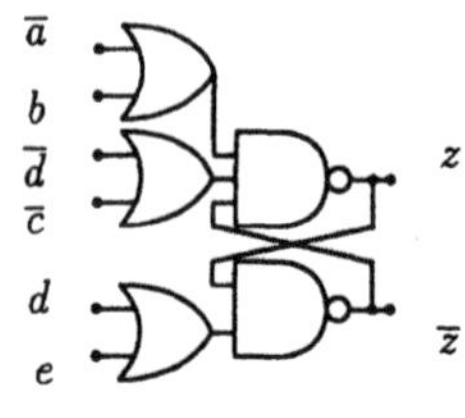


(b)

$$a \wedge \neg b \vee d \wedge c \rightarrow z \uparrow$$
$$\neg d \wedge \neg e \rightarrow z \downarrow$$

(c)

Figure 2: Dual rail implementation of control signals. (a) Transistor schematic, (b) logical diagram, and (c) production rules

in the design of the first GaAs microprocessor. However, it resulted in a circuit with a large power consumption (4 Watts) and modest performance (70 MIPS).

For the second GaAs processor, a different approach is used. Each signal is implemented with a dual-rail encoding, always generating both positive and negative sense. An advantage of this approach is that no inverters are necessary to generate the boolean expressions for the production rules. A drawback is that combinational gates require extra circuitry to generate dual-rail outputs, as do signals coming in from the datapath.

Figure 2 shows how a specific set of production rules are implemented in dual-rail. Note the feed-back transistor on the outputs of the individual nor gates. These transistors have the same function as the one on the "squeeze" [Brown 92] buffer, allowing the use of much stronger pull-up transistors.

# 4 RESULTS

Using these techniques, several circuits have been designed, fabricated and tested on the HGAAS II and HGAAS III processes offered by Vitesse Semiconductors. Among them are two small RAM's, a register file, and two microprocessors.

The first microprocessor design uses the circuits described in [Tierno 92]. The main concern during the design was to get around parameter variation problems and noise margin considerations. This was achieved, but at the expense of power and performance. At 70 MIPS, the processor consumes 4 W.

Several new ideas and circuits were incorporated into the second design, many of them presented in this paper. All of those circuits were fabricated and tested successfully.

This design was extensively simulated with Hspice. The expected performance of this design was about 200 MIPS with a dissipation of 2 W. Power and speed predictions have been very accurate so far using the Hspice models. However, in this case the measured performance is only 100 MIPS. The causes are still under investigation. There is some evidence of fabrication problems, and underestimation of the parasitic capacitances as extracted by the MAGIC layout program.

Another factor affecting performance is pad delay. So far, we have used ECL levels on the outside, to be able to interface to standard parts and simplify prototyping. Pad delays are in the order of 1ns, mostly spent in level conversion; also, the padframe uses a considerable amount of power —close to $1A$ worst case for the processor. This delay and power can be greatly reduced by designing matched pad drivers and receivers in a system composed exclusively of GaAs parts. It would certainly be a requirement in the interface with cache memory.

# 5 CONCLUSIONS

In the course of this project, we have designed a number of very different GaAs circuits. Most of them had very strict requirements, to overcome the limitations of GaAs. No general solution is given to synthesize all logic circuits in a design as big as a microprocessor. Instead, we found specific solutions to implement completion trees, control circuits, PLA's, registers, etc.

Porting the design of the original CMOS microprocessor was almost as easy as expected. A few changes were necessary because of the complexity of register cells in the first GaAs version. These changes were carried over to the second processor to speed-up the redesign.

In general, we are satisfied with our solutions to the problem of designing asynchronous circuits in GaAs, though we were expecting better performance. The MESFET is far from being an ideal switch, and, as a result, gates with good gain, noise, delay, and power characteristics end up being quite complex. An overhead in delay and power is paid to make those gates more reliable, which offsets the raw performance gain of the GaAs MESFET over the Si MOSFET.

More importantly, integration levels in GaAs are very limited compared to CMOS. An important source of performance in digital circuits is parallelism, which can more than make up for the difference in raw speed. Circuits that can make efficient use of parallelism, like microprocessors with wide datapaths, or memories, are greatly penalized by reduced integration levels.

There are, of course, inherently sequential problems, like serial communication channels. When the problem is well suited for GaAs, we can apply the tools of asynchronous design, and carry over to GaAs all the advantages of asynchronous circuits.

## ACKNOWLEDGMENTS

We are indebted to Marcel van der Goot for his help in generating high quality software support. Acknowledgment is due to Steve Burns and Pieter Hazewindus, for their participation in the original design, and many very useful comments and conversations, and H. Peter Hofstee for his collaboration in the design of asynchronous memories.

The research described in this paper was sponsored by the Advanced Research Projects Agency, ARPA Order number 6202, and monitored by the Office of Naval Research under contract number N00014-87-K-0745.

## REFERENCES

Alain J. Martin. Compiling communicating processes into delay-insensitive VLSI circuits. *Distributed Computing*, 1(4):226–234, 1986.

Alain J. Martin, Steven M. Burns, T. K. Lee, Drazen Borkovic, and Pieter J. Hazewindus. The design of an asynchronous microprocessor. In Charles L. Seitz, editor, *Advanced Research in VLSI: Proceedings of the Decennial Caltech Conference on VLSI*, pages 351–373. MIT Press, 1989.

Alain J. Martin. Synthesis of asynchronous VLSI circuits. In J. Straunstrup, editor, *Formal Methods for VLSI Design*, pages 237–283. North-Holland, 1990.

S. I. Long and S. E. Butner. *Gallium Arsenide Digital Integrated Circuit Design*. McGraw-Hill, New York, 1990.

A. J. Martin. Asynchronous datapaths and the design of an asynchronous adder. *Formal Methods in System Design*, 1(1):117–137, 1992.

Richard Brown. Private communication, 1992.

J. A. Tierno. Designing asynchronous circuits in Gallium Arsenide. CS-TR-92-19, California Institute of Technology, 1992. Master's Thesis.

PART TWELVE

# Communication Busses

# Optimizing the Communication Overheads during the Allocation of Global Memories and Busses

*J. Müller, R. Kumar*
*Automation of Circuit Design, Forschungszentrum Informatik (FZI)*
*Haid-und-Neu-Str. 10-14, D 76131 Karlsruhe, Germany*
*Tel: ++49/721/9654412, Fax: ++49/721/9654459*
*email: {jmueller|kumar}@fzi.de*

**Abstract**

One of the important aspects in synthesizing from multi-process specifications is to deal with the existence of complex data between the processes. In this paper we focus on the allocation of memory elements and busses for realizing such data exchanges. The synthesis approach minimizes the communication delays, taking the area demand for the needed communication structure into consideration. The global data are clustered according to a distance factor which is derived by an analysis of the specification. This clustering information is then used to appropriately partition the global data into subsets which can be assigned to a cost minimal memory and bus structure.

**Keywords**

system-level synthesis, communication synthesis

## 1 INTRODUCTION

Most existing commercial and academic VHDL based synthesis tools do not take a system view point of the overall behaviour and hence concentrate on the synthesis of single process specifications. Although some tools synthesize multi-process VHDL descriptions, they do not adequately handle the synchronization aspects between the various processes, thus leading to implementations with hazards, in the presence of multiple clocks. A second major problem in synthesizing multi-process descriptions is the consideration of complex data which can be more efficiently implemented using register files or RAMs, as compared to single registers or latches. Our approach towards multi-process synthesis considers these descriptions from a system point of view, i.e. the global synchronization of the processes and the global storage of the data are accounted for. However in this paper, we mainly concentrate on the memory/bus mapping aspects of the multi-process synthesis.

Two approaches which also consider the above-mentioned problems separately are worth mentioning. In (Berthet a. o., 1992) the possibility of synthesizing VHDL arrays are discussed. They map each array onto one generic RAM primitive, which is implemented using a RAM generator, however without further exploitation of the design space. In (Narayan and Gajski, 1994) the bus assignment of communicating processes is examined. The number of communication channels (busses) and the data transfers which are to be mapped onto them is assumed. Based on this, they then determine the bitwidths of the busses and the possible protocols. Our approach takes a more global view of the problem by simultaneously considering the trade-offs between the sizes, number and type of memories and also the number and bitwidths of the necessary busses. Furthermore, the communication delays for memory/bus accesses are also incorporated.

This paper is organized as follows: after a brief overview of the overall synthesis system, the parameters influencing the optimization of the memory/bus allocation are discussed. This is followed by the partitioning concepts and finally some experimental results are presented.

## 2 OVERVIEW OF THE SYNTHESIS SYSTEM

A behavioural, multi-process VHDL specification is transformed into a structural VHDL description of a shared-memory system architecture (figure 1) and a set of behavioural descriptions. The structural description defines the interconnection between the memories, processors, and the global controller. In order to ensure correct communication, specialized interface components are automatically inserted for the memories, processors and busses. These components are available as a predesigned library of generic VHDL models and are instantiated appropriately. In addition, each process in the input specification is transformed into a behavioural processor description with the suitable bus ports. Furthermore, a behavioural description of the global controller is also generated. The synthesis of the behavioural descriptions can then be performed by using commercial design tools like the Synopsys Design Analyser.

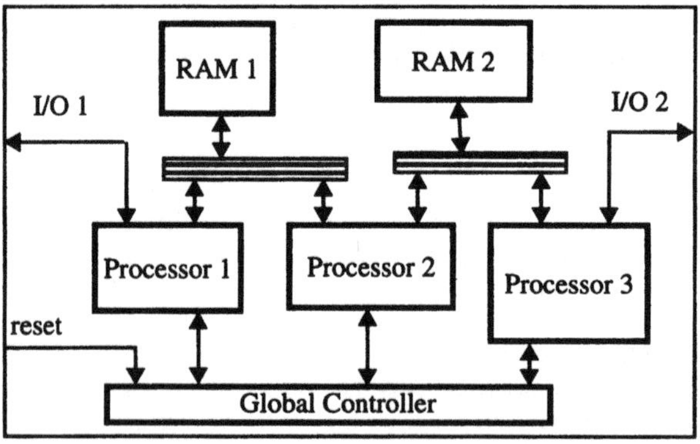


**Figure 1** Example for a generated system target architecture

In our approach we distinguish between synchronization and data exchanges. The synchronization of the processors is performed by the global controller, and the data which resides in the global memories is exchanged between the processors via the busses.

In order to define our VHDL subset for the input specification we follow the recommendations given by Camposano a. o. (1991). The communication between different processes and the environment can be specified only by means of global signals. The synchronization signals that are used for the synchronization of the processes are those signals which are used at least in one *wait statement* and additionally, they must be of type BIT. The other signals in the input specification are used for the data exchange. These data signals are assigned to several RAM modules or global registers.

In this paper, we present an approach to determine a nearly optimal communication structure for the data exchange which takes the area and the temporal communication overheads into account.

# 3 FACTORS INFLUENCING THE GENERATION OF THE COMMUNICATION STRUCTURE

The overall problem that we are solving can be stated as follows:

*given a multi-process VHDL description comprising the global signals for synchronization and data transfers, find a communication structure (i.e. the memories and the busses) such that the costs and access conflicts are minimized.*

There are two aspects that are to be considered during the generation of the communication structure, namely, the costs and the occurrence of the conflicts. The costs that influence the optimization of the memory and bus mapping of the data signals are the overall area and the access times. The occurrence of the conflicts is analyzed by using a Petri-net model of the communication among the VHDL processes. A brief description of the costs and the conflict analysis is given below.

## 3.1 Area costs

The area costs can be divided into three parts: the area needed for the individual processors and the global controller, the storage of the global signals and the interconnection between the components of the system. In the context of our task, we need to consider only the area costs for the storage elements and the interconnections and hence the costs to implement the processors and controller are ignored. The total area costs for the storage and the interconnections can be lumped together and considered as communication costs. The area cost for a RAM module depends on the number of stored signals and the bitwidth of the module. The interconnection costs depend on the number and the size of the interface components of the memory modules, which are needed to perform read, write or read/write accesses together with the bus arbiter which is used to resolve concurrent bus accesses among the processors.

### 3.2 Time costs

The time cost function has to give a relative measure for the communication overhead caused by a certain system structure. The factors influencing the cost are:

- the access protocol determined by the type of memory element which stores a signal
- the number of accesses to a signal
- if a set of signals are assigned to one memory element, then a sequence of accesses is needed
- if the bitwidth of the signal is greater than the bitwidth of the memory then more than one access is required

It is to be noted that, all data accesses to data which are implemented in different memory elements can be performed in parallel. Therefore, the total cost function is determined by the maximum access delay of all RAM modules and registers.

### 3.3 Conflict Analysis

If more than one signal is assigned to one memory element then the probability of concurrent accesses to the set of signals is needed to evaluate the effect of serialization delays. To rate the probabilities of access conflicts the synchronization scheme of the processors must be analyzed.

In (Müller and Krämer. 1993) a Petri net based approach to examine the system behaviour was described. A Petri net model is used to detect which parts of the different processes can work concurrently. The markings of the reachability graph are regarded as the set of all possible states of the system. Data accesses take place during the transitions between the successive system states. By a Markov process the probabilities of the system states can be calculated which imply the probabilities of concurrent accesses to each pair of data signals. We have improved this approach by some features. The Petri net model was modified into a timed Petri net with fixed firing delays. Here the desired clock frequencies of the processors are taken into account.

## 4 SYNTHESIS OF THE COMMUNICATION STRUCTURE

The task now is to map disjoint sets of data signals onto memories which implies the overall system structure, while minimizing the costs and access conflicts. The complexity of this problem is large and hence we developed an approach which consists of two phases: first a set of cluster trees which mirrors the similarity of the data signals is created, using an abstract measure which is derived from the communicating processes in the input description. In the second phase, these cluster trees are partitioned. Each partition then corresponds to a set of disjoint subsets of the data signals, where each subset has been assigned to a specific type of memory element. These memory elements implicitly determine the kind of interface and bus components thus leading to a candidate system structure. Therefore, concrete cost functions taken from a library, can be used in this phase.

The cluster trees are created using a concept called *distances* between the signals. The distance between two signals is large if the probability of access conflicts between these signals is high, and it is small if the signals are accessed by many common processes. The distance between two signal $s$ and $q$ can be calculated by:

$$d_{s,q} = \omega \cdot p_{s,q} + \frac{a_s \cdot a_q}{a_{com}^2 + 1} \tag{1}$$

$p_{s,q}$ is probability of access conflicts between the signals $s$ and $q$,
$a_s$ and $a_q$ are the number of processes accessing the signals $s$ and $q$,
$a_{com}$ is the number of processes accessing both signals,
$\omega$ is a weighting factor to normalize both terms.

The algorithm to build a cluster tree, based on the distances (eqn. 1), was described in (Krämer and Müller, 1992). The leaf nodes of such a cluster tree represent the individual data signals (figure 2). The cluster tree is constructed in such a way that signals with small distances are collected in common subtrees. The shape of a cluster tree depends on the weighting factor $\omega$. In order to explore a large design space, all significant cluster trees resulting between a maximum and minimum $\omega$ ($\omega$-range) are created, by using a divide-and-conquer algorithm. The significant cluster trees are created by recursively halving the $\omega$-range, as long as the resulting cluster trees are different or a minimum $\omega$-range is reached.

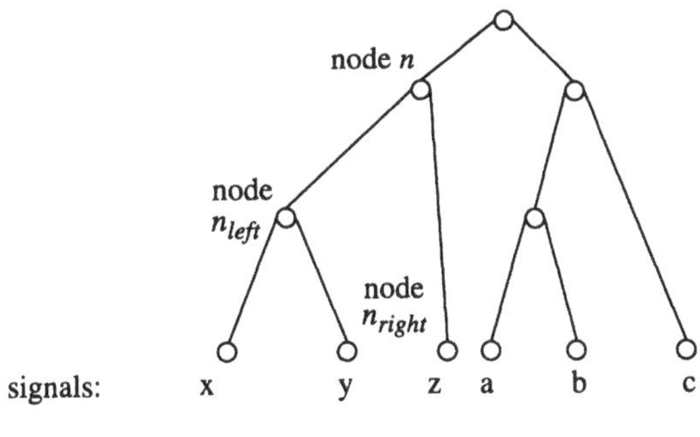


**Figure 2** Example of a cluster tree

The cluster tree is recursively traversed from the leaf to the root and appropriate cuts are made in the tree, for creating the different partitions. The following objective function is used to compare two partitions:

$$Obj(\gamma, T, A) = \gamma \cdot \ln T + \ln A \tag{2}$$

$T$ represents the costs for the communication delay and $A$ represents the costs for the area demand. The parameter $\gamma$ is a user defined factor which biasses a faster or a smaller design. By setting $\gamma$ as 2 we get the well known $AT^2$ factor for optimization.

The cluster tree is cut at node $n$ and declared as leaf node if the cost of implementing its subset $S_n$ of signals in one memory module is less than the cost for the separate implementations of the signal subsets on the left and the right successor subtrees, i.e.:

$$Obj(\gamma, T_n, A_n) < Obj(\gamma, max(T_{n_{left}}, T_{n_{right}}), A_{n_{left}} + A_{n_{right}}) \quad (3)$$

It is to be noted in eqn. 3, that the parameters $T$ and $A$ consider the type of memory, i.e. register or RAM and the corresponding bitwidths into account. The formulas for $T$ and $A$ are complex and hence beyond the scope of this paper.

The cluster tree is cut as high as possible so that the overall objective function is minimized. The final cuts result in the memory and bus mapping of the signals.

## 5 Examples and Results

To illustrate the partitioning approach we have used a 5 process example which computes the determinant of a 3x3 matrix. One process reads the data of the input matrix row by row, three processes calculate the 3 sub-determinants, and the last process sums the sub-determinants and writes the result to the output. The synchronization of the processes was specified so that the input process can restart as soon as the calculation of the sub-determinants has finished, which results in an asynchronous two stage pipeline. The parameters for the bus cost functions were got by instantiating the generic library of interface components built using a commercial standard component library. For the RAM costs we have assumed a RAM generator.

By modifying the weighting parameter $\omega$ between 0 and 50 we have got 12 different cluster trees. To find the best partitions, for a given $\gamma$ all the cluster trees were examined. In our experiments, $\gamma$ was varied between 0.01 to 100 in order to obtain different realizations ranging from a single RAM solution to separate registers for each signal. Table 1, shows the best 4 results obtained with $\gamma = 2$.

**Table 1** Best Partitions for the example with $\gamma = 2$

| *RAM 15 bit* | *RAM 30 bit* | *Registers* | *T* | *A (gates)* | *Obj(γ,T,A)* |
|---|---|---|---|---|---|
| 1 (9 words) | 0 | 3 (30 bit) | 0.29 | 10281.0 | 6.731 |
| 1 (9 words) | 1 (4 words) | 0 | 0.33 | 9205.3 | 6.881 |
| 0 | 1 (12 words) | 0 | 0.33 | 9534.5 | 6.916 |
| 1 (15 words) | 0 | 0 | 0.43 | 6268.5 | 7.035 |

The best solution uses one 15 bit RAM with 9 words and three 30 bit registers. In the second solution the registers are replaced by a 30 bit RAM with 4 words which is the minimum word number, dictated by the RAM generator. The last result uses only one 15 bit RAM, where each 30 bit data needs 2 words.

## 6 REFERENCES

Berthet, C., Rampon, J. and Sponga, L. (1992) Synthesis of VHDL Arrays on RAM Cells, Proc. of EURO-VHDL, pp. 726-731

Camposano, R., Saunders, L. F., Tabert R. M. (1991) VHDL as Input for High-Level Synthesis, IEEE Design &Test of Computers, pp. 43-49

Krämer, H., Müller, J. (1992) Assignment of Global Memory Elements for Multi-Process VHDL Specifications, Proc. of ICCAD, pp. 496-501

Müller, J. , Krämer, H. (1993) Analysis of Multi-Process VHDL Specification with a Petri Net Model, accepted at EURO-VHDL

Narayan, S. , Gajski, D. D. (1994) Synthesis of System-Level Bus Interfaces, Proc. of EDAC, pp. 395-397

Narayan, S. , Gajski, D. D. (1994) Protocol Generation for Communication Channels, Proc. of 31th DAC

## 7 BIOGRAPHY

*Jens Müller* received the diploma degree in computer science from the technical university of Dresden in 1992. Then he started to work at the computer science research center (FZI) at Karlsruhe. His fields are the high-level synthesis and system-level synthesis. His special interests are the synthesis of the communication structures of complex systems, the exploration of different target architectures and performance estimation at system-level.

*Ramayya Kumar* completed his Ph.D. at the Indian Institute of Technology (IIT) Delhi in 1986. He worked as a scientific staff in the Department of Electrical Engineering at IIT Delhi between 1980-1986. From 1986-1988 he worked at the corporate research centre of Siemens in Munich, where he was involved with the development of the concepts for the CALLAS System. During 1989-1993, he worked at the University of Karlsruhe where he founded the hardware verification group. Since 1993 he heads the design automation group at the Computer Research Institute (FZI) Karlsruhe. His areas of interest are formal methods in hardware, synthesis and CAD frameworks, in which he has published over 40 papers.

36

# Optimal and Robust Scheduling of Communications in Bus Architectures

*Elof Frank and Thomas Lengauer*
*German National Research Centre for Computer Science*
*D- 53754 St. Augustin, Germany, email:* `elof.frank@gmd.de`

**Abstract**

Bus interconnect architectures in their most general form consist of several functional units with associated memory accessible to each functional unit. Global buses communicate data values from one functional unit to another. The time at which certain values are communicated affects the size of the local memories and the number of buses that are needed. We address the problem of scheduling communications in a partitioned bus architecture and present a formal approach for minimizing the number of buses and registers. This approach is robust in the sense, that it can incorporate resource (register and bus) constraints and also regard mutual exclusive communications. It is also adaptable to a large variety of different architectures.

**Keywords**

interconnect and register allocation, communication graph scheduling, bus architectures

## 1 INTRODUCTION

A common approach to high-level synthesis involves scheduling a data flow graph and allocating functional units and then proceeds to allocate wires resp. buses and storage. In bus-based architectures, there is a direct correlation between the time for transporting values over the buses and the number of buses and registers needed. If only one main memory (RAM, register file) is available for all the variables, the time of communication is fixed to the execution time of the operation using this value. In the fetch phase of its communication time, the value will be loaded onto a bus and will then be directly fed into the port of the appropriate functional unit. Therefore, the number of buses needed is given by the maximum number of distinct values consumed by all functional units at each time step. The number of registers is given by the maximum number of variables, which are alive at any time step. A variable is *alive* from its time of production until its time latest time of consumption. Several authors use the concept of a centralized memory architecture to validate their data flow graph scheduling and estimate interconnect and storage cost ([TS86, McF86, DN89, HLH91]). Even for simple examples, the number of buses needed easily exceeds technology-dependent constraints.

In contrast, having local memory (registers, register file, distributed RAM's) in each functional unit allows for the flexibility of scheduling communications at advantageous time steps. Our target architecture is a partitioned bus architecture, which has been studied before in [Ewe90, MCdP91, MRS91]. In this architecture (see figure 1a) a functional unit consists of an ALU and a register file. Each communication of a variable from a register to an ALU or vice versa has to be transferred over a global bus. These global buses can be dynamically partitioned by activating one or several switches on the bus. Each bus can now transfer more than one value, as long as the routes of the communications do not overlap. Partitioned bus architectures result in small-area designs as well as in reduced bus lengths and therefore in shorter fetch cycle delays. The time, at which a value is transferred, is variable within a given production and consumption time frame. The transfer time for each value influences the number of buses and registers in each functional unit. Therefore, the problem of scheduling communications is to determine for each value of the data flow a transfer time, such that a cost function in terms of the number of buses and registers is minimized.

We present a formal approach to the problem of scheduling communications. The formulation is based on integer programming. It is robust in the sense, that resource constraints on the number of buses and the number of registers in each functional unit can be incorporated. Furthermore, this formulation can be customized to a variety of other bus architectures, such as non-partitioned bus architectures and two-dimensional bus architectures. Similar to data-flow scheduling, mutual exclusive communications can be scheduled simultaneously but only use one bus. We can incorporate mutual exclusive communications into our formulation.

The rest of this paper is organized as follows. In the next section we discuss previous scheduling approaches to similar bus architectures. In section 3 we define the problem of scheduling communications. In section 4 we present our integer programming formulation. In section 5 we present our results and compare them to a non-partitioned bus architecture with central memory. We only sketch briefly the extension to mutual exclusive communications.

# 2 RELATED RESEARCH

Several authors synthesize towards bus architectures with local memory and have presented methodologies and heuristics to solve the problem of communication scheduling. In the Cathedral-II system [DRSC86] the synthesis process is directed towards a multi-processor architecture, in which the processors are connected by a restricted set of customized buses. The freedom of scheduling communications is not exploited. Every value is stored in the source functional unit and transferred over a bus when it is needed, i.e. as late as possible. Lanneer et al. [LCGD94] present a general data routing paradigm. In their architecture a value can be scheduled not only at different time steps but also along different global routes. A global route is an interconnection between storage devices. They present an efficient heuristic to the problem based on a force-directed scheduling approach[PK89]. Duncan and Hendry [DH93] synthesize towards an architecture, in which the buses are not partitioned. The authors solve the problem of communication scheduling using a greedy approach. Our formulation can be adopted to their target architecture. Monteiro et al. [MRS91] partition the buses of a given bus oriented data path and thereby achieve a speed enhancement. Ewering [Ewe90] schedules communications on a partitioned bus architecture by enumerating all possible scheduling times for a communication as well as all possible bus assignments of these communications. He then solves a maximum independent set problem to decide which communications do not conflict each other. For larger designs, he uses a heuristic to increase the efficiency of the algorithm.

In this paper we present a formal approach to solving the problem of scheduling communications optimally. Optimality is measured in terms of the number of buses and registers needed to realize the schedule. We do not solve the problem of register assignment and bus assignment here. These problems can be solved in a later phase by general high-level synthesis techniques, such as the left edge algorithm [KP87] or a weighted clique partitioning approach [TS86].

# 3 DEFINITIONS AND PROBLEM FORMULATION

We assume, that the tasks of data-flow scheduling, module allocation and operation assignment have been performed. The resulting data structure is a *communication graph* $G = (V, E)$, where the node set $V$ represents the ALUs and the edge set $E$ represents values of the data flow, that are produced and consumed by different ALUs at different times (see figure 1c). An edge $e = (s, t_1, \ldots, t_k)$ represents a value, that is produced at source node $s$ and consumed at target nodes $t_1, \ldots, t_k$. Let $k$ be the number of targets of an edge $e$. Each edge $e$ is labeled with a $k+1$-tupel $(\text{asap}, \text{alap}_1, \ldots, \text{alap}_k)$. A value, which is produced at the end of the execute phase – assuming a two phase (fetch, execute) clocking style – is stored in a local register and is available for communication in the following fetch phase (asap) at the earliest. Each target node $t_i$ consumes this value at time $\text{alap}_i$. We call a pair $(s, t_i)$ of an edge $e$ a *communication requirement* and the interval $[\text{asap}, \text{alap}_i]$ the *mobility range* for this communication requirement. Note, that a node can appear multiple times among the target nodes. We assume, that all input variables are available in specific registers before the computation starts, but, for the sake of brevity, we will not describe how we distribute the input values onto the registers. Figure 1b) and c) show an instance of a scheduled data flow graph and the resulting communication graph, given by a module allocation using one adder ADD, one subtracter SUB and one multiplier MUL.

Furthermore, we assume, that a linear order on the functional units – a placement of the communication graph within the architecture – is given. Duncan and Hendry [DH93] as well as Ewering [Ewe90] solve a hamiltonian path problem on the communication graph to compute a linear placement. Let $\pi_i$ be the position of node $v_i$ within the data path from the left. The route of a communication requirement $(s, t_i)$ is given as the interval $[\pi_s, \pi_{t_i}]$. For example, the linear placement of the communication graph in figure 1c) is as depicted from left to right.

The problem of scheduling communications is to compute, for each communication requirement, a transfer time $\tau(s, t_i)$, such that $\tau(s, t_i) \in [\text{asap}, \text{alap}_i]$. Depending on these transfer times and the given placement, resources buses and registers are allocated, to meet the following requirements:

- Each communication requirement $(s, t_i)$ occupies a bus at time $\tau(s, t_i)$.
- Obviously, two simultaneous transfers of the same value (the same edge $e$) only occupy one bus.
- Two communication requirements $(s, t_i)$ and $(s', t_j)$ of different values only occupy one bus track, if their routes do not overlap, that is if $[\pi_s, \pi_{t_i}] \cap [\pi_{s'}, \pi_{t_j}] = \emptyset$.
- Each value occupies a register in its source functional unit $s$, as long as at least one communication requirement $(s, t_i)$ has not yet been transferred.
- Each value occupies a register in a target functional unit $t_i$, as soon as the corresponding communication requirement $(s, t_i)$ is transferred and until this value is consumed at $t_i$. If the transfer time is as late as possible, the value is directly fed into the port of the ALU and does not need to be stored in a local register. Otherwise, a communication requirement will appear once again as a local

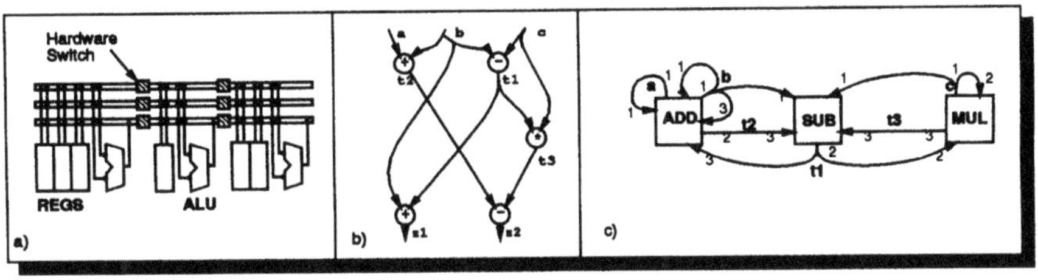


**Figure 1** a) partitioned bus architecture b) scheduled DFG c) ordered communication graph. The edges of the communication graph are labeled with their asap and alap times close to the producing and consuming terminal nodes.

communication at consumption time on the target bus segment only and thereby, it will be fed into the port of the ALU.

Let $B$ denote the number of buses and $R$ denote the number of registers needed. We minimize the weighted sum $\alpha B + (1-\alpha)R$, where $\alpha \in [0,1]$. Depending on the application, the priority is to be given to either the registers or the buses.

Figure 2 shows two possible schedules of the communication graph in figure 1c). For each time step, the communication graph and only those communication requirements that are transferred at this time are shown. The number of buses is the maximum number of distinct communication requirements connected to any node at any time. For the first schedule, we need three buses and for the second schedule we only need two buses. Note, that in the second example local communications are introduced for the variables $t1$ and $t2$ at time step three, although they were transported globally at time step two. The variable lifetimes in each functional unit are shown below the communication schedules and in the first case four registers are needed, whereas in the second case six registers are needed. For the same example and a non-partitioned bus architecture with only one central register file four buses and four registers would have been needed.

# 4 INTEGER PROGRAM FORMULATION

Let the communication edges be enumerated $e_i = (s_i, t_{i_1}, \ldots, t_{i_{k_i}})$. Then a communication requirement $(s_i, t_{i_j})$ corresponds to edge $e_i$. Let $c_{i,j,\tau} = 1$, iff communication requirement $(s_i, t_{i_j})$ is scheduled at time $\tau$, else $c_{i,j,\tau} = 0$. The first set of constraints guarantees, that each communication requirement is scheduled exactly once.

$$\sum_{\tau=\mathrm{asap}_i}^{\mathrm{alap}_{i_j}} c_{i,j,\tau} = 1, \qquad \forall (s_i, t_{i_j}) \tag{1}$$

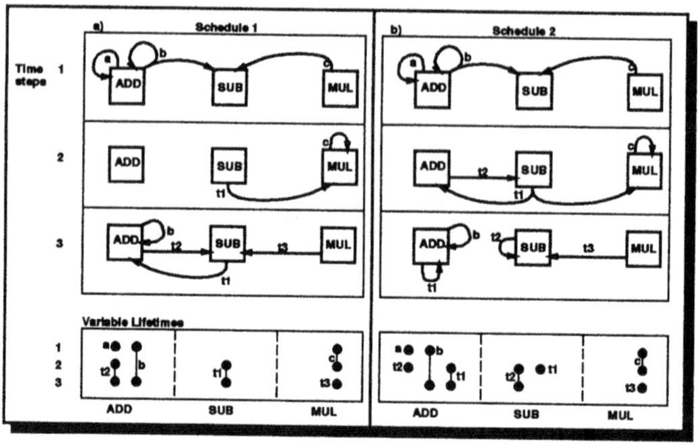


**Figure 2** Two different schedules resulting in a) 3 buses and 4 registers and b) 2 buses and 6 registers

Let $b_{i,l,\tau} = 1$, iff any communication requirement of edge $e_i$ is scheduled at time $\tau$ and originates from, passes or ends at a functional unit $v$, for which $\pi_v = l$ (equation 2). If a communication requirement is scheduled earlier than $\text{alap}_{i_j}$, then one extra local communication on the target bus segment is needed in order to feed this value into the port of an ALU (equation 3).

$$b_{i,l,\tau} \geq \{c_{i,j,\tau}\} \qquad 1 \leq l \leq |V|, \quad \forall\tau, \quad \forall(s_i, t_{i_j}) : \text{asap}_i \leq \tau \leq \text{alap}_{i_j},\ l \in [\pi_i, \pi_{i_j}] \tag{2}$$

$$b_{i,l,\tau} \geq \{1 - c_{i,j,\tau}\} \quad 1 \leq l \leq |V|, \quad \forall\tau, \quad \forall(s_i, t_{i_j}) : \qquad \tau = \text{alap}_{i_j},\ l = \pi_{i_j} \tag{3}$$

The number of buses needed is bounded from below by the number of distinct communications, that use a bus segment at the same location $l$ of the data path and at the same time $\tau$.

$$B \geq \sum_{1 \leq i \leq |E|} b_{i,l,\tau} \qquad 1 \leq l \leq |V|, \quad \forall\tau \tag{4}$$

$s_{i,j,\tau} = 1$, iff communication requirement $(s_i, t_{i_j})$ has not been scheduled before or at time $\tau$. $t_{i,j,\tau} = 1$, iff communication requirement $(s_i, t_{i_j})$ has been scheduled before or at time $\tau$. Obviously,

$$s_{i,j,\text{asap}_i} = 1 \qquad \forall(s_i, t_{i_j}) \tag{5}$$

$$t_{i,j,\text{asap}_i} = c_{i,j,\text{asap}_i} \qquad \forall(s_i, t_{i_j}) \tag{6}$$

$$s_{i,j,\tau+1} = s_{i,j,\tau} - c_{i,j,\tau} \qquad \forall(s_i, t_{i_j}), \quad \forall\tau : \text{asap}_i \leq \tau < \text{alap}_{i_j} \tag{7}$$

$$t_{i,j,\tau+1} = t_{i,j,\tau} + c_{i,j,\tau+1} \qquad \forall(s_i, t_{i_j}), \quad \forall\tau : \text{asap}_i \leq \tau < alap_{i_j} - 1 \tag{8}$$

| 1 | 2 | 3 | 4 | 5 | 6 | 7 |
|---|---|---|---|---|---|---|
| name | control steps | ALUs | B | R | B' | R' |
| diffeq | 8 | 5 | 3 | 8 | 6 | 7 |
| dct | 7 | 18 | 6 | 34 | 12 | 24 |
| dct | 7 | 18 | 8 | 33 | 12 | 24 |
| ellipf | 15 | 5 | 4 | 21 | 6 | 17 |

**Table 1** Results for different benchmarks

$S_{i,\tau} = 1$, iff some communication requirement of edge $e_i$ is scheduled at or after time $\tau$. In this case a register is needed for this communication in the source functional unit at time $\tau$.

$$S_{i,\tau} \geq s_{i,j,\tau} \quad \forall \tau, \forall (s_i, t_{i_j}),\ \text{asap}_i \leq \tau \leq \text{alap}_{i_j} \tag{9}$$

$T_{i,k,\tau} = 1$, iff any communication requirement $(s_i, t_{i_j})$ of edge $e_i$, for which $t_{i_j} = v_k \in V$, is scheduled before or at time $\tau$ and thus a register is needed at node $v$. Let $T(e_i)$ be the set of distinct target nodes of edge $e_i$.

$$T_{i,k,\tau} \geq t_{i,j,\tau} \quad \forall \tau, \quad \forall v_k \in V, \quad \forall (s_i, t_{i_j}) : t_{i_j} = v_k,\ \text{asap}_i \leq \tau \leq \text{alap}_{i_j} \tag{10}$$

Finally, the number of registers $R_k$ in a functional unit $v_k \in V$ is bounded from below by the number of values, that are alive at any time $\tau$ in $v_k$. $R$ is then the overall number of registers.

$$R_k \geq \{ \sum_{e_i=(v_k,\ldots)\in E} S_{i,\tau} + \sum_{\substack{e_i \in E \\ v_k \in T(e_i)}} T_{i,k,\tau} \} \quad \forall v_k \in V, \forall \tau \tag{11}$$

$$R = \sum_{v_k \in V} R_k \tag{12}$$

Obviously, if the designer wants to specify a limit on the number of buses or the number of registers, an appropriate constraint can easily be incorporated into the design.

# 5 RESULTS AND CONCLUSIONS

In table 1 we present the results for the benchmarks *Differential Equation Solver* (diffeq), $5^{th}$ *Order Elliptical Filter* (ellipf) [HLS94], and *Discrete Cosine Transform* (dct). Columns two and three denote the number of ALUs and the number of time steps, which result from data flow graph scheduling and module allocation. For the resulting communication graph, we solve the integer program using the CPLEX mixed integer program solver [CPL92]. The resulting number of buses and registers is given in columns four and five. Column six and seven give the resource utilization, if a central register file and non-partitioned global buses would have been used, as for example in [HLH91].

As can be seen from table 1, the number of buses is less than in a non-partitioned bus architecture with a central register file, at the cost of an increased number of registers. In rows two and three there are two different schedules for the *Discrete Cosine Transform*, which result in different register and bus allocations. CPU-time for the examples varies from a few seconds to approx. 2 hours for the bigger DCT examples, in which the global time frame for the computation is increased. Yet, as the number of buses is critical in many designs, this extra time and the flexibility of analyzing different schedules is by far acceptable. This is especially true, as we solve the problem of communication scheduling optimally. In contrast to a heuristic approach, if a limit on the number of buses is imposed, we can be sure to meet this constraint, if this is possible at all. The IP-formulation can be extended to regard mutual exclusive communications. For this, the bus and register constraints (equations 4 and 11) are set up only for those communications, which are not mutually exclusive. Mutually exclusive communications are counted in separate constraints, even if they are scheduled simultaneously.

*Acknowledgement* This work has been supported by the German Ministry for Education and Research (BMBF) under grant 01-IS-102

# REFERENCES

[CPL92] CPLEX Optimization, Inc. *CPLEX*, 1992.

[DH93] A. A. Duncan and D.C. Hendry. DSP Datapath Synthesis Eliminating Global Interconnect. *Proceedings of the EURODAC Conference 1993*, pages 46 – 51, 1993.

[DN89] S. Devadas and R. Newton. Algorithms for Hardware Allocation in Data Path Synthesis. *IEEE Transactions on Computer-Aided Design of Integrated Circuits and Systems*, CAD-8(7):768–781, 1989.

[DRSC86] H. DeMan, J. Rabaey, P. Six, and L. Claesen. Cathedral-II: A Silicon Compiler for Digital Signal Processing. *IEEE Design and Test*, 3(6):13–25, 1986.

[Ewe90] C. Ewering. Automatic High Level Synthesis of Partitioned Busses. *Proceedings of the International Conference on Computer Aided Design*, pages 304–307, 1990.

[HLH91] C. T. Hwang, J. H. Lee, and Y. C. Hsu. A Formal Approach to the Scheduling Problem in High Level Synthesis. *IEEE Transactions on Computer-Aided Design*, 10(4):464–475, 1991.

[HLS94] *The 7$^{th}$ High-Level Synthesis Workshop Benchmarks*, 1994.

[KP87] F. J. Kurdahi and A.C. Parker. REAL: A Program for REgister ALlocation. *Proceedings of the 24th Design Automation Conference*, pages 210–215, 1987.

[LCGD94] D. Lanneer, M. Cornero, G. Goossens, and H. DeMan. Data Routing: A Paradigm for Efficient Data-Path Synthesis and Code Generation. *7th High-Level Synthesis Symposium*, pages 17–22, 1994.

[MCdP91] A. Mignotte and M. Crastes de Paulet. Resource Assignment with Different Target Architectures. *Proceedings of the International Conference on Computer Design: VLSI in Computers*, pages 172–177, 1991.

[McF86] M. C. McFarland. Using Bottom-Up Techniques in the Synthesis of Digital Hardware from Abstract Behavioral Descriptions. *23rd Design Automation Conference*, pages 474–480, 1986.

[MRS91] F. Monteiro, B. Rouzeyre, and G. Sagnes. High Level Synthesis: A Data Path Partitioning Method Dedicated to Speed Enhancement. *Proceedings of the European Design Automation Conference*, pages 123–128, 1991.

[PK89] P. G. Paulin and J. P. Knight. Scheduling and Binding Algorithms for High-Level Synthesis. *26th Design Automation Conference*, pages 1–6, 1989.

[TS86] C.-J. Tseng and D.P. Siewiorek. Automated Synthesis of Data Paths in Digital Systems. *IEEE Transactions on Computer-Aided Design of Integrated Circuits and Systems*, 5(3):375–395, 1986.

PART THIRTEEN

# System Level Synthesis

37

# Cosynthesis In CASTLE

*J. Wilberg[+], R. Camposano [*], M. Langevin[+], P. Plöger[+], T. Vierhaus[+]*
*[+] GMD-SET,*
*Schloß Birlinghoven, D-53757 Sankt Augustin, Germany,*
*{wilberg, langevin, ploeger, vierhaus}@gmd.de*
*[*] Synopsys Inc.,*
*700 East Middlefield Road, Mountain View, CA 94043-4033, USA,*
*raul@synopsys.com*

**Abstract**

The CASTLE cosynthesis environment is described. These tools generate both a synthesizable VHDL processor description and a corresponding compiler back-end from a single processor specification on a block-diagram level. The designer can enter this block-diagram via CASTLE's Schematic Entry. The cosynthesis system offers new possibilities for optimizing the design on a system-level, since the designer can take into account the complex interaction between hardware, software, and algorithms. This is illustrated by comparing different IDCT (Inverse Discrete Cosine Transform) algorithms on different processors.

**Keywords**

Cosynthesis, retargetable code generation, video processing

## 1 INTRODUCTION

The success of the RISC concept showed the necessity of codesigning the processor hardware and the compiler (Cocke, Markstein, 1990). Advanced synthesis systems offer the possibility to generate complex processors from abstract specifications (Camposano, Wolf, 1991). Programming these processors is typically a difficult and time consuming task. Due to the complexity of the current processor generation (instruction level parallelism, pipelining), programming on an assembler level is only feasible for small programs. Designing a compiler is a complex project on its own (Ackland, 1994). Therefore changes in the processor structure are often prohibitive since they require a major redesign of both the processor and the compiler. This situation hinders efficient design space exploration which is necessary when developing complex systems.

A solution to this problem is a cosynthesis system which generates both, a synthesizable hardware description and a compiler back-end from a single system specification (Camposano, Wilberg)(Wilberg, et al., 1994b). This is our definition of cosynthesis. In this case the task of keeping the processor structure and the compiler consistent is handled by the synthesis system. Therefore the designer can concentrate on specifying the main system structure by performing a design space exploration.

The paper describes the tools used for the cosynthesis in CASTLE (Codesign and Architecture driven Synthesis TooL Environment). The tools in CASTLE are only loosely coupled which allows to configure different design flows for various application domains. This provides more flexibility for the synthesis but, as detailed below, this approach also poses new demands on the inter-tool communication.

Retargetable code generation for digital signal processing applications is also considered in Cathedral's CHESS system (Van Praet, et al. 1994) and in the FlexWare system (Liem, et al., 1994). The PEAS-1 (Alomary, 1993) system uses the GNU C compiler to generate code for RISC-like scalar processors. Scheduling with respect to VLIW (Very Long Instruction Word) architectures is addressed, e.g., in the Mimola (Marwedel, 1993) system. Codesign of hardware and software components is considered in Vulcan (Ernst, et al., 1993) and Cosyma (Gupta, DeMicheli, 1993). Our major contribution is in the development of a codesign methodology to derive a well-balanced processor for a given application and a given target architecture template. The paper concentrates on the cosynthesis phase of this methodology, i.e., the generation of a VHDL processor description and a compiler back-end from a single schematic.

The remainder of the paper is organized as follows. The next section gives a general overview of the cosynthesis approach used in the CASTLE system. Section 3 describes the VHDL-libraries of generic processor components which form the basis for the processor cosynthesis. Section 4 describes the underlying data structure of the schematic editor and the main steps for generating a VHDL processor description and the corresponding compiler back-end. Section 5 presents first results obtained with the tool suite in CASTLE, while Section 6 concludes the presentation.

## 2 OVERVIEW OF THE TOOLS

The CASTLE workbench approach (Theißinger, et al., 1994)(Wilberg, et al., 1994a) allows to configure the synthesis tools to form a design flow. The design flow can be tailored to a specific application domain. A design flow for cosynthesis of high-performance processors is shown in Figure 1. It allows to derive a well-adapted processor in short time. The methodology facilitates the exploration of many different processor architectures from which the best one can be selected for the final hardware implementation.

The current implementation of our methodology uses the retargetable compiler and the simulator of the MOVE system (Hoogerbrugge, Corporaal, 1994) developed at the University of Delft. A VLIW architecture template is used that permits to specify many architectural parameters such as the number and latency of functional units, the number of register file inputs and outputs, the number of busses and their interconnections with the components.

To explore this large design space, the design flow starts with a requirement analysis of the considered application domain. This permits to determine, e.g., the most frequent operations and data transfers, and the critical procedures. Then, during the design exploration phase, the performance characteristics of the processor are determined by exploring relaxed architectures for which only some architectural parameters are constrained.

Based on the results of the above analysis, the designer specifies the main structure of the processor. The structure is entered into CASTLE's schematic entry on a block-diagram level.

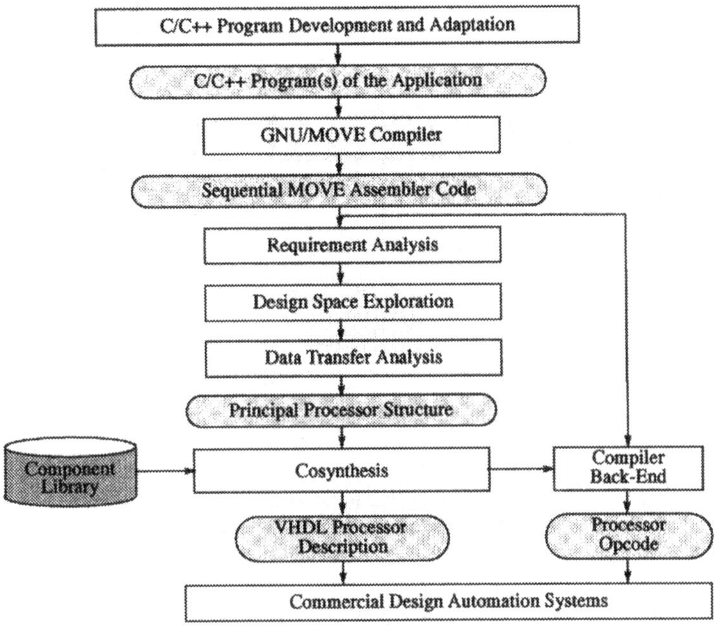


**Figure 1** Cosynthesis as part of a Hardware/Software codesign flow.

The schematic entry provides a comfortable interface for the designer *and* it is used to manage the cosynthesis as depicted in Figure 2. A single schematic is sufficient to derive suitable formats for the hardware synthesis and, on the other hand, for code generation.

Therefore, the schematic entry is the central element which ensures the consistency between the processor hardware and the compiler generation. The backbone of the approach is formed by a library of processor components as described in the next section.

## 3 LIBRARY OF GENERIC PROCESSOR COMPONENTS

The library contains generic processor components such as adders, registers, sequencers or multifunctional units. They are characterized by generics like word width, number of stages, etc. This is a compact way of describing large classes of components, for example, register files with different word width, different number of words, and different number of input and output ports are represented by just a single library element. VHDL is used for implementing the library. This supports hardware reuse and simplifies the extension of the library, since adesigner can add new elements to the library by including their VHDL descriptions. Furthermore, VHDL allows to separate the interface declarations (component package) and the internal realization (entity/architecture file). This permits to optimize the internal realization of

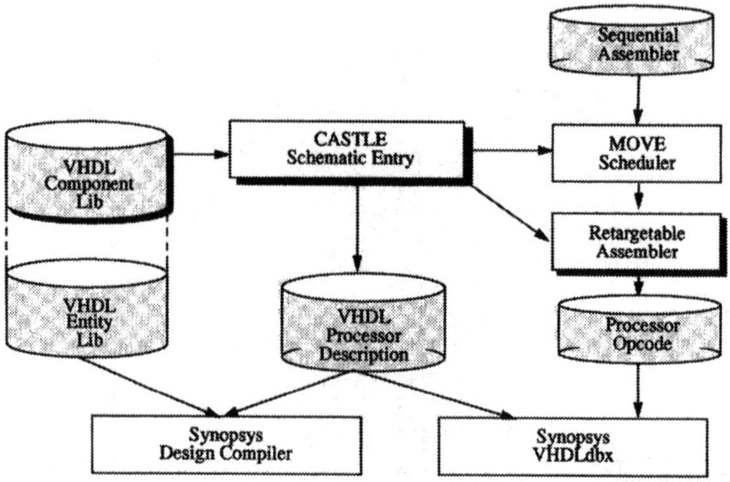


**Figure 2** Details on cosynthesis environment. Boxes denote tools. Shaded symbols are described in detail.

each processor component without effecting the overall processor description. In addition to this behavioral description, further information is included in the library to guide the simultaneous generation of hardware and software:

- The component ports are grouped into different categories (dataIn, dataOut, control, flags, etc.) where each group is handled differently during the netlist generation.
- The result latencies, supported assembler instructions, etc. of the units are stored.
- Programs for encoding the supported assembler instructions are stored with each unit.

An example from the component library is shown in Figure 3. The generic VHDL code is augmented with lines having a comment delimiter in front followed by the key term 'CASTLE'. At the beginning of the package some global attributes like '`hidden`', '`broadcast`' or '`flags`' are set. They will guide the schematic entry to treat some pins of the components in a special way. E.g., signals with '`broadcast`' attribute will be propagated to all instantiated components and a port will be added for them at the designs top level entity. The wires '`clk`' or '`reset`' will be suppressed when the schematics is displayed to the designer.

In the component declaration of the multiply/shifter type ('`MPY_ShifterType`') the latency, the supported operations, and the commands for the machine level encoding of this components are given. These commands will be interpreted when the back-end is generated. Finally, the ports are grouped by their functionality.

From a mere hardware point of view our component library contains similar elements like Synopsys' Design Ware (Synopsys, 1994) library. The special annotation extends their approach in a natural way[1] though. It is easy to maintain this VHDL library file or even to substitute it. It has furthermore the advantage of keeping all the relevant information in one file

```
package libComponents is
--CASTLE default hidden generics: N Log_N;
--CASTLE default control: control cin WriteEnable;
--CASTLE default broadcast: clk reset;
--CASTLE default flags: cout;
...
component MPY_ShifterType
-- CASTLE result latency: {Output, Stages};
-- CASTLE MOVE operation: mul shl shr shru;
-- CASTLE begin operationEncoding:
--       mul  { set contrWord(Control) 1;}
--       shl  { set contrWord(Control) 2;}
--       shr  { set contrWord(Control) 4;}
--       shru { set contrWord(Control) 8;}
-- end CASTLE;
  generic(Log_N: integer := 4;
          Stages: integer := 3);
  port   (
    In1,In2:in  std_logic_vector(2**Log_N-1 downto 0);
    Control:in  std_logic_vector(3 downto 0);
    Clk: in std_logic_vector(0 downto 0);
    Output:out std_logic_vector(2**Log_N-1 downto 0));
end component;
...
end libComponents;
```

**Figure 3** A component example from the VHDL library

only. Since the schematic entry will also base its graphical representation on it, a parser ('libComp' in Table 5) extracts a format which can be used to initialize the SIR/CASTLE schematic entry tool accordingly.

# 4 THE SCHEMATIC ENTRY IMPLEMENTATION

This section focuses on the data structure that is used to generate both the VHDL processor netlist and the files for the retargetable code generation. The interactive generation of the schematics demands a lot of screen manipulations on one hand. The code generation in the compiler back-end is facilitated when regular expression matching is employed on the other hand. Furthermore, the building blocks for a processor datapath show much regularity such that an object oriented approach is appropriate. Therefore we use an object oriented extension of the tool communication language TCL (Ousterhout, 1994)(Mc Lennan) for the implementation that provides all the required features.

Three main collection classes form the basis for representing the processor structure. The base class '`NetCircuit`', which will be called circuit manager in sequel, manages a collection of instances of the root class '`BaseUnit`' or one of its derivations, respectively. It models the network topology and can dump it as a structural VHDL netlist. The '`BaseCircuit`' derivation supports all methods which are necessary for the interactive manipulation and display of the underlying netlist, while the '`CompCircuit`' class encapsulates the output of the encoding during the back-end code generation.

1. The front end of CASTLE (Steinhausen, et al., 1993) will also comprise a full 93 VHDL parser which will enable us to use the 'attribute' statement for specifying this information in future.

In Figure 4, an instance of class 'NetCircuit' is shown. It has a list of contained functional units, lists of processor input ports, output ports, etc. In addition, there are a number of methods for creating new units, new connections, and so on. Therefore the circuit manager provides an ideal way for representing architecture templates. For example, the method 'makeUnit' can be overloaded differently for each derivation of a specialized circuit manager. Tools like the schematic entry can call this function without having to know the way a circuit manager inserts the units. Currently, a circuit manager for VLIW processors is used. In the future different circuit managers will be provided to represent other architecture templates like micro controller circuits, etc.

## 4.1 Generation of VHDL Processor Descriptions

After invoking the schematic tool, the user can specify a processor datapath at the block diagram level. He can instantiate functional units contained in the VHDL component library. A listbox window pops up for the actual selection. The values of some component generics may be changed. E.g., the user may change the number of 'Stages' for the 'MPY_ShifterType', but not the 'Log_N' value since this was declared 'hidden' at the beginning of the library. The placement of new components is performed in a simple left to right order, which suffices for the VLIW case. This may be overloaded when deriving manager classes for other architecture templates. Then components can be wired. Multiplexors for multiple fanin ports are automatically inserted. Also, when a feedback from an output port of a unit to any input port unit is placed, the schematic tool automatically adds a forwarding unit. After netlist specification, the result can be dumped into a structural VHDL format. The broadcast signals will be inserted automatically. For the actual hardware synthesis on RT level we currently use the Synopsys Design Compiler. The next section describes how a compiler back-end is generated.

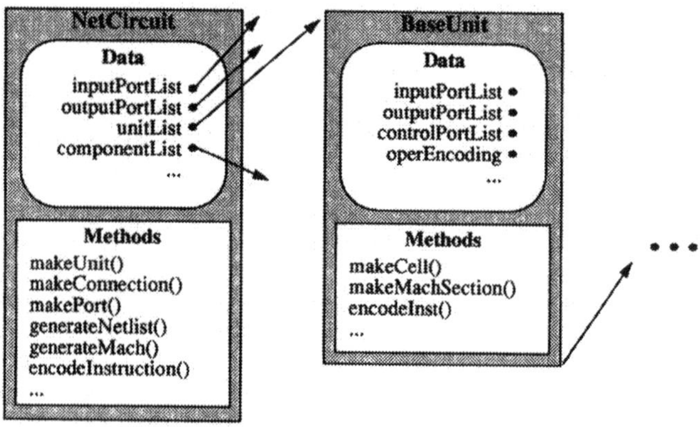


**Figure 4** Circuit manager as toplevel interface to an object-oriented processor representation.

## 4.2 Compiler Back-End Generation

Efficient code generation is especially important when designing complex systems. Even highly optimized full-custom chip sets, like AT&T's AVP4000 video compression, require a significant amount of microcode (45k lines in case of the AVP). Therefore the CASTLE system generates a compiler along with the processor hardware.

The GNU/Move compiler is used as a front-end. It translates a C/C++ program into the sequential MOVE assembler code. All front-end optimizations of the GNU gcc/g++ are performed. For our purposes the sequential MOVE assembler code can be considered as a RISC-like 3-operand code for a load/store architecture. No particular processor topology is assumed when the sequential assembler code is generated.

The retargetable compiler back-end translates this code into the opcode of the designed VLIW processor. The instruction word of a VLIW processor is completely orthogonal (Rau, Fisher, 1993), i.e., each unit is controlled by a separate bitfield in the instruction word (the instruction word of the processor is formed by concatenating all these bitfields). Only static scheduling is used, i.e., the operations to be executed in parallel are determined at compile-time. The compiler back-end can be subdivided into two parts:

1. The **scheduler** which generates the parallel assembler code (the scheduler of the Move system is used). It has two inputs: (1) the sequential assembler code generated by the front-end and the machine description file, which is produced by the schematic entry. The scheduler checks the dependencies among the operations and the resource constraints imposed by the structure of the specified processor. Based on this information it parallelizes the assembler code and augments it with labels that specify the allocation and binding of each operation. The result is a text file which lists the assembler operations that can be executed in parallel and the units on which each of these operations is going to be executed. The label names were chosen by the schematic entry and stored in the machine description file. Hence, these labels can be used in the retargetable assembler to retrieve the binding information for the operation encoding.
2. The **retargetable assembler** translates the parallel assembler code into the binary instruction format of the specified VLIW processor. It uses the same object-oriented processor structure which is also used for the VHDL code generation. The encoding of the operations is done in the following way: First, the control word of each unit object is initialized to the default value zero, i.e., a nop-operation is assigned to each functional unit. Next the circuit manager iterates over all operations in an instruction.

   For each operation the allocation and binding labels are parsed, and the unit is determined that should execute the operation. The operation is forwarded to that unit object. The operation encoding within the unit object is based on an AWK (Aho, et al., 1988) like program which is stored in the component library for each unit type (Figure 3). The encoder is implemented as switch over regular expressions. Each branch consists of a statement of the following form:

   *<regular expression matching the operation>* **{**
           *<Tcl statements to generate the encoded instruction>***}**

   If for example a '`shl`' operation should be encoded by a unit of type '`MPY_shifterType`', the matching encoding statement is

   ```
   shl  { set contrWord(Control) 2;}
   ```

The statements in the braces will be executed to encode the operation. In this case the part of the control word connected to the port `'Control'` of the functional unit will be set to the binary value of 2 (`'0010'`). The value is automatically expanded to the correct bitsize of the port, even if this size depends on generic parameters of the unit. Finally, the individual instruction words, stored in each unit, are concatenated to form the complete instruction word of the VLIW processor.

Separating the scheduler and the retargetable assembler, as described in this section, allows to include post-scheduling optimizations (e.g., local microcode optimization (Langevin, et al., 1995), software pipelining (Lam, 1989)) into the compiler back-end.

# 5 RESULTS

This section applies our cosynthesis methodology to compare different implementations of an IDCT on different processors. The IDCT is one of the most computationally intensive functions in an MPEG-1 decoder. The following four different implementations of the algorithm are used in this experiment:

1. matrix_dct: 8x8 matrix multiplication.
2. matrix_butterfly_dct: two 4x4 matrix multiplications and one butterfly layer (Uramoto, et al., 1992).
3. j_rev_dct: the fast IDCT implemented in the *mpeg_play* program (Patel, et al., 1993).
4. j_rev_dctOpt: the optimized IDCT version of the *mpeg_play* program.

The instruction mix of these procedures, if embedded into the *mpeg_play* program, is shown in Table 1. The *flower.mpg* sequence was used as data input to the *mpeg_play* program. The different procedures exploit the symmetric properties of the IDCT kernel to reduce the number of operations. In addition, the *j_rev_dctOpt* procedure also exploits the special properties of the MPEG-1 standard, by first checking for zero coefficients before calculating the transformation. This gives a further reduction of the average operation count at the cost of a more irregular control flow.

For each of the above implementations the effects of some architectural parameters on the performance of the application are evaluated. Figure 5 shows the results of this analysis for the *j_rev_dctOpt* implementation when varying the number of some unit types. The analysis was performed by compiling the application for relaxed architectures where only one unit type is constraint at a time.

This first phase of analysis permits a fast reduction of the design space. For instance, a maximum of 4 adders and two instances of each other unit type of Figure 5 seems sufficient to achieve a near optimal performance, given the available compiler and target architecture. In a second phase, to obtain a more accurate estimation of the achievable performance, an analysis constraining many parameters at a time can be performed.

At this point, we have a good idea of the potential trade-off between the performance and the hardware resource requirements. Final architectures are derived by fixing all the parameters, including the interconnection networks. For this purpose, a data transfer analysis is performed. Figure 6 shows the results for the *j_rev_dctOpt*. In this case a result of an ALU operation is

**Table 1** Instruction mix results for different implementations of the IDCT. The flower.mpg sequence was used as data input where a macroblock consists of 16x16 pixel for the luminace component and 8x8 pixel for each chrominance component.

| *DCT Function* | *Operations / Macroblock* | *Assembler Instructions* | *ADD, SUB COMP, LOGIC* | *MULT, SHIFT* | *LOAD, STORE* | *REST* |
|---|---|---|---|---|---|---|
| matrix_dct | 15641.1 | 170 | 44.77% | 20.45% | 34.78% | 0.00% |
| matrix_butterfly_dct | 9133.3 | 187 | 45.75% | 22.75% | 31.50% | 0.00% |
| j_rev_dct | 3141.7 | 234 | 52.76% | 19.40% | 20.30% | 7.54% |
| j_rev_dctOpt | 2084.7 | 869 | 53.78% | 17.01% | 19.79% | 9.43% |

**Figure 5** Performance in number of clock cycles for relaxed processors when processing the *j_rev_dctOpt* function. The flower.mpg sequence was used as data input.

most often used as input for a succeeding ALU operation. The traffic to the register file is reduced if this data transfer is supported by a direct hardware connection from the ALU output to the ALU input.

For the experiments reported here, four different processors were specified. Table 2 shows the results when synthesizing the VHDL library components with the Synopsys Design Compiler. A wordwidth of 32-bit is used for all components. Different generic parameters were chosen for the register file configuration. These components were used to construct the processors.

The resulting gate count of the processors is shown in Table 3.The sizes of the register files are changed: P1 and P2 have 16 word (w) register files, P3 and P4 have 32 w register files. P1 has the smallest number of arithmetic units and P4 has the highest number of units. The flexibility of the interconnection network gradually increases from P1 to P4. A more efficient implementation of the processors might be achieved if demanding components like the register files or the multipliers are replaced by special full-custom blocks.

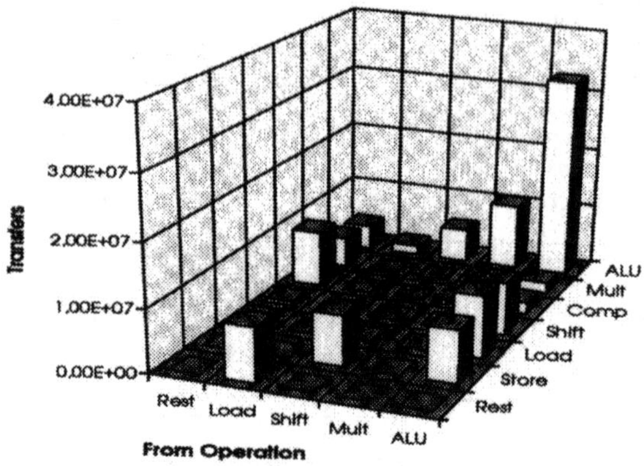


**Figure 6** Data transfers between different operations for the *j_rev_dctOpt*.

**Table 2** Complexity of library components when synthesized from the LSI10k librray.

| *Library Element (Parameters)* | *Gate Count / Instance* | *Library Element (Parameters)* | *Gate Count / Instance* |
|---|---|---|---|
| RegFile (16w, 2in, 4out) | 8996 | ALU | 1237 |
| RegFile (16w, 3in, 6out) | 13374 | ALU_Shift | 2297 |
| RegFile (32w, 1in, 2out) | 12277 | Compare | 389 |
| RegFile (32w, 2in, 4out) | 16661 | MPY | 17505 |
| RegFile (32w, 3in, 6out) | 26233 | ShiftLeftRight | 1176 |

**Table 3** Complexity of different processors used in the codesign example.

| *Component Type* | *P1* | *P2* | *P3* | *P4* |
|---|---|---|---|---|
| Register | 8996 | 13726 | 16661 | 27641 |
| Arithmetic | 20418 | 21637 | 21637 | 23934 |
| Interconnection | 2080 | 4050 | 4050 | 5860 |
| Total (incl. glue) | 31499 | 39426 | 42391 | 57435 |

The speed up obtained for the different IDCT implementations on the different processors is shown in Table 4. The results show the importance of a codesign of hardware, software, and algorithms. For example, the algorithmic improvements resulted in a speed up of 5, the hardware improvements obtained a speed up of 3, and only the combined optimization in the codesign achieved a speed up of 15. These optimization can not be carried out separately. For example, reducing the number of operations by additional checks, as in *j_rev_dctOpt* is a good idea for a scalar-like processor (P1), but the introduced dependencies hinder efficient parallelization in case of a higher degree of instruction level parallelism (P4). In summary, only the codesign can achieve the best results with regard to the overall system.

Table 5 summarizes the complexity of some CASTLE tools used in the cosynthesis environment

**Table 4** Average number of cycles per marcoblock for the different algorithms and different processors. The flower.mpg sequence was used as data input.

| *DCT Function* | *Cycles/Macroblock* | | | | *Speed Up* | | | |
|---|---|---|---|---|---|---|---|---|
| | *P1* | *P2* | *P3* | *P4* | *P1* | *P2* | *P3* | *P4* |
| matrix_dct | 23760 | 20935 | 12468 | 8610 | 1 | 1.13 | 1.91 | 2.76 |
| matrix_butterfly_dct | 14692 | 13886 | 7589 | 5572 | 1.62 | 1.71 | 3.13 | 4.26 |
| j_rev_dct | 5807 | 4822 | 2134 | 1548 | 4.09 | 4.93 | 11.13 | 15.34 |
| j_rev_dctOpt | 4340 | 3739 | 1804 | 1615 | 5.47 | 6.36 | 13.17 | 14.17 |

**Table 5** Code sizes of different CASTLE programs.

| *Program* | *Language* | *Lines of Code* |
|---|---|---|
| Analysis Environment | C++, Scripts | 3712 |
| CDFGView | itcl, C++, GAWK | 1386 |
| Schematic Entry | itcl | 4645 |
| Retargetable Assembler | itcl, Perl | 2868 |
| libComp | GAWK | 1107 |
| Library Component Package | VHDL | 661 |
| Library entity/architecture file | VHDL | 3396 |
| Library Testbenches | VHDL | 3826 |

## 6 CONCLUSION

The paper described the tools in the CASTLE cosynthesis environment. A single object-oriented processor description is used to generate both a synthesizable VHDL description of the processor and a corresponding compiler back-end. That is, the consistency of the processor hardware and the compiler is automatically handled by the cosynthesis system. This approach offers new possibilities for design space explorations which take into account the complex interaction between hardware, software, and algorithms.

Future work must address the cosimulation which is very time consuming for the complex image processing tasks and the problems of test generation for complex application-oriented processors.

## 7 REFERENCES


B. Ackland (1994) "The Role of VLSI in Multimedia", IEEE J. Solid-State Circuits, vol. 29, no. 4, pp. 381-388, April.

A.V. Aho, B.W. Kernighan, P.J. Weinberger (1988) *The AWK Programming Language*, Reading: Addison-Weseley.

A. Alomary, et al. (1993) "PEAS-I: A Hardware/Software Co-design System for ASIPs", Euro DAC, Hamburg, pp. 2-7.

R. Camposano, W. Wolf (eds.) (1991) *High-Level VLSI Synthesis*, Kluwer Academic Publishers, Boston.

R. Camposano, J. Wilberg: "Embedded System Design", to be published.

J. Cocke, V. Markstein (1990) "The evolution of RISC technology at IBM", IBM J. Research and Development, vol. 34, no. 1, pp. 4-11, Jan.

*Design Ware Databook*, Synopsys® Inc. (1994), vers. 3.1a, March.

R. Ernst, J. Henkel and Th. Benner (1993) "HW/SW cosynthesis for microcontrollers", IEEE Design & Test, pp. 64-75, Dec.

R.K. Gupta, G. De Micheli (1993) "Hardware-Software Cosynthesis for Digital Systems", IEEE Design & Test, vol. 10, no. 3, pp. 29-41, Sept.

J. Hoogerbrugge, H. Corporaal (1994) "Transport-Triggering vs. Operation-Triggering", Compiler Construction Conference.

M. S. Lam (1989) *Systolic Optimizing Compiler*, Kluwer Academic Publishers, Dordrecht, Netherlands.

M. Langevin, E. Cerny, J. Wilberg, H.T. Vierhaus (1995) "Local Microcode Generation in System Design ", G. Goossens, P. Marwedel (eds.): *Code Generation for Embedded Processors*, Kluwer Academic Publishers, to appear.

M.J. Mc Lennan: "[incr Tcl] – Object-Oriented Programming in Tcl", AT&T Bell Laboratories, Allentown, PA 18103, michael.mclennan@att.com.

C. Liem, T. May, P. Paulin (1994) "Instruction-Set Matching and Selection for DSP and ASIP Code Generation", Proc. of the European Design and Test Conf. (ED&TC), pp. 31-37, Paris.

P. Marwedel (1993) "Tree-Based Mapping of Algorithms to Predefined Structures", Proc. of the Int. Conf. on CAD (ICCAD), pp. 586-593, Santa Clara CA.

J. K. Ousterhout (1994) *Tcl and the Tk Toolkit*, Addison-Wesley Publishing Company, Reading, MA.

K. Patel, B.C. Smith, L.A. Rowe (1993) "Performance of a Software MPEG Video Decoder", Proc. 1st ACM Int. Conf. on Multimedia, Anaheim, CA.

B. R. Rau, J. A. Fisher (1993) "Instruction-Level Parallel Processing: History, Overview, and Perspective", J. Supercomputing, vol.7, no. 2, pp. 9 - 50, May.

U. Steinhausen, et al. (1993) "System-Synthesis Using Hardware/Software Codesign", Int. Workshop on Hardware-Software Co-Design, Cambridge, MA, Oct. 7-8.

M. Theißinger, P. Stravers, H. Veit (1994) "An Interactive Environment for HW/SW Co-Design", 3rd Int. Workshop on Hardware/Software Codesign, Sept. 22-23, Grenoble, pp. 203-209.

S.-I. Uramoto, et al. (1992) "A 100-MHz 2-D Discrete Cosine Transform Core Processor", IEEE J. Solid-State Circuits, vol. 27, no. 4, pp. 492-498, April.

J. Van Praet, G. Goossens, D. Lanneer, H. DeMan (1994) "Instruction Set Definition and Instruction Selection for ASIPs", 7th Int. Symp. High-Level Synthesis, Niagara-on-the-lake, Ontario, Canda, pp. 11-16, May 18-20.

J. Wilberg, R. Camposano, U. Westerholz, U. Steinhausen (1994a) "Design of an Embedded Video Compression System – A Quantitative Approach", Int. Conf. Computer Design, Cambridge, MA, pp. 428-431, Oct. 10-12.

J. Wilberg, R. Camposano, W. Rosenstiel (1994b) "Design Flow for Hardware/Software Cosynthesis of a Video Compression System", 3rd Int. Workshop on Hardware/Software Codesign, Sept. 22-23, Grenoble, pp. 73-80.

# 38

# Optimization of Heterogeneous Multiprocessors for Complex Image Processing Applications

*M. Schwiegershausen, P. Pirsch*
*University of Hannover*
*Laboratorium für Informationstechnologie,*
*Schneiderberg 32, 30167 Hannover, Germany.*
*Telephone: +49-511-762-5050. Fax: +49-511-762-5051.*
*email:* schwiege@mst.uni-hannover.de

**Abstract**

This paper presents a new approach for the derivation of heterogeneous multiprocessor systems and its implementation as CAD tool. These heterogeneous systems, consisting of application-specific as well as programmable processors, are highly suitable for performing composite schemes of image processing algorithms under real time constraints. By means of a formal optimization approach, using mixed integer linear programming, it becomes possible to derive efficient, real time hardware, being optimal in terms of area expense and throughput rate.

**Keywords**

Heterogeneous multiprocessors, optimization, mixed integer linear programming.

## 1 INTRODUCTION

Today's image compression and coding schemes, like JPEG, MPEG, etc. have to be processed under real time constraints with high demands on computational and throughput rates. This can only be achieved by massive application of parallel processing and pipelining as provided by multiprocessor systems. These schemes can be splitted into several subtasks with different requirements leading to different architectures appropriate for the individual tasks. This results in a heterogeneous multiprocessor system for the implementation of the complete scheme. Heterogeneous systems in turn, may be composed of different, application-specific processors for implementing low and medium level tasks with high computational demands and programmable processors to provide the flexibility required by medium and high level tasks (see Figure 1). Nevertheless, the large multitude of suitable processors for each single algorithm makes it impossible for the human designer to choose manually the best combination. Consequently, a formal approach for the optimization is mandatory.

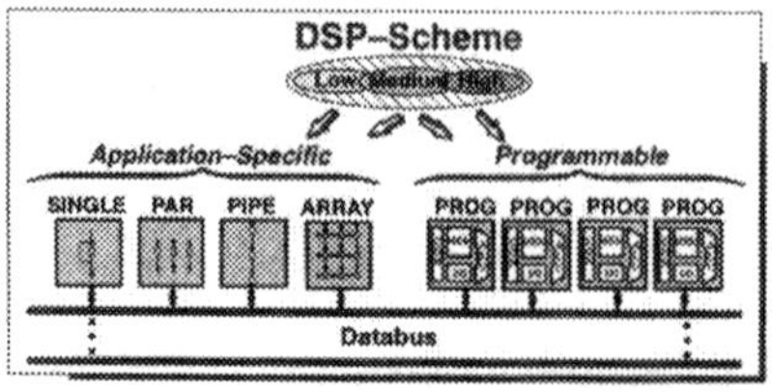


**Figure 1** Heterogeneous multiprocessor.

This optimization includes the mapping of the composite scheme of algorithms onto the multiprocessor system. There has been extensive research on the mapping problem, mainly in the field of high-level synthesis, where methodologies for mapping one single algorithm onto a dedicated datapath (Rosenstiel and Krämer, 1991) or onto an array of processing elements like systolic array processors (Kung and Jean, 1988) were tackled. Furthermore, synthesis scripts like *CATHEDRAL II, 2nd* (Rabbaey *et al.*, 1988) were developed, mainly addressing medium throughput applications. Finally, in (Bachmann *et al.*, 1993) an approach for domain-specific multiprocessor systems is presented but in opposite to our work it aims at calculating feasible schedules without considering the area expense as well as the throughput rate. Therefore, these approaches have to be extended for the derivation of heterogeneous multiprocessor systems, suitable for performing composite schemes of image processing algorithms under real time constraints. So, in this paper a new system level design methodology is presented, including a formal approach for the optimization of such heterogeneous systems.

This paper is organized as follows: Section 2 introduces our system level design methodology including the optimization. Section 3 shows first experimental results. Section 4 gives some hints on the implementation of the methodology as CAD tool. Finally, concluding remarks are made in Section 5.

## 2 DESIGN METHODOLOGY

The derivation of heterogeneous mutiprocessor systems for the execution of composite schemes of image processing algorithms can be divided into three steps, the *decomposition* of the whole scheme into single tasks or algorithms, the *assignment* of all possible processors for each single task (design space exploration), and finally the *optimization* of the overall system, thus selecting the best combination of application-specific and programmable processors (allocation) as well as deriving the most suitable order of execution (scheduling), see Figure 2. In the proceeding, the three design steps of our methodology are explained in more detail, with emphasis on the *optimization*:

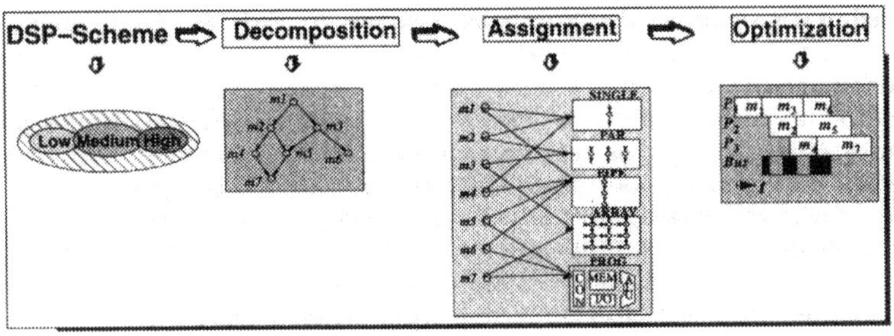


**Figure 2** Design methodology.

## 2.1 Decomposition

In the first step the overall image processing scheme has to be decomposed into single tasks or image processing algorithms. Thereby, the dependencies between the different tasks can be modelled by means of a directed, acyclic graph $G_T = (V, E)$, whereas each vertex $v_i \in V$ represents exactly one task $m_i$, as shown in the box *decomposition* of Figure 2.

## 2.2 Assignment

During this design step it is determined for each task which assignments of the single task's algorithm to suitable architectures of a given processor library is feasible. In order to restrict the large design space to a finite set of parametrizable, architectural alternatives, a processor library with five different types for each image processing algorithm is currently supported. This processor library comprises a single processor with dedicated data path (`SINGLE`) performing all operations sequentially, several such dedicated processors in parallel (`PAR`) or connected in pipeline mode (`PIPE`), and an array of dedicated processors (`ARRAY`) providing both, pipelining and parallel operation mode. Additionally, a flexible and programmable processor (`PROG`) preferably suitable for performing the medium level tasks is provided.

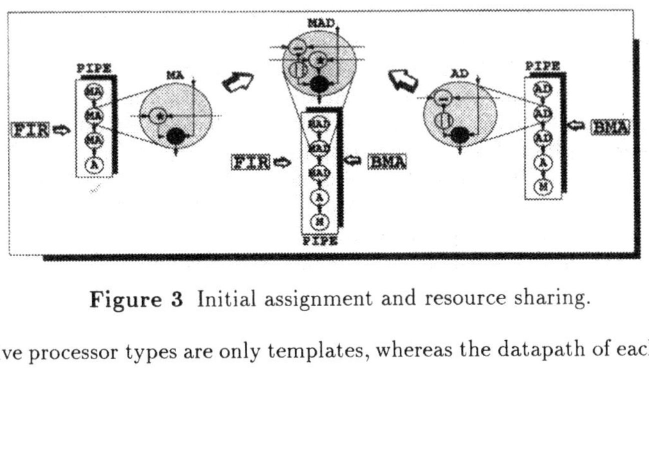


**Figure 3** Initial assignment and resource sharing.

These five processor types are only templates, whereas the datapath of each of the types is adapted to the operations that have to be performed for each task. For example, a PIPE processor's datapath performing a filter task like FIR differs from a PIPE processor's datapath performing a motion estimation task like BMA. Figure 3 shows the initial assigment for these two algorithms, in case a processor of type PIPE is selected. After application of resource sharing, the two processor pipelines can be combined to a third processor of same type, able to perform both algorithms, thus leading to a higher processor utilization. However, if it is advantageous to select the combined processor of type PIPE or the two single processors, also of type PIPE, with respect to the efficiency of the overall system can not be answered on the fly. On the contrary, this leads to a combinatorial optimization problem, as described next in the third step of the design methodology:

## 2.3 Optimization

Finally, it is necessary to select from the large amount of feasible assignments derived so far, the one which leads to an overall multiprocessor system being optimal in terms of throughput rate and area expense. This optimization problem can be formulated by means of mixed integer linear programming (MILP). Although the idea to tackle scheduling, allocation, and binding by MILP, is not essentially new, the presented approach is extended in comparison to previous works (Prakash and Parker, 1992) by considering a multitude of different application–specific processors as well as regarding overlapped and periodic execution.

Our MILP model, as presented in (Schwiegershausen et. al., 1994) consists of binary decision variables, integer valued as well as real valued timing and area variables. A set of restrictions guarantees unique processor– and bus–selection as well as non overlapping processor– and bus–usage. The objective function of the MILP takes into consideration the computation time (latency) $T$, the computation period $P$ as well as number $Y$ and area $A$ of processors and buses. By means of individual weights $c_T, c_P, \ldots, c_B$ different constraints of the designer can be considered.

# 3 EXPERIMENTAL RESULTS

In order to validate the design methodology, a typical complex image processing application was chosen: Low bit rate coding according to the hybrid videocodec scheme based on CCITT recommendation H.261 (CCITT, 1990). It is used for data reduction necessary to transmit video data on a line with $p$ x 64 kbit/s ($1 \leq p \leq 30$). The application consists of several computational extensive regular, low level tasks like motion estimation, i.e. block matching algorithm (BMA), discrete cosine transform and its inverse (DCT, IDCT), and FIR filtering (LF) as well as data dependent, irregular medium level tasks like quantization and its inverse (Q, IQ) and variable / runlength coding (VLC, RLC) combined to coding (COD), see Figure 4a). The corresponding task graph is sketched in Figure 4b); the possible assignments of the $M = 12$ tasks to $N = 19$ processors, assuming four different processor types (SEQ, PAR, PIPE, ARRAY) for the low level tasks and a programmable processor (PROG) for the medium level coding task can be seen in Figure 4c).

The MILP model (147 variables, 336 constraints) could be solved by *branch-and-bound* within a few cpu–minutes. For this purpose, the LP–Solver of the NAG * Fortran–Library was adapted to these needs.

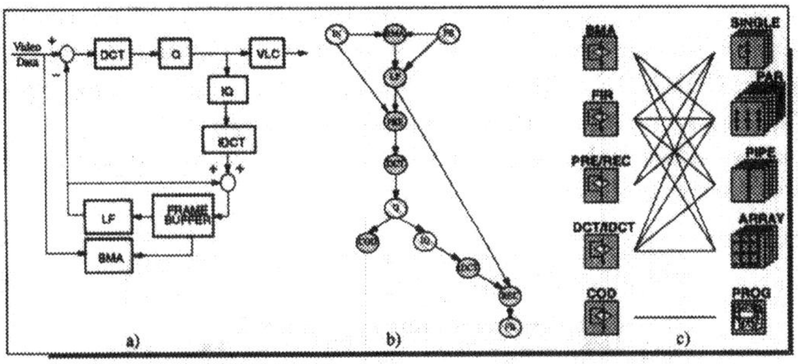


**Figure 4** a) Video encoder algorithm according to H.261 b) Task graph according to H.261 c) Possible assignments of tasks to processors.

According to the designers goal or priority order, different solutions for an efficient heterogeneous multiprocessor system can be derived. The four solutions with different priorities like minimum area expense or minimum computation period according to the four columns of the table in Figure 5a) are additionally sketched in the area– and time–plane:

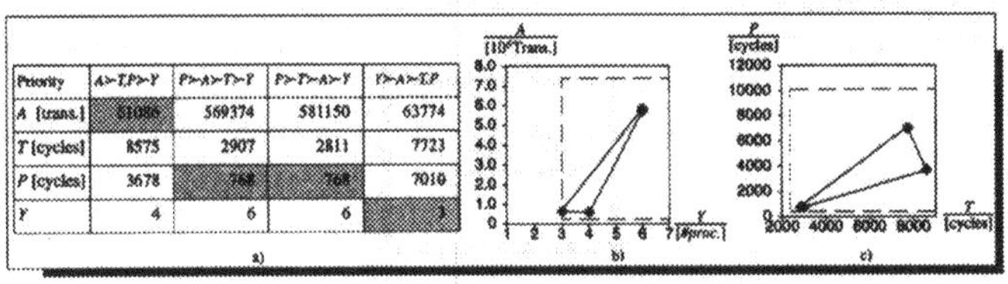

| Priority | A≻T,P≻Y | P≻A≻T≻Y | P≻T≻A≻Y | Y≻A≻T,P |
|---|---|---|---|---|
| A [trans.] | 51086 | 569374 | 581150 | 63774 |
| T [cycles] | 8575 | 2907 | 2811 | 7723 |
| P [cycles] | 3678 | 768 | 768 | 7010 |
| Y | 4 | 6 | 6 | 3 |



**Figure 5** a) Performance and area expense b) Design space concerning $A, Y$ c) Design space concerning $P, T$.

The possible design spaces according to the supported processor library and the real time constraint of the (video) application are marked by the dashed rectangles, showing the possible solutions concerning area expense $A$ and number of processors $Y$, Figure 5b) as well as concerning computation period $P$ and latency $T$, Figure 5c). For example, the upper edge of the $P, T$ plane reflects the real time constraint ($P \leq 10101$). Clearly, any feasible solution must reside inside the dashed design regions. Thus, depending on the designer's main goal either area or time optimal solutions can be derived.

*The Numerical Algorithms Group Limited

# 4 IMPLEMENTATION

The presented design methodology including the optimization approach using mixed integer linear programming is part of a system–level synthesis aid for the derivation of heterogeneous multiprocessors. It is currently under development as a protoype CAD implementation in COMMON LISP/CLOS † on Sparc Stations including a graphical user interface (GUI) based on CLUE/CLX ‡. An overview of the CAD system is given in Figure 6 with a specialized task graph as central database, connected to the algorithm– as well as to the processor–library.

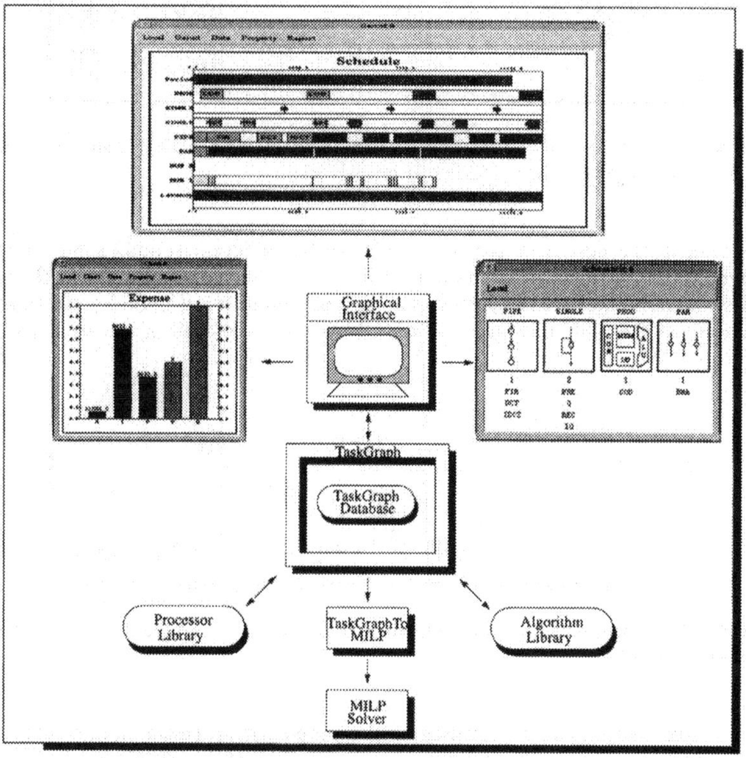


**Figure 6** Overview of the CAD tool.

All feasible assignments are derived, and the transformation to a MILP formulation is performed, and finally the (MI)LP solver is invoked. The results of the optimization can be visualized, using the graphical user interface.

†CLOS Specification, ANSI, X3J13

‡Common Lisp User Interface Environment/X

# 5 CONCLUSION

In this paper, a methodology for the design of heterogeneous multiprocessor system is presented. In order to find the best combination of application-specific and programmable processors for these heterogeneous systems, an optimization using mixed integer linear programming is included in the methodology. Thereby, it becomes possible to derive an overall system which is optimal with respect to a designer's priority or main goal like minimum area expense or maximum throughput.

# 6 ACKNOWLEDGEMENTS

This work is supported by the Deutsche Forschungsgemeinschaft (Pi-169/5).

# REFERENCES

Bachmann, A., Schöbinger, M., and Thiele, L., 1993, Synthesis for Domain-Specific Multiprocessor Systems including Memory Design, in *VLSI Signal Processing VI* (ed. L.D.J. Eggermont *et al.*), IEEE Press, New York, 417–25.

CCITT Study Group XV, 1990, Recommendation H.261, Video Codec for Audiovisual Services at p x 64 kbit/s, Report R37, Geneva.

Kung, S.Y., and Jean, S.N., 1988, A VLSI Array Compiler System (VACS) for Array Design, in *VLSI Signal Processing III* (ed. R. W. Brodersen and H.S. Moscovitz), IEEE Press, New York, 495–508.

Prakash, S., and Parker, A.C., 1992, Synthesis of Application Specific Heterogeneous Multiprocessor Systems, *Journal of Parallel and Distributed Computing*, **16**, 338–51.

Rabbaey, J., de Man, H., Vanhoof, J., Goosens, G., and Cathoor, F., 1988, CATHEDRAL II: A Synthesis System for Multiprocessor DSP Systems, in *Silicon Compilation*, Addison-Wesley, 311–60.

Rosenstiel, W., and Krämer, H., 1991, Scheduling and Assignment in High Level Synthesis, in *High Level VLSI Synthesis* (ed. R. Composano and W. Wolf), Kluwer Academic Publishers, 355–82.

Schwiegershausen, M., Schönfeld, M., Pirsch, P., 1994, Mapping complex image processing algorithms onto heterogeneous multiprocessors regarding architecture dependent parameters, in *Algorithms and Parallel VLSI Architectures* (ed. M. Moonen), Elsevier Science.

# Self-test with deterministic test pattern generators*

*Arno Kunzmann and Sven Kriebel*
*Forschungszentrum Informatik (FZI)*
*Haid-und-Neustrasse 10-14*
*D-76131 Karlsruhe, Germany*
*Phone (+49) 721 9654456 Telefax (+49) 721 9654409*
*Email kunzmann@fzi.de*

**Abstract**

This paper addresses the problem of generating deterministic test patterns via a cost-efficient self-test circuitry. The proposed solution is based on primitive linear feedback shift registers (LFSR) and a mapping of their outputs on the inputs of the circuit under test. By construction, an exhaustive enumeration of all LFSR states ensures a complete stuck-at fault coverage. For the computation of the mapping function two approaches will be discussed that are based on an tuned ATPG algorithm or alternatively on a given deterministic test pattern set. Experimental data with over 30 benchmark circuits underline the efficiency of these approaches; with only a few exceptions a LFSR of length 20 is sufficient to guarantee complete fault coverage.

**Keywords**

Self-test, ATPG, deterministic test patterns, pseudo-exhaustive testing.

## 1 INTRODUCTION

In the field of microelectronic CAD the technology push has a strong impact on future testing strategies. With the increasing number of transistors per area unit, in sub-micron technologies, clock frequencies are expected to increase greatly, and simultaneously the wiring delays will have a growing influence on critical path delays. These developments require cost intensive test equipment which handle high speed circuits with an increasing number of external I/Os, or the usage of innovative self-test techniques, which are the focus of this paper.

Various self-test techniques have been proposed to reduce the costs of a VLSI production test. An essential task of all these techniques is to specify hardware for test pattern generation that ensures the required fault coverage is achieved at a low cost. For this purpose, in the past,

* This work is partly supported by the ESPRIT project LINK (6855)

several approaches have been pre-sented tuned on (weighted) random test (e.g., (Waicukauski, 1988), (Wunderlich, 1988)), deterministic test (e.g., (Dähn, 1983), (Kunzmann, 1994)) or pseudo-exhaustive patterns (e.g., (McCluskey, 1981), (McCluskey, 1984)). Using linear feedback shift registers (LFSR) both conventional and optimized random test patterns can be generated very inexpensively, but the necessary number of test vectors and thus the test application time can grow quickly and prevent its application. Another option, self-test techniques with deterministic test patterns require only a few test patterns. Therefore, the test application time is not critical, but the resulting test hardware effort remains very cost intensive in general. The application of pseudo-exhaustive test techniques requires that each output depends on a restricted number of inputs to enable the realization of cost-efficient test hardware typically by counters or LFSRs. The enumeration of all input combinations covers almost all combinational faults including all detectable stuck-at faults, and this kind of test pattern enumeration makes an explicit ATPG algorithm overflow. The main disadvantage of pseudo-exhaustive testing is that, in general, the number of influencing inputs has to be limited by adding specific design-for-testability hardware (e.g., (McCluskey, 1981), (Wunderlich, 1988)). The required segmentation yields increased production costs, and several additional test sessions may be necessary.

While for the test response analysis a LFSR is assumed, the main focus of this paper will be on the development of cost-efficient test generation hardware. The basic idea of the proposed approach can be characterized by the following question: How can a LFSR of minimal width, including a mapping of its outputs to one or more inputs of the circuit being tested, be computed such that an exhaustive enumeration of all test patterns does not exceed the available test application time while ensuring a complete fault coverage ?

One answer is based on an extended ATPG algorithm which decides which inputs can always be stimulated by the same value. Obviously, these groups of inputs can be stimulated by a common LFSR output. Experimental results show that the achievable low widths of the required LFSR or counter enable the enumeration of all test patterns, and thus a low-cost self-test solution can be realized. Besides the ATPG based option, a static approach will also be discussed. Based on a deterministic test pattern set and with greatly reduced CPU-time requirements this alternative approach does not demand any modifications of the ATPG algorithm, and provides satisfactory solutions.

The paper is organized as follows: in Section 2, a motivation will be given as to why purely considering the circuit structure has some limitations, which can be overcome, however, if functional information is also considered. Then, two basic approaches will describe how functional independent input pairs and groups can be identified. In Section 4, the basic self-test hardware structure is described. Finally, some experimental data will be given to show the efficiency of the self-test hardware.

## 2 MOTIVATION AND BASIC DEFINITIONS

The following two definitions are helpful to describe the state-of-the art technology which uses structural dependencies to reduce the number of inputs that have to be exhaustively enumerated:

**Definition 1:** *Dependence Matrix*
For a given circuit N with n primary inputs and m primary outputs the dependence matrix D(N) has n columns and m rows. An entry is set to "*", if and only if the corresponding input influences the corresponding output. Otherwise, there is no entry value.

**Definition 2:** *Structural Independence*
An input pair (i,k) is said to be *structurally independent* if the corresponding dependence matrix contains no row with an entry "*" in both columns i and k.

Figure 1 shows the dependence matrix of a sample circuit with five symbolic logic gates. As indicated by the depicted 4-bit counter, two input pairs can be set to the same value due to the structural independence. In this case, instead of 64 test vectors, the enumeration of 16 patterns is sufficient.

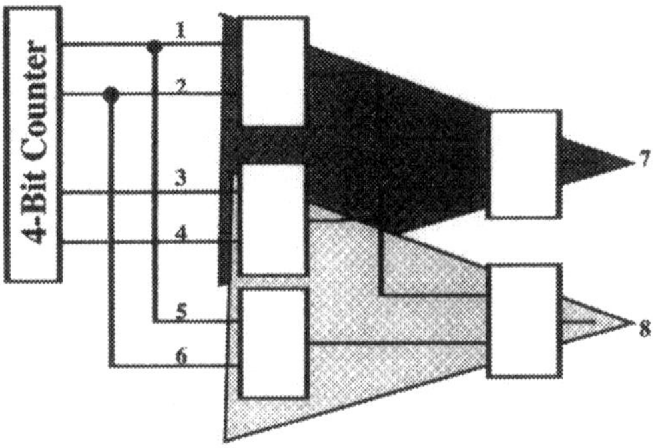


```
        123456
1:(7)   ****
2:(8)     ****
```

(a) (b)

**Figure 1** (a) Structural independencies for a sample circuit and (b) the corresponding dependence matrix.

In order to increase the number of compatible input pairs, thereby reducing the number of test patterns, the following two case studies will show how functional circuit information can be used for further optimizations.

In Figure 2, a small sample circuit consisting of three logic gates is shown. According to the criteria described above and depicted in Figure 2(a),the initial circuit specification shows no structural independence that would allow the connection of an input pair. But presuming the stuck-at fault model, the proposed input combinations in Figures 2(b) and 2(c) obviously allow all faults to be detected. This statement can easily be verified by enumerating all possible eight and four input combinations, respectively. The resulting gate assignments ensure that all stuck-at faults can be both stimulated and propagated to the primary input. In this example, instead of 16 test patterns only four test patterns are needed to guarantee complete stuck-at fault coverage.

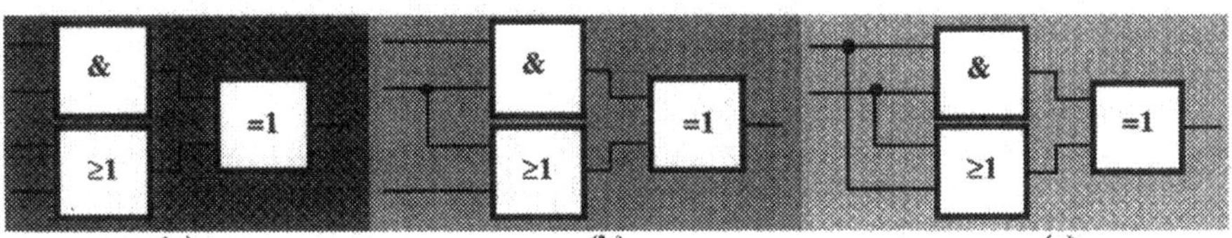


(a) (b) (c)

**Figure 2** Input combinations due to functional independencies.

A second observation to motivate the development of a new test method is based on a partly unspecified deterministic test pattern set. In Figure 3(a), the dependence matrix of the benchmark circuit c17 is given, and the second primary output is influenced by all four primary inputs. Therefore, no structure-based simplifications are possible. In Figure 3(b) a test set is given, which has been generated by the automatic test pattern generator (ATPG) SPROUT9v (Kunzmann, 1990) with the specific option of preserving partly unspecified patterns. If both unspecified values of the last given test pattern are both set to the same logical value, then the first and last bit position of all test vectors are equal. In this case these two inputs can be substituted with a single common input. The original fault coverage is guaranteed due to this procedure, and the necessary number of test patterns can be cut in half.

```
        1234
1:(8)   ***
2:(9)   ****
```

(a)

```
    1234
1:  0110
2:  0x10
3:  1001
4:  1011
5:  1111
6:  x01x
```

(b)

**Figure 3** (a) Dependence matrix and (b) test vector set of benchmark circuit c17.

## 3 BASIC APPROACH

In the following subsections two basic approaches will be discussed in order to determine the described functional independencies. The first approach uses a tuned ATPG algorithm in order to decide whether the selected input pair is functionally independent. The alternative second approach is based on a given set of deterministic test patterns with partly unspecified values which increase the chance of combining input pairs.

### 3.1 ATPG based approach

The principal observations and situations described above prove the existence of functional independencies. A general rule and simultaneously an essential boundary condition for identifying these independencies is to guarantee the original fault coverage. Therefore, pairs of inputs can only be combined if it is sure that the fault coverage will not be reduced.

Based on these assumptions the definition of functional independence can be formulated as follows:

**Definition 3:** Functional Independence
Let FC denote the target fault coverage of the circuit being tested. An input pair (i,k) is said to be functionally independent if a set of test patterns TP can be found that fulfills the following two requirements

(R1) There exists no test pattern p with $v_p(i) \neq v_p(k)$ {$v_p(j)$ denotes the value of pattern p at bit position j, $p \in TP$}.

(R2) Fault simulation with pattern set TP yields the target fault coverage FC.

This definition emphasizes the fact that the boundary condition for the input minimization procedure is the achievable fault coverage, i.e. the number of detectable faults as the result of a conventional deterministic test pattern generation. Consequently, the number of redundant faults must not increase. Also, no restriction on the basic fault is made. It should be stated, that by definition structural independencies are also functional independencies, since the fault coverage is not decreased if no common output can be influenced.

It should also be noted that in (McCluskey, 1984), besides the structure dependence, the so-called "functional dependence" is introduced. But here, an output is functionally dependent on an input w, if the output function f strictly requires the input variable w. According to this definition, functional independence means that w can be set to the constant value of 0 or 1 without changing the output function and therefore, the corresponding input fault is not detectable. This redundancy also enables the combination of this input with any other input without changing the correct output function. Unfortunately, the efficiency of using functional dependence is not satisfactory, as already expected in (McCluskey, 1984): in the over 30 analyzed benchmark circuits no functionally independent input could be found.

The following description for computing functional independencies is based the ATPG algorithm SPROUT9v (an enumeration algorithm similar to FAN and SOCRATES) which is extended to offer some specific features for CPU-time reduction. Here the stuck-at fault model is presumed.

In the beginning, the basic fault coverage including aborted and redundant faults has to be determined. Inputs can only be combined if the original fault coverage is achieved, i.e. no additional aborted or redundant faults may occur. Therefore, the target fault sets will contain the complete set of detectable faults.

Basically, the ATPG algorithm has to be executed with all possible combinations of input pairs to reach the described goal. But due to the quadratic increase with the number of inputs this procedure would result in excessive computation times. Therefore, several strategies were developed and implemented to reduce the necessary computation time. These strategies include

- **Initial computation of structure-based independencies**
The computation of functional independencies is always performed after the determination of structure-based input combinations to reduce the effective number of inputs that have to be enumerated.

- **Ordering of the fault sets**
Only faults have to be examined that can be influenced by connecting the input pair currently under consideration.

- **Stopping of test generation when an aborted or undetectable fault occurs**
Assuming a fault set that has to be completely covered, the ATPG algorithm can stop as soon as at least one single aborted fault (e.g., due to a given maximum number of backtrackings) or fault is classified as undetectable. In all these cases, further computations can be suppressed. This feature requires the modification of the ATPG algorithm.

- **Fault simulation with adapted deterministic test pattern sets**
A further reduction can be achieved if previously computed test sets are used for fault simulation. This requires a modification of these test sets, such that the values of input pairs currently under evaluation are equal. All test vectors that do not fulfil this requirement have to be doubled: the corresponding bit positions of the first vector are set to "0", in the second vector these values will be replaced by "1" (see Figure 4). This strategy can either be realized by an extra conversion program or integrated into the ATPG algorithm.

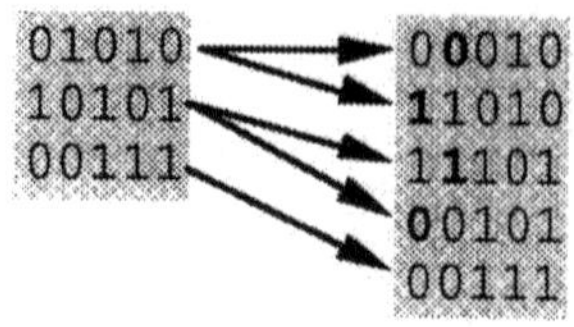


**Figure 4** Test set adaptation, assuming a coupling of inputs 1 & 2.

The new concept for a method of computing input pairs that are compatible due to functional dependencies, is given in the following figure.

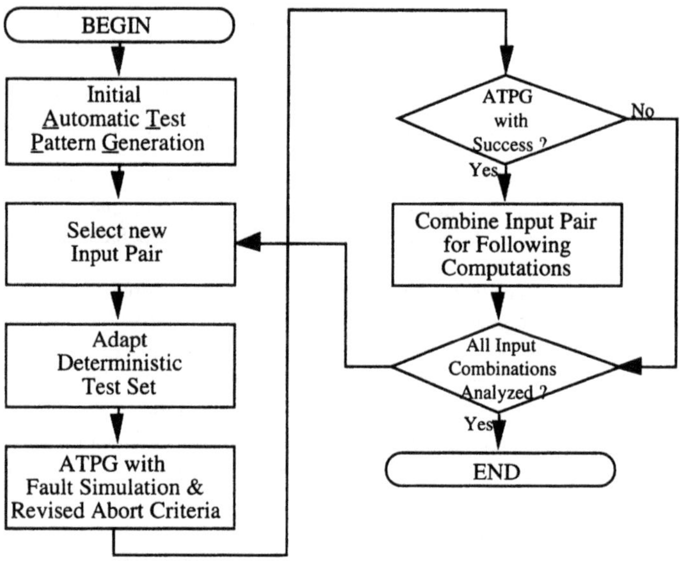


**Figure 5** Computation of functional independence.

According to the given definition of functional independence, the procedure starts with a conventional deterministic test pattern generation to compute the basic fault sets and the test pattern set for the subsequent phases. For all possible input combinations, the test patterns which are used for fault simulation must be adapted to speed up the ATPG process. If no aborted or additional redundant faults occur, the ATPG phase was successful and the current input pair can be combined. For circuits with n primary inputs, n*(n-1)/2 iterations are necessary to enumerate all possible input combinations. But fortunately, in practice this does not result in an n*(n-1)/2 times increased computation time compared with a conventional ATPG. Due to both the test pattern adaptation and the abort criteria it turned out that this factor is on average 2.9. Assuming that first structure dependencies are determined, mid-sized circuits (e.g., all circuits listed in the result table) can be treated within a 5 hour CPU-time limit.

## 3.2 Test pattern based approach

Besides the described ATPG based input grouping algorithm, a static method has also been developed based on a fixed set of deterministic test patterns which guarantees complete fault coverage. This approach only requires the results of a conventional ATPG tool, and there are no modifications mandatory, as described in the previous section. The possible input groupings are derived from the test pattern set by examining functional coherencies between any pair of bit positions for all patterns. These coherencies will represent structural or functional independencies of the circuit inputs.

Since unspecified test patterns increase the probability of detection of coherencies, the ATPG tool should be able to generate partially determined test patterns, i.e. bit positions are set to a fixed value only if this is a necessary condition for detecting the fault. These Don't Care (DC) values can be associated with an arbitrary value to enable an input grouping.

In this test pattern based approach the above given definition of functional independence is extended to also admit Boolean functions with several inputs, besides the trivial setting of inputs to the same value. The definition of this extension is tuned on the given pattern set and will allow further reductions of the effective test pattern width.

**Definition 3.1** *Generalized functional independence*
Let TP be a deterministic test pattern set of width w. A bit position k ($k \leq w$) is called "functionally independent" from a set of bit positions BP ($k \notin BP$), if there exists a Boolean function f with |BP| inputs that are compatible to all pattern specific input values of BP and to the given value of output k.

With each selected input set that allows the definition of a compatible Boolean function, one input can be eliminated by adding the corresponding logic gates. Furthermore, with each eliminated input the test pattern set has to be checked for adaptation: according to the Boolean function both forward and backward implications that are typically used in ATPG algorithms, have to be executed. In case of executable implications, DC values are set to the forced value. With a decreasing number of DCs the probability of finding new input groupings is also decreased. Therefore, the proposed approach follows the goal of first grouping those inputs which will preserve the maximum number of DCs (Greedy Algorithm). Figure 6 shows the basic effect of this selection strategy; while in the upper branch by combining input 2 and 3, the effective input number can only be reduced to three. The preferred alternative solution requires only two effective inputs since two input pairs can be grouped.

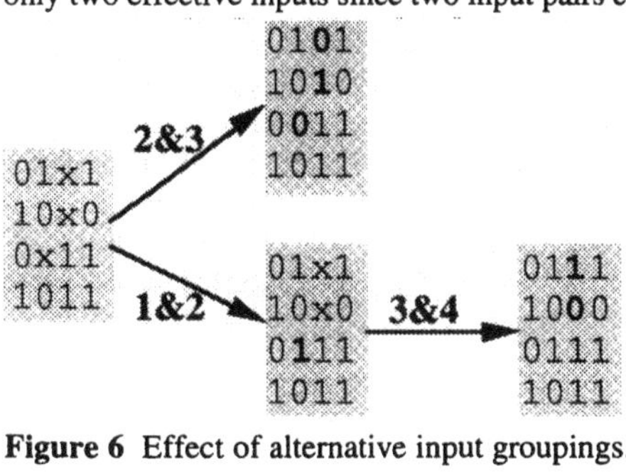


**Figure 6** Effect of alternative input groupings.

To reach the goal of preserving a large number of DCs the selection strategy is followed to group n inputs only if n-1 inputs can not be grouped. The following two advantages can be observed:

1. The number of bit positions and thereby the relevant number of DCs is kept as small as possible. Therefore, the probability of eliminating DCs decreases.
2. With growing input number the computational complexity increases. Therefore, preferring input groupings with fewer inputs results in a significant gain in computation time.

Groupable inputs are found by enumerating all possible input combinations and selecting the most cost-effective solution. To reduce the required computation time, we used the heuristic to sort the inputs of an increasing number of DCs. In this case, input groupings with a small DC number are analyzed first, and the number of DCs that are forced to take a fixed value is decreased.

Another method of reducing the required computation time is to sort the given test pattern set. Once a combination of an input k and a set of inputs BP is selected, a Boolean function f must be determined such that k = f(BP). Therefore, each test pattern is tested until there is either a contradiction with respect to the selected function for all test patterns have already been considered. In the latter case the inputs can be grouped, otherwise a new input combination has to be considered until all combinations have been enumerated. Obviously, it is desirable to detect contradictions as soon as possible. Since with an increasing number of DCs the probability of contradictions decreases, the pattern set is sorted by an increasing DC number.

There exist several strategies to further decrease the resulting test pattern width, e.g. by ignoring test patterns which cause contradictions in finding a permissible Boolean function and thereby reducing the initial fault coverage, or by trying to substitute an "incompatible" test pattern with an alternative one. Both optimization strategies are currently under development.

## 4 SELF TEST HARDWARE

In a direct comparison with conventional pseudo-exhaustive test techniques, the proposed approach uses a very similar self-test hardware structure. Moreover, since only a single test session is required, the core of the self-test hardware can be based on one counter of the appropriate length.

The test hardware structure of an external self-test is depicted in Figure 7, where the test pattern generator feeds in both the indicated scan path and the primary inputs of the circuit being tested. To distinguish between test and system mode, each primary input and most flipflop outputs have to be multiplexed, controlled by the given test mode signal. As shown in Figure 7, all grouped inputs that can be stimulated with the same values (groups A and B with 3 and 2 inputs, respectively), and only one flipflop per group has to be set to the target value. Therefore, the indicated scan path includes only two flipflops instead of five. This procedure helps to reduce the test hardware overhead by a shortened scan length and to shorten the test application time. The test responses can be analyzed by a conventional feedback shift register of appropriate width.

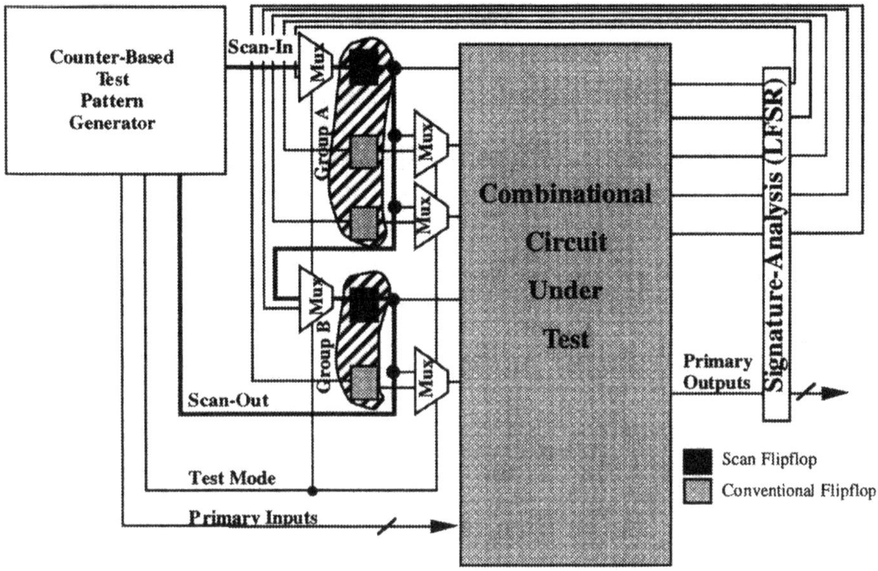


**Figure 7** Self-test hardware.

# 5 EXPERIMENTAL RESULTS

The described procedure has been evaluated by the ISCAS benchmark circuits((Brglez, 1985), (Brglez, 1989)). For the analyzed sequential circuits, a complete scan design is presumed. All experiments use the results of the ATPG tool SPROUT9v, where the stuck-at fault model is presumed (Kunzmann, 1990).

The following table shows the initial number of primary inputs and the resulting number of relevant inputs, if structural and functional independencies are identified and the corresponding input pairs are replaced by common inputs. This is indicated by the two phases (A) and, based on these interim results, phase (B). Due to the large number of inputs resulting from phase (A), additional random patterns have been applied to circuit c7552 before computing functional independencies.

Assuming an upper limit of 20 inputs that can be exhaustively enumerated, only two benchmark circuits c2670 and s838 exceed this limit (bold text) when using the ATPG based approach. For all other 30 circuits, a pseudo-exhaustive test can be performed with less than or equal to $2^{20}$ test vectors. In all cases the number of relevant inputs can be significantly decreased. As depicted in Table 1, the resulting total number of compatible inputs due to functional independencies is 714. Since this number expresses additional input combinations, the relative increase compared with only using structural circuit information exceeds 150%.

By construction, the original fault coverage is kept unchanged, no test hardware is necessary and a single test session is sufficient, since all primary outputs can be tested simultaneously. Assuming that all structural dependencies are computed first, the overall CPU-time could be kept below 5 hours for each analyzed circuit (SUN Sparc 2).

**Table 1** Experimental results for the ATPG and test pattern based approach

| | | *ATPG Based* | | | | *Test Pattern Based* | | | |
|---|---|---|---|---|---|---|---|---|---|
| | | *(A) Structural Level* | | *(B) Functional Level* | | *Boolean Function with* | | | |
| *Circuit* | *Primary Inputs* | *Indep. Inputs* | *Relevant Inputs* | *Indep. inputs* | *Relevant Inputs* | *n ∈ {1,2,3,4}* *n=1* | *n=2* | *n=3* | *n=4* |
| **c432** | 36 | 0 | 36 | 23 | 13 | 16 | 15 | 14 | 13 |
| **c499** | 41 | 0 | 41 | 32 | 9 | **41** | **34** | **33** | **32** |
| **c880** | 60 | 15 | 45 | 26 | 19 | **27** | **22** | 19 | 18 |
| **c1355** | 41 | 0 | 41 | 31 | 10 | **41** | **35** | **33** | **33** |
| **c1908** | 33 | 0 | 33 | 19 | 14 | **31** | **31** | **28** | **25** |
| **c2670*** | 233 | 111 | 122 | 98 | **24** | **41** | **39** | **37** | **27** |
| **c3540** | 50 | 0 | 50 | 32 | 18 | **29** | **27** | **26** | **23** |
| **c5315** | 178 | 109 | 69 | 54 | 15 | **32** | **25** | **24** | **23** |
| **c6288** | 32 | 0 | 32 | 26 | 6 | **32** | **30** | **28** | **25** |
| **c7552*** | 207 | 13 | 194 | 179 | 15 | **65** | **59** | **57** | **49** |
| **s27** | 7 | 1 | 6 | 3 | 3 | 4 | 4 | 3 | 3 |
| **s208** | 19 | 1 | 18 | 6 | 12 | 11 | 10 | 8 | 8 |
| **s298** | 17 | 9 | 8 | 1 | 7 | 7 | 7 | 6 | 6 |
| **s344** | 24 | 11 | 13 | 5 | 8 | 9 | 8 | 7 | 7 |
| **s349** | 24 | 11 | 13 | 7 | 6 | 9 | 8 | 7 | 7 |
| **s382** | 24 | 10 | 14 | 6 | 8 | 9 | 8 | 7 | 7 |
| **s386** | 13 | 1 | 12 | 1 | 11 | 11 | 10 | 10 | 10 |
| **s420** | 35 | 1 | 34 | 14 | 20 | **24** | 15 | 12 | 11 |
| **s444** | 24 | 10 | 14 | 5 | 9 | 10 | 8 | 8 | 7 |
| **s510** | 25 | 5 | 20 | 10 | 10 | 9 | 8 | 8 | 8 |
| **s526** | 24 | 10 | 14 | 2 | 12 | 14 | 13 | 13 | 13 |
| **s641** | 54 | 27 | 27 | 8 | 19 | 17 | 12 | 11 | 10 |
| **s713** | 54 | 27 | 27 | 10 | 17 | 17 | 12 | 11 | 10 |
| **s820** | 23 | 2 | 21 | 8 | 13 | 13 | 12 | 12 | 11 |
| **s832** | 23 | 2 | 21 | 8 | 13 | 13 | 12 | 12 | 12 |
| **s838*** | 67 | 1 | 66 | 34 | **32** | **50** | **25** | **21** | 19 |
| **s953** | 45 | 26 | 19 | 3 | 16 | 16 | 15 | 13 | 12 |
| **s1196** | 32 | 9 | 23 | 6 | 17 | 17 | 15 | 14 | 14 |
| **s1238** | 32 | 9 | 23 | 8 | 15 | 17 | 15 | 14 | 14 |
| **s1423** | 91 | 31 | 60 | 45 | 15 | **34** | **22** | 19 | 19 |
| **s1488** | 14 | 0 | 14 | 2 | 12 | 13 | 11 | 10 | 10 |
| **s1494** | 14 | 0 | 14 | 2 | 12 | 13 | 11 | 10 | 10 |
| **TOTAL** | **1596** | **452** | **1144** | **714** | **430** | **692** | **578** | **535** | **496** |

* additional application of 10,000 random test patterns before computing compatible inputs

In Table 1 the last four columns represent the results of the pattern based approach that requires an average CPU-time of about one tenth of the ATPG based method. These columns correspond to an optional realization of Boolean functions with between 1 and 4 inputs and contain the effective number of inputs, i.e. the necessary LFSR or counter width. While the results for the s-benchmark circuits are very positive , even if the input number of the Boolean functions is restricted to one or two, for only two c-benchmark circuits the relevant input number could be reduced to 20. This is mainly due to the fixed initial deterministic test pattern set. But the achieved effective input numbers emphasize that with relatively low CPU-time effort a substantial reduction can be obtained. Therefore, it is planned to use these results, where necessary, as an interim solution for the ATPG based approach, to reduce the overall CPU-time requirements. A more detailed discussion of the experimental results with respect to the test pattern based approach can be found in (Kriebel, 1994).

## CONCLUSIONS

A new self-test approach was presented that uses a conventional LFSR or alternatively a counter for the test pattern generation. This technique requires a mapping of the LFSR outputs to the circuit inputs, where, in general, one output is connected with several circuit inputs. While classic methods in the field of pseudo-exhaustive test use only structural information to find adequate input groupings, in this paper functional independencies were also introduced. For the computation of functional independent input pairs or groups, both an ATPG and a test pattern based approach were presented. While the first method uses an extended ATPG algorithm to verify the compatibility of a selected input pair, the second is purely based on a given set of deterministic test patterns without requiring any changes to the ATPG algorithm and, as a further advantage distinctly reducing the CPU-time. The pattern based approach also offers the option of including dependencies of more than one input by introducing the ability to determine compatible Boolean functions.

Based on the selected stuck-at fault model, experimental data show that the proposed grouping of both structural and functionally independent input pairs is highly efficient in minimizing the necessary LFSR width. With a reasonable CPU-time, the required LFSR width can be reduced to under 21 for almost all benchmark circuits, which ensures an affordable test application time at very low self-test hardware overhead and additional production cost.

## REFERENCES

Amazonas, J.R. and Strum, M. (1989) Pseudoexhaustive Test Techniques: A New Algorithm to Partition Combinational Circuits, 1st European Test Conference, Paris, 392-7.

Brglez, F. and Fujiwara, H. (1985) A neutral netlist of 10 combinational benchmark circuits and a target simulator in Fortran, Int. Symposium on Circuits And Systems (ISCAS-85), 695-8.

Brglez, F.; Bryan, D. and Kozminski, H. (1989) Combinational Profiles of sequential benchmark circuits, Int. Symposium on Circuits And Systems (ISCAS-89), 1929-34.

Hellebrand, S. and Wunderlich, H.-J. (1990) Tools and Devices Supporting the Pseudo-Exhaustive Test, Proc. 1st European Design Automation Conference, Glasgow.

Hellebrand, S.; Wunderlich, H.-J. and Haberl, O.F. (1990) Generating Pseudo-Exhaustive Vectors for External Testing, Proc. International Test Conference, 670-9.

Kriebel, S. (1994) Erzeugung deterministischer Testmuster für den Selbsttest digitaler Schaltungen unter Verwendung von Zählern minimaler Breite, Diploma Thesis, University of Karlsruhe.

Kunzmann, A. and Wunderlich, H.-J. (1990) An Analytical Approach to the Partial Scan Design, JETTA, Vol. 1, Kluwer, Boston, 163-74.

Kunzmann, A. (1992) Generation of Deterministic Test Patterns by Minimal Basic Test Sets; Proceedings EuroDAC, Hamburg.

Kunzmann, A. (1993) FPGA based Self-Test with Deterministic Test Patterns; in "Lecture Notes on Computer Science", Springer-Verlag, 174-82.

McCluskey, E.J. and Bozorgui-Nesbat, S. (1981) Design for Autonomous Test, IEEE Transactions on Circuits and Systems, Vol CAS-28, No. 11, 1070-8.

McCluskey, E.J. (1984) Verification Testing - A Pseudoexhaustive Test Technique, IEEE Transactions on Computers, Vol C-33, No. 6, 541-6.

Daehn, W. (1983) "Deterministische Testmustergeneratoren für den Selbsttest von integrierten Schaltungen", Ph.D. Thesis, University of Hannover.

Schulz, M. H. and Auth, E. (1988) SOCRATES: Advanced Automatic Test Pattern Generation and Redundancy Identification Techniques; FTCS-18, Tokyo, 30-5.

Seroussi, G. and Bshouty, N.H. (1988) Vector Sets for Exhaustive Testing of Logic Circuits, IEEE Transactions on Information Theory, Vol. 34, No. 3, 513-22.

Waicukauski, J.A. et al. (1988) WRP: A Method for Generating Weighted Random Patterns, IBM Journal of Research and Development, Vol. 33, No. 2, 149-61.

Wunderlich, H.-J. (1987) On Computing Optimized Input Probabilities for Random Tests, Proc. 24th Design Automation Conference, Miami Beach, 392-8.

Wunderlich, H.-J. (1988) Multiple Distributions for Biased Random Test Patterns, Proc. International Test Conference, Washington, 236-44.

## BIOGRAPHY

Arno B. Kunzmann received the diploma degree and the Dr. rer. nat. (Ph.D.) degree in computer science from the University of Karlsruhe, in 1982 and 1989, respectively. Since 1990 he joins the Computer Science Research Center Karlsruhe (FZI). Here he is head of the department ,Microelectronic System Design. His research activities include design for testability methods, especially self-test strategies, and algorithms for deterministic test pattern generation.

Sven Kriebel received his diploma degree in computer science from the University of Karlsruhe in 1995. His current research activities are concentrated on ATPG algorithms and self-test strategies.

# INDEX OF CONTRIBUTORS

# KEYWORD INDEX

GPSR Compliance
The European Union's (EU) General Product Safety Regulation (GPSR) is a set of rules that requires consumer products to be safe and our obligations to ensure this.

If you have any concerns about our products, you can contact us on

ProductSafety@springernature.com

In case Publisher is established outside the EU, the EU authorized representative is:

Springer Nature Customer Service Center GmbH
Europaplatz 3
69115 Heidelberg, Germany

www.ingramcontent.com/pod-product-compliance
Lightning Source LLC
Chambersburg PA
CBHW070743130926
51338CB00014B/22
*9780412726903*